FUNCTIONAL ANATOMY OF THE EXTREMITIES

Publication Number 855

AMERICAN LECTURE SERIES®

A Monograph in

The BANNERSTONE DIVISION *of*
AMERICAN LECTURES IN ORTHOPAEDIC SURGERY

Edited by

J. ROBERT CLOSE, M.D.

Director of Research
Orthopaedic Hospital
Los Angeles, California

FUNCTIONAL ANATOMY OF THE EXTREMITIES

Some Electronic and Kinematic Methods of Study

By

J. ROBERT CLOSE, M.D.

Director of Research
Orthopaedic Hospital
Los Angeles, California

With a Foreword by

HENRY J. RALSTON, Ph.D.

Research Physiologist, Biomechanics Laboratory,
Department of Orthopaedic Surgery, School of Medicine,
University of California, San Francisco.

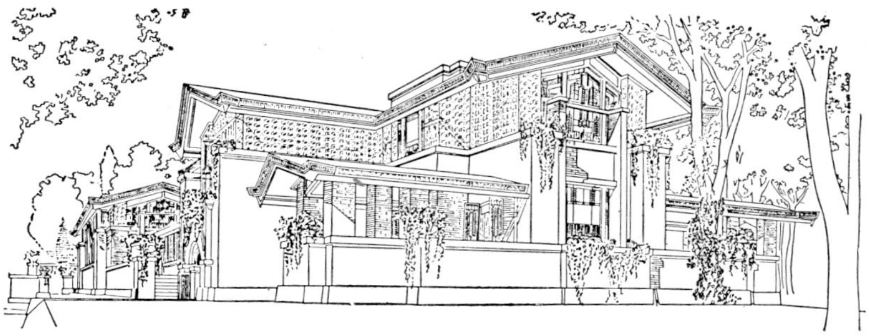

CHARLES C THOMAS • PUBLISHER

Springfield • Illinois • U.S.A.

Published and Distributed Throughout the World by
CHARLES C THOMAS • PUBLISHER
BANNERSTONE HOUSE
301–327 East Lawrence Avenue, Springfield, Illinois, U.S.A.

ISBN 0–398–02525–8

Library of Congress Catalog Card Number: 72–75912

Printed in the United States of America
BB-14

To

MARY ISABEL CLOSE

FOREWORD

During the last half-century enormous progress has been made in our knowledge of neuromuscular physiology. Much of that progress was made possible by the development of electronic techniques.

In an earlier monograph, "Motor Function in the Lower Extremity: Analyses by Electronic Instrumentation," Dr. Close created an unique text, of great value not only to the practicing orthopaedist but also to anyone interested in pursuing experimental studies in basic neuromuscular physiology. Of particular importance was Dr. Close's critical description and analysis of electromyographic techniques and the pitfalls to be avoided in the interpretation of electromyographic data.

In the earlier book, only muscles of the lower extremity were considered from the orthopaedic standpoint. The value of the work has now been greatly increased by incorporation of material dealing with the upper extremity.

The present writer is a physiologist, not an orthopaedist, but his long association with Dr. Close leads him to feel sure that this new book will be as important for the practicing orthopaedist as it certainly is for the experimental physiologist.

HENRY J. RALSTON, PH.D.
Research Physiologist, Biomechanics Laboratory,
Department of Orthopaedic Surgery, School of Medicine,
University of California, San Francisco.

PREFACE

The orthopaedic surgeon is concerned with fractures, reconstructive procedures involving the neuromusculoskeletal system, and the organization and supervision of programs of rehabilitation. His thinking and techniques are not based entirely upon his knowledge of previously established procedures. Since each problem that he encounters has never before been quite duplicated, he must often rely upon his knowledge of function of body parts in order to implement necessary departure from tried methods. To be totally dependent upon previously accumulated knowledge of the experience of others is to defeat the possibility of advances that he may add to accepted skills. One is in the best position to act creatively and with imagination if he is aware of the basic neuromusculoskeletal principles of function. These include the nature of muscle contraction, patterns of neuromuscular behavior and the geometry of the skeletal system and the types and magnitudes of forces to which it is normally subject.

It is the purpose of this book to describe some of these basic considerations and to point out how awareness of them may expand our thinking when confronted with the problems of the orthopaedic surgeon.

This monograph is based principally upon neuromuscular and musculoskeletal studies begun by the author and associates in 1955. The first study was an analysis of the muscles of the normal lower extremity, followed by similar analyses of the phasic activity of muscles before and after tendon transfer in the postpoliomyelitis patient. Over a period of several years, the muscles of approximately 150 patients with tendon transfers in the lower extremity were studied.

Although the crippling effects following poliomyelitis are now generally seldom found among the people of this country, the medical centers of this region are concerned with treatment of considerable numbers of such patients from outside the United States. A number of well-organized programs for this type of reconstructive and rehabilitative treatment are found among the medical institutions of this area.

Along with the 150 postpoliomyelitis patients where tendon transfer had been performed, the muscles of the lower extremity of the normal subject were also studied. The phasic activity of the muscles of the lower extremity was easily recorded, with the action potentials displayed by a cathode ray oscilloscope, the output of which produced an excellent sound track for the 16 mm sound camera. This resulted in an automatically synchronized audiovisual display of the subject during locomotion and other activities and of the action potentials from the muscles involved.

Analyses of the action potentials of skeletal muscle were then undertaken using the electronic counter to demonstrate action potential counts during muscular contraction, with simultaneous display of the total electrical activity produced, using a specially designed absolute-value integrator. The ratio of these two values was continuously demonstrated on the x-y oscilloscope, this representing the average value of the action potential. It was found that this determination was useful in order to establish average action potential values for the differently constituted muscles of the body and for those of patients with various neuromuscular disorders.

These objectives led us into a thorough exploration of most of the available techniques and equipment. Many types of recording equipment were tested and compared. Electrodes of many types and sizes were evaluated, and for the most part, dual nylon-coated copper wires measuring 75μ in diameter were used. Methods were developed for making synchronous recordings from several muscles. The technique was later used for multichannel recordings of the muscles of the forearm and hand in normal subjects and patients where tendon transfer had been performed.

Due to the irregular spatial distribution of the many muscle fibres of the motor unit, which is thought to be divided into subunits, as described by Buchtal,[1] it was concluded that the single action potential seen on the usual recording was influenced by adjacent and perhaps interdigitating muscle fibers or subunits of closely associated motor units. Such recordings are also influenced to an unknown extent by neighboring motor units. We therefore set out to establish a means of direct observation of, and direct recording from, the solitary motor unit. We have been able to accomplish this in experimental animals—the rabbit and cat—by single axon stimulation after microdissection of the proximal branch of the nerve to the anterior tibial muscle. The characteristics of the resulting action potential were studied in order to learn the requirements of recording equipment. Thus the amplitude, duration and frequency spectrum of this single oscillation or waveform became known to us. The waveform of the single motor unit was subjected to complete Fourier analysis. The tension of the single motor unit was later recorded in synchrony with the action potential of the single motor unit. The relation between these two waveforms, that of the single motor unit action potential and its tension curve, synchronously recorded under various conditions, was found to be a profitable area of study. It was thought that this test—single axon stimulation with simultaneous muscle action potential recording—could well add valuable information to the conventional surgical muscle biopsy. It is quite feasible to arrange apparatus with sterile probes to conduct the test in the modern hospital operating room. An alternative is removal of a small neuromuscular specimen for *in vitro* study. Such procedures might well be expected to add considerably to our information concerning neuromuscular diseases.

A knowledge of the skeletal framework, the geometry of its joints, is essential if one is to gain understanding of the functions of muscles which act within the ranges and planes of motion of the various joint systems. Three-dimensional analyses of the motions of several joints were undertaken. For example, the range of motion of the subtalar joint and the axis about which motion takes place were studied extensively, as described in Chapter III. The importance of this axis, about which the calcaneus rotates with respect to the talus, is evident when one considers at just what point in the dorsum of the foot transferred tendons are to be placed. Knowledge of the axis is of paramount importance in understanding the position in which the bone graft for subtalar arthrodesis is to be placed.

Invaluable contributions were made to the study, some over a period of several years, by the following individuals.

Dr. Roy Frank, physicist, contributed the section found in Chapter II, "The Differential Effect and Distortion Caused by Coaxial Shielding in Coaxial Electromyographic Electrodes." The discussion found in Chapter VII, "Determination of Optimum Recording Frequency Limits by the Application of the Fourier Transform to the Isolated Triangular Spike," was also contributed by Dr. Frank, who later, by computer analysis, produced a meticulously calculated frequency spectrum of the action potential of the single motor unit.

For many years Mr. Frank N. Todd of the Biomechanics Laboratory of the University of California, Berkeley, prepared most of the illustrations and took part in the experimental work. Mr. Todd made invaluable contributions over a period of several years to the studies of the phasic activity of the muscles of the lower extremity before and after tendon transfer (Chap. VI).

Mr. Alfred Benjamin, R.B.P., A.R.P.S., Director of the Audio-Visual Department, Orthopaedic Hospital, Los Angeles, provided expert counsel and supervised the reproduction of the photographic displays of Chapter VI.

Of particular value during the past few years was the daily assistance of Mrs. Caroline C. Kidd, M.A. Mrs. Kidd meticulously checked the bibliography, prepared many of the illustrations of Chapter VI, typed a substantial part of the manuscript, and regularly took part in the more recent experimental work. She constantly searched the literature pertinent to the studies described herein. The magnitude of her effort can hardly be overestimated. For her contribution, the author and this institution are exceedingly grateful.

Of great value was the careful reading and evaluation of the manuscript of the first edition by Professor H.J. Ralston, physiologist, the University of California School of Medicine and the College of Physicians and Surgeons, San Francisco. Doctor Ralston, whose particular interest is the physiology of muscle, has long been closely associated with the Biomechanics Laboratory of the University of California School of Medicine.

The preparation of the manuscript was facilitated and made an enjoyable endeavor by the continuous assistance and stimulating cooperation of Mr. Payne Thomas of Charles C Thomas, Publisher.

To each of the above, and to many other dedicated workers in this field, the author is deeply indebted for their countless contributions which have made this work possible.

J.R.C.

ACKNOWLEDGMENTS

I am grateful for aid provided by the following institutions:

The Orthopaedic Hospital Medical Center, Los Angeles.

The National Institutes of Health, Research Grant NB 02300.

Rehabilitation Research and Training Center, The University of Southern California, RT-18.

The Orthopaedic Research and Education Foundation.

CONTENTS

FUNCTIONAL ANATOMY
OF THE EXTREMITIES

Chapter I

INTRODUCTION

This monograph is based on analyses and techniques that we have employed in the study of neuromuscular and musculoskeletal function in the lower extremity. It is written for orthopaedic surgeons, physiatrists, neurologists, physical therapists and others engaged in the surgical and rehabilitative aspects of orthopaedic surgery. It is hoped that the orthopaedic surgeon, in particular, will find useful basic information in the description of the techniques employed in these studies. Where basic studies have yet no known practical application, they may assist us in understanding the methods of nature—methods ingeniously employed by the neuromusculoskeletal system to produce complex movements.

The great anatomical importance of muscle in the study of the central nervous system is appreciated when one realizes that the end result of all motor nervous system activity is finally expressed through muscle—the terminal of the neuromuscular system. Action potentials have been thought to be produced even by thinking.[49,58,127,128] Jacobson and Max many years ago asked subjects to think of activities which would involve certain muscles to which electrodes had been applied. Using the most sensitive recording equipment, however, we have failed to demonstrate action potentials from the muscle into which electrodes have been inserted when the subject is asked to concentrate on an activity requiring the use of the muscle. "Miniature end plate potentials" have been observed by Katz[107] in inactive muscle, using intracellular electrodes. According to Jacobson,[49] motor activity is even an essential condition for thinking. This statement has not been substantiated by the most refined techniques which employ calibrated, cathode-ray oscilloscopes with very sensitive preamplifiers.

One can say that the sum total of cortical motor activity is represented peripherally through the selected, regulated, graded activity of the motor units of the muscles.[103] The motor unit or final common path consists of the anterior horn cell, its axon, and the group of muscle fibres which its branches supply. Physical activity in muscle produces electrical potential change, and the resulting action potentials can be expressed as visual or auditory records. The aggregate action potentials may be counted electronically. One can imagine the value of representative counts simultaneously recorded throughout the muscles of the body. Such a series of continuous counts would be a numerical representation of composite motor cortical activity, though recorded peripherally. Although such electronic counts are taken from small portions of muscle examined, it can be shown that adjacent portions of the muscle have a comparable, sometimes very similar, and often virtually identical, pattern of electrical activity.[41] This has been repeatedly demonstrated by the dual-channel oscilloscope (Figs. 1 and 2). Certain muscles thus appear to have a predominant rhythm or frequency (rate) of motor activity.

It is the orthopaedic surgeon who must concern himself with muscle, perhaps more than the physician of any other category. He repairs, separates, transfers and even divides more muscle tissue than does the physician of any other specialty. The action and dynamics of muscle are therefore of the greatest concern to him. The final effect of his work may be known to him only if he understands general patterns

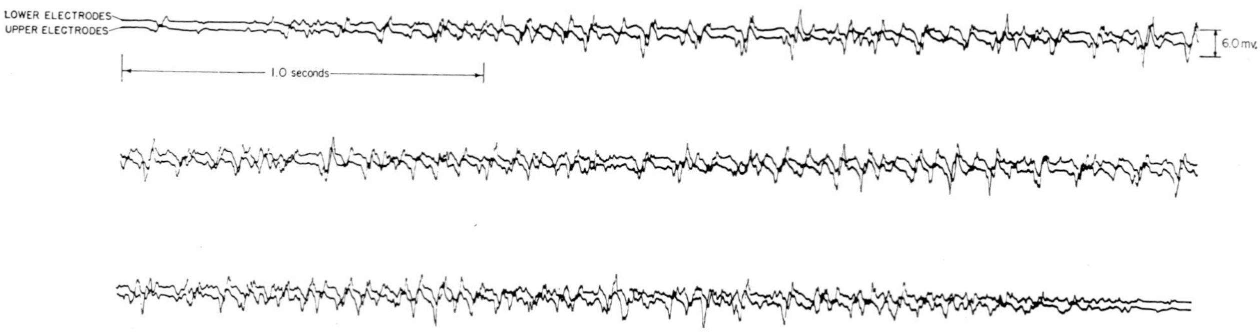

Figure 1. Synchronization of adjacent portions of normal anterior tibial muscle during voluntary contraction. Recorded by differential method. (From *Journal of Bone and Joint Surgery,* October, 1960.)

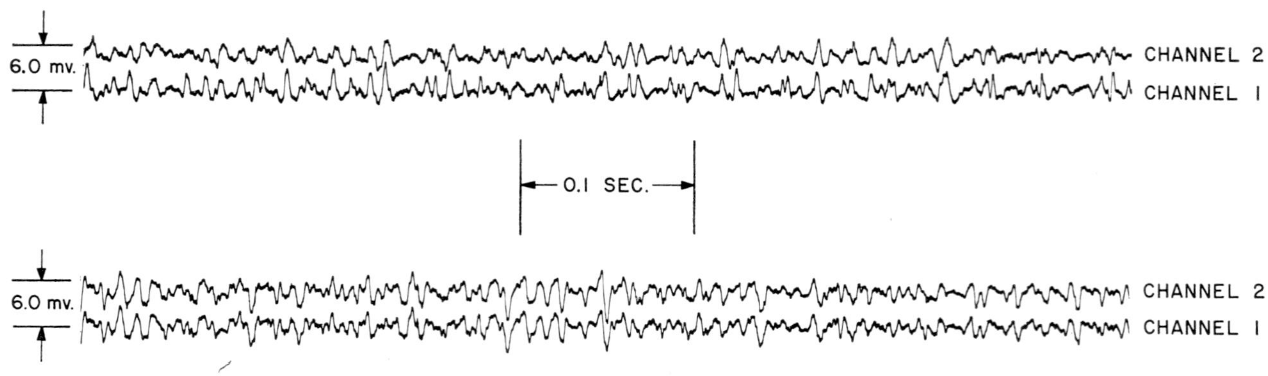

Figure 2. Synchronization in two adjacent portions of the anterior tibial muscle; dual internal electrodes, differential amplification; voluntary dorsiflexion against resistance.

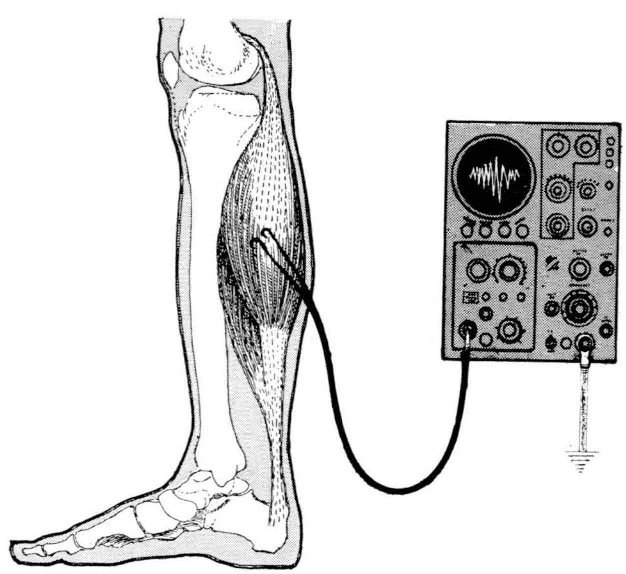

Figure 3. A differential method for recording from muscle. Dual-wire internal electrodes are led to the differential preamplifier of wide, adjustable frequency range. (From Instructional Course Lectures, American Academy of Orthopaedic Surgeons, 1959.)

of neuromuscular behavior. In certain surgical procedures, it becomes of the greatest concern to the orthopaedic surgeon to know the effect of destruction or denervation of a portion of the contractile elements of a muscle. To the orthopaedic surgeon, the relative "coarseness" or delicacy of each muscle is of the greatest importance. If a portion of a muscle is denervated by trauma or surgery, assessment of the resulting loss may be evaluated if one appreciates the nature of its components, such as the number of fibres in the motor units of which the muscle is composed. Attention is called to the fact that muscle accounts for 40 to 45 percent of the normal body weight.

Muscle is also of great importance to the orthopaedic surgeon in his consideration of the etiology and treatment of such deformities as scoliosis, clubfoot, flatfoot, valgus foot, pronated foot and the many poliomyelitis residual deformities such as the externally rotated tibia which may result when there is a strong, unopposed or weakly opposed biceps femoris muscle in the thigh. It has been pointed out that in scoliosis we must consider other factors than the osseous, ligamentous and joint systems of the spine.[123] This is best exemplified by the scoliosis resulting from muscular imbalance or contractures following poliomyelitis. Whether such muscular imbalance is present in other types of scoliosis has yet to be determined. Properly standardized and calibrated comparative electromyographs may provide further insight regarding such deformities. The necessity of multichannel, carefully calibrated, synchronized recording equipment is obvious. This equipment should have suitable frequency-recording range to adapt to the frequencies of the action potentials of contracting motor units (Figs. 3, 8, 20, 21). Although the "EEG-EMG" multichannel pen writers may have certain limited applications, dependence upon this very low frequency, "convenient" apparatus can result in serious disservice to those attempting to evaluate action potentials of muscle on a precise basis over the complete frequency range of this phenomenon. It must be granted that the lumbering ink writer, if motivated by preamplifiers of acceptable design, will "scan" and average (after a fashion) the real action potentials of muscle. Even on high-grade apparatus, many of the action potentials during strong contractions to a varying degree become coincident and grouped together. This resulting record may also represent an average, but one of infinitely superior fidelity and accuracy (Figs. 4, 4A).

Through an understanding of the relative behavioral importance of normal muscle groups, one gains further insight into the abnormal conditions often found in cerebral palsy. For example, analyses of the action of the adductors and the axis about which they act (Figs. 38, 39, 40) assist greatly in understanding the "scissors gait" and planning its treatment.[35] One should not consider adductor tenotomy or obturator neurectomy without first assessing the contractile state of the glutei as well as the adductors.

One of the best examples of the importance of knowledge of basic muscular neurophysiology in orthopaedic surgery has been brought out to us in studies of tendon transfer. We at one time believed that muscle could be transferred in any manner surgically logical. Using relatively simple modern electronic techniques, one can easily assay the resulting neuromuscular activity following tendon transfer. We have thus been enlightened regarding the neurophysiology of muscles under consideration for transfer.[40]

It should be noted that muscular-contraction timing during locomotion (phasic activity) in the paralytic patient may differ from the normal. The paralytic may use his available remaining musculature in a different manner from the normal individual for whom we have derived "normal patterns of phasic activity during walking."[70,140] The degree of variation from the normal pattern depends upon the extent of paralysis. An example is the iliopsoas muscle which we have found to be completely inactive in the normal subject during walking (Fig. 91). The paralytic (Fig. 129) may make considerable use of the iliopsoas during walking. It is quite possible that the normal subject elects to

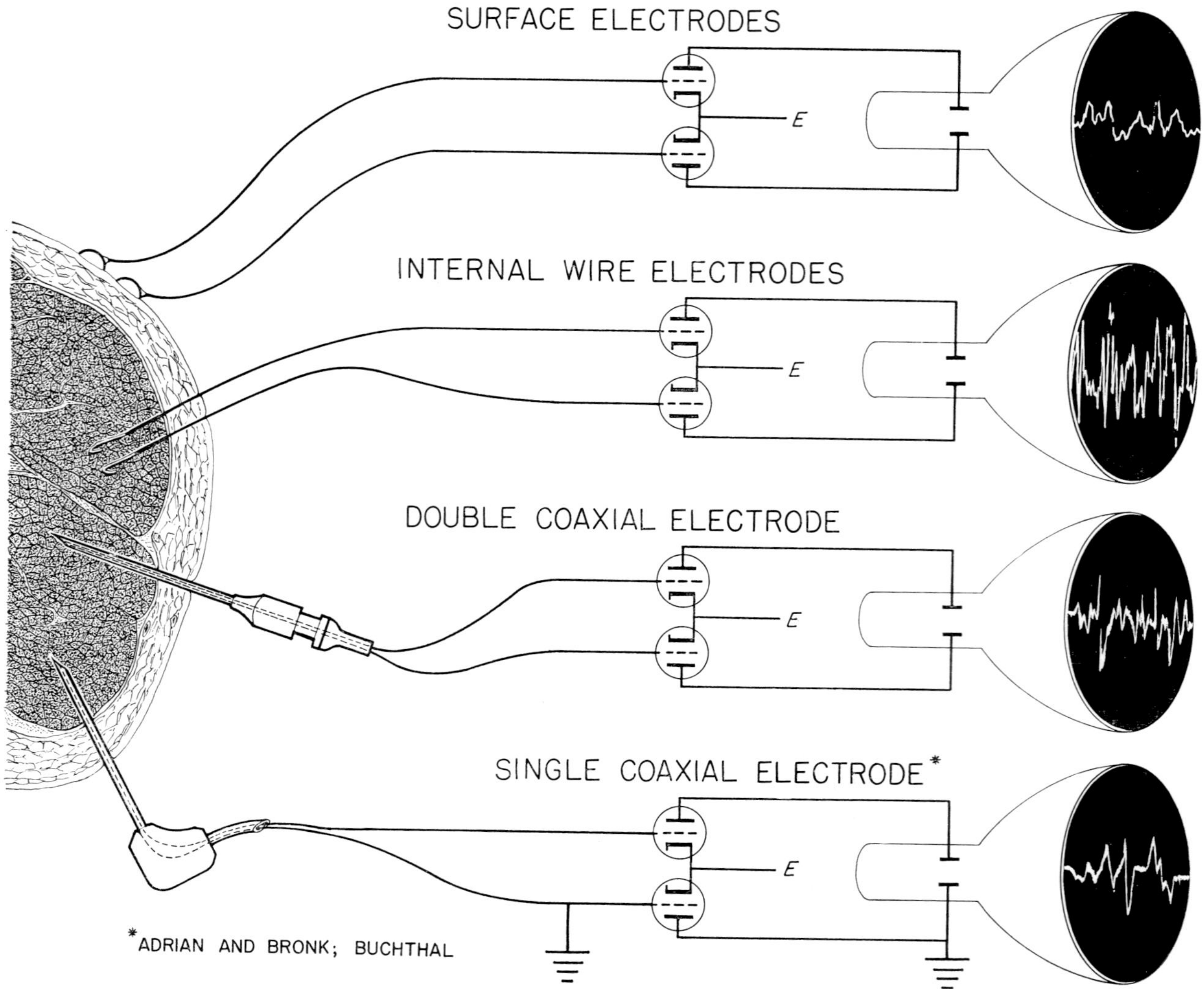

Figure 4. Four methods for recording action potentials from muscle, with sample records.

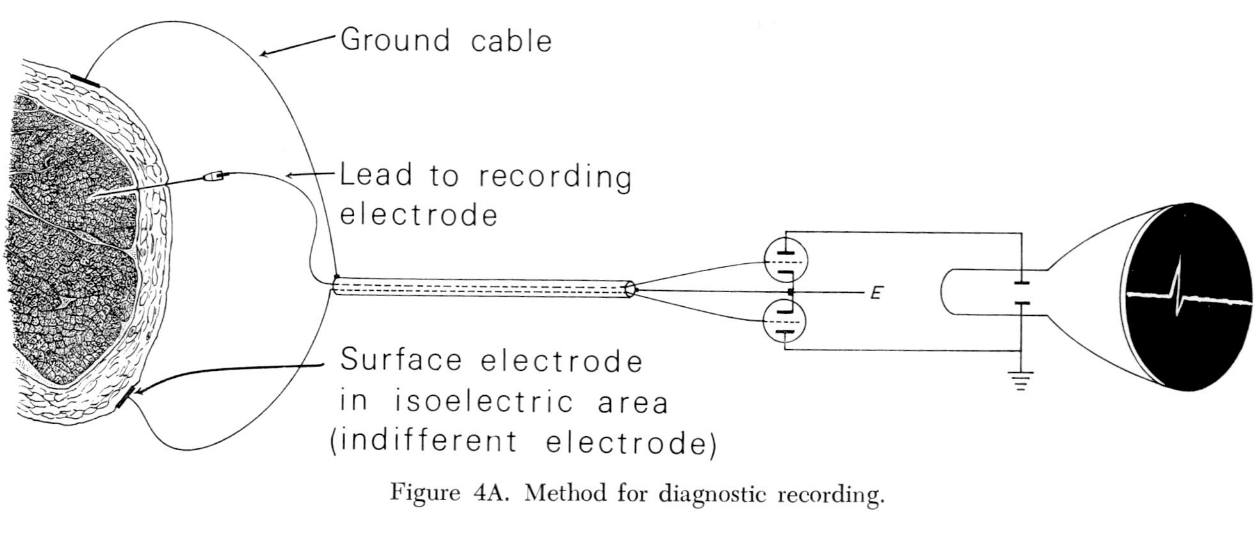

Figure 4A. Method for diagnostic recording.

employ other hip flexors such as the rectus femoris and that the paralytic with weak or absent rectus femoris or sartorious musculature may resort to the iliopsoas as a hip flexor. An analysis of such a pattern of iliopsoas behavior is shown in Figure 129.

Thus patterns of phasic activity in the paralytic patient are not necessarily comparable to the normal. In most of the cases that we have selected for presentation, however, where tendon transfer has been done, there was a minimal amount of paralysis. The muscles of these subjects behaved very much according to the normal pattern. Such a selection of subjects for study was necessary in order to relate phasic activity in the paralytic, before or after transfer, to the consistent pattern of phasic activity of the normal subject. The normal subject is our only beginning reference.

The electromyographic technique is now widely accepted as a most efficient method for recording the presence of activity in muscle. Representing the passage of many impulses down the fibres, it closely coincides with the essential function or activity in muscle. Since the electromyogram used in the study of muscle function is usually recorded from relatively large probes compared to the size of muscle fibres, it represents an average of the electrical activity of thousands of fibres. During strong contractions, the identity of the individual motor unit action potential is averaged and lost. Appreciation of the details of the tracing therefore requires considerable analysis. From consideration of the time relation of the action potential to the activity of the single motor unit, we learn that contraction follows the action potention almost immediately (Chap. VII).

The electromyogram makes possible the following studies or techniques:

1. The timing of muscular activity during such actions as walking (phasic activity).
2. Analyses of voluntary contraction enabling one to ascertain the relative importance of each muscle as it makes its contribution to various motions about joint systems. Such studies may be included in the records of the patterns of central nervous system activity and organization, using multichannel equipment. Multichannel equipment makes possible an excellent display of reciprocal activity.
3. Studies of the patterns of action potential oscillograms in normal and pathological conditions.

Since the precise and complete histology of the single motor unit is not yet completely understood, we are not always justified in attributing single action potential spikes to individual motor units. During weak contractions, the EMG spikes may often represent single motor unit action potentials (Fig. 160). As the strength of the contraction increases, the density of action potentials likewise increases—the normal interference pattern. This leads to confusion in processes of counting or integrating. The normal interference pattern (Fig. 5) obscures completely the single motor unit action potential. This does not mean that the record has lost its value as an index of neuromuscular activity. It reliably represents the general level of activity, and the action potentials can still be counted and integrated. The resulting data are of considerable significance[41] (Chap. III).

If it were possible to record from a solitary motor unit where influence from adjacent or interdigitating motor units could be definitely eliminated, very significant information could be obtained. A method for demonstrating this single quantum of peripheral neuromuscular activity[149] is described in Chapter VII. Such a study makes possible an appreciation of aggregate series of potentials. By measuring motor unit action potential characteristics (amplitude, duration and rate), one learns the requirements of the instruments, such as frequency response and sensitivity, necessary to best record action potentials from muscle.

Since it can be demonstrated that motor units behave synchronously to a considerable degree (Figs. 1, 2), it may be found that the ideal record of the single motor unit is actually a miniature of the single spikes conventionally recorded from normal contracting skeletal muscles.

In consideration of the action of the indi-

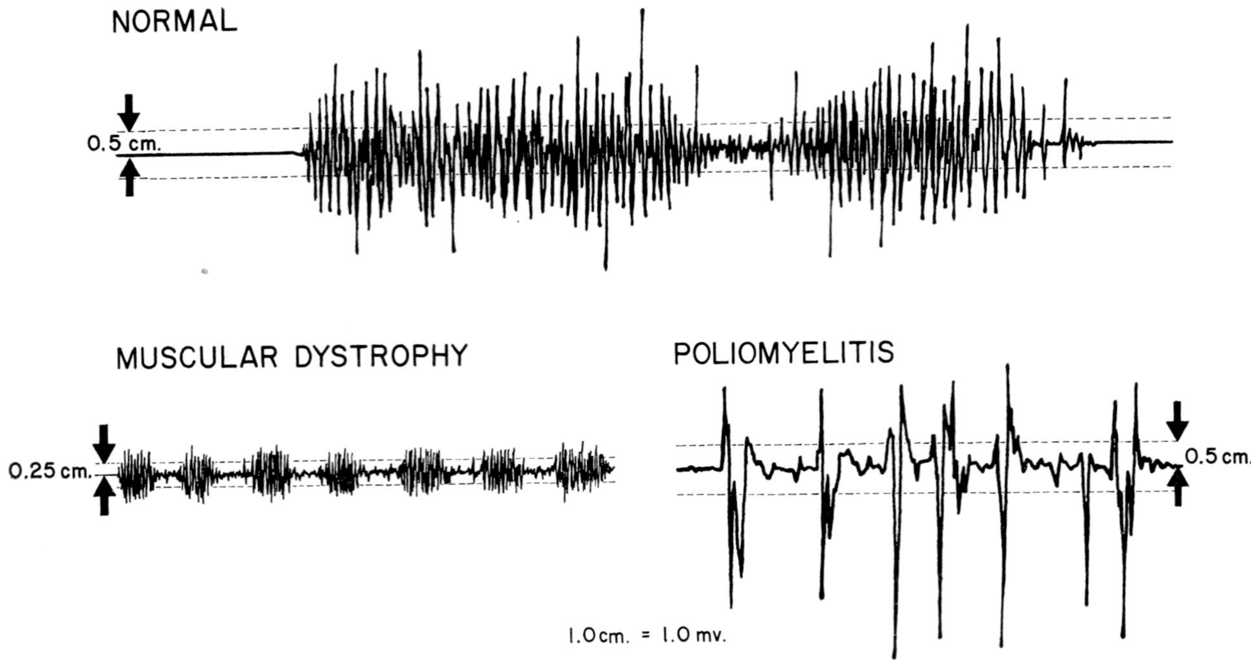

Figure 5. Sensitivity settings used for electronic counting of action potentials in normal and pathological muscle. Typical records of action potentials of three types of muscles —normal, muscular dystrophy, poliomyelitis (after maximum recovery).

vidual muscle, the knowledge of axes of parts and centers of rotation of the joints of the skeleton is essential. Many muscles and their tendons can be thought of as acting about centers of rotation or axes. For example, one can best understand the action of the iliopsoas muscle by considering the center of rotation of the hip joint (Fig. 37). Also, one is at a loss to understand the action of the adductor musculature (Figs. 38, 39, 40) inserting posteriorly on the femur along the linea aspera without knowledge of the mechanical axis of the lower extremity (Fig. 38). The adductors, for the most part, actually pass anterior to the mechanical axis. This group of muscles demonstrates beautifully the observations of Duchenne[56] and Jackson[116] that single muscles have no function as such. The adductors, strangely, behave differently on electrical stimulation than on voluntary activity, regarding their ability to move the lower extremity about its mechanical axis of rotation. This fact is further evidence that we have no independent action of single muscles but that each contributes its action with the actions of other muscles to perform a variety of motions. Each movement calls upon all various available muscles to contribute most efficiently to the final motion required.

Another excellent example of axes which influence the actions of muscle is the well-known compromise axis of the subtalar joint (Figs. 33, 34, 35), knowledge of which is essential in any analysis of the actions of the muscles of the leg (Chap. III). The axis is of obvious importance in consideration of muscle tendon transfers which pass across or in relation to the ankle and subtalar joints (see Chap. III). Such studies which explain the methods employed by muscles to achieve their various types of motion are typical of the great value of functional anatomy to orthopaedic surgery. One may profitably employ the sound motion picture camera, with the sound track consisting of a synchronized auditory electromyogram of the muscle concerned. This technique is easily applied to any of the muscles of the extremities. It makes possible analyses of the participation of the various muscles during movements which can be observed as the auditory electromyogram is heard. This audiovisual medium is almost indispensable for study of the action of such muscles as the iliopsoas, tensor fascia lata and adductors. Many reviews of the final records may be readily made, leaving little doubt concerning the relative participation of each muscle in the activities of the extremities. It has been our practice to make repeated sound motion picture studies of such muscles, both in normal and pathological states, particularly where surgical procedures involving these muscles are contemplated. The postoperative comprehensive data complete the study.

The study of locomotion is thus concerned with

1. The participation of various muscles, their temporal activity and the degree of this activity.
2. The axes or centers of rotation about which muscles are able to produce movement of a body segment.
3. Forces such as gravity, acting on the body and its segments, and acceleration of movement of parts which the muscle may have to oppose or act in conjunction with in order to accomplish its purpose.
4. Deficiencies of movement due to pathological processes in the muscle.

In the consideration of individual muscular action, one must remember that each muscle contributes only a part of the resulting composite movement. We are obliged nevertheless to study the actions of single muscles or small groups of muscles. We are often led into considerations of functional parts of the muscle. We find ourselves dividing the muscle into fasciculi, motor units, single fibres and ultimately to considerations of the molecular components of the contractile fibre—the complex myosin and actin molecules. A great deal of investigation has been done on these complex protein structures and their interrelated, elusive method of producing muscular contraction. Szent-Gyorgi,[163] in describing theories of muscular action and the complex structure of myosin and actin molecules along with the concept of sliding filaments, concludes that only a better understanding of the events within the muscle fibre can enable us to cope with its disorders such as dystrophy. It is his feeling that the secret of muscular contraction conceals most of the mysteries of cellular function not yet understood.

In general, muscular activity is the result of integrated cortical patterns. Occasionally, it is quite possible that we call upon a single muscle to produce certain movements. Examples of such muscles may be the extensor indicis proprius or the long extensor or the long flexor of the thumb. Even such apparently isolated muscle action as the examples mentioned have yet to be convincingly demonstrated as such by multichannel recording systems. Multichannel systems, as described in Chapter VI, demonstrate the participation of muscles in group actions. At the present time, with our own equipment, we are able to record from as many as twelve muscles simultaneously. Such a procedure, however, requires the insertion of an impractical number of electrodes into the muscles concerned. This may affect normal body motion. It is our practice, therefore, to record from no more than six muscles at one time. Since each channel must represent the action potentials of an individual muscle, exclusive of the influence of adjacent muscles, the differential mode must be employed. Two small wire electrodes must be meticulously inserted into each muscle (Fig. 6). The position of the electrodes may definitely be ascertained by electrical stimulation.

Although a significant transmembrane direct current potential (—90 mv) is recorded when a single tiny glass-enclosed electrode is inserted directly into the muscle fibre under microscopic control, it is along the fibre membranes of contracting muscle that rapidly fluctuating, propagated action potentials occur. This potential change, associated with differential premeability of the membrane to potassium, sodium, calcium, chloride, and other ions, is most conveniently recorded by alternating-current methods. The activity of the whole muscle is reliably represented by records from two internal metallic electrodes inserted into the muscle belly (Fig. 6). The recording is not depen-

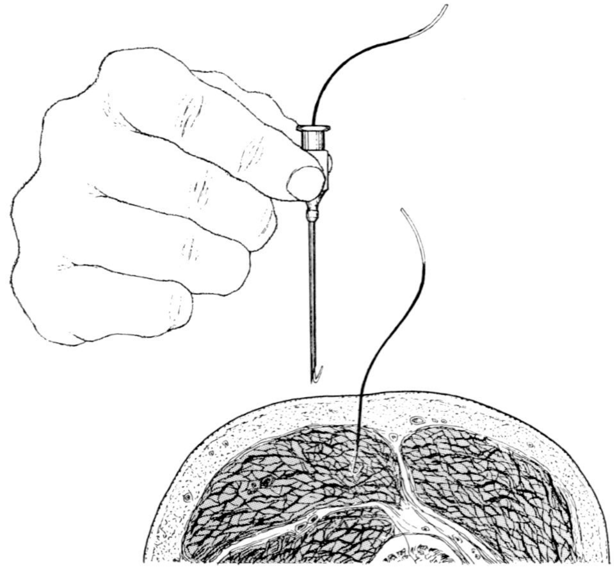

Figure 6. Insertion of dual internal wire electrodes, diameter 0.003 in., for differential method. (From Instructional Course Lectures. American Academy of Orthopaedic Surgeons, 1959.)

dent on, or significantly influenced by, proximity of the electrodes to the muscle's motor point. The recording is most conveniently made by differential amplification. Undoubtedly these electrodes (which may penetrate a few fibres, surround, or come into the immediate proximity of others) can be relied upon to lead off action potentials to the oscilloscope quite characteristic of the muscle's state of activity. When the normal muscle relaxes, electrical activity—a potential change—ceases.

It is our practice at the present time to use two internal wire electrodes inserted at adjacent sites of muscle in order to obtain the most satisfactory isolated and dependable recording. This method employs the customary and reliable differential amplifier. It is quite possible, however, to use the single-ended method employing one electrode and ground plate, and in many applications we have found this to be a satisfactory method. This method can be used reliably only when the ground plate can be placed directly over the muscle tested. This applies to muscle function—not diagnostic tests. If the single electrode is remote from the ground plate placed elsewhere on the extremity, the record is less reliable, since intervening muscles or other tissue may contribute to the record. For example, when the recording electrode is placed in the anterior tibial muscle and the ground plate placed over the calf musculature, some electrical activity is demonstrated on plantar flexion, when the anterior tibial muscle is known to be silent. Plantar flexion produces no visible activity when two internal electrodes are inserted into the anterior tibial muscle. Only the dual internal probes (differential technique) have the unique quality of confining the recorded action potential to the single muscle or part of muscle from which isolated records are required (Fig. 83).

Many other types of applicators or probes have been used. The surface electrode has many uses[105] and does not require penetration of the muscle. It has, however, certain deficiencies.[15,16] This method usually fails to confine the recording to an individual muscle. It is totally unsuitable for deep muscles or small muscles. The signal reaching the surface electrode requires higher amplifier gain setting. This increase in gain also causes correspondingly more "noise" to be recorded. Gain setting varies with the amount of subcutaneous tissue between the electrodes and the underlying muscle. This makes comparison of recordings taken from various subjects or from the different muscles of the same subject meaningless. The low-frequency records of this method may be due to the distance of the electrodes from the actual source of the action potential (Figs. 31 and 53). Perhaps an improvement is represented by the plastic suction cup[16] containing an electrode, or by adherent surface electrodes. Such applicators can be closely applied to the surface. Used in pairs, this method has the potentiality of producing a representative record of a large portion of muscle.

One can imagine the ideal situation where one might make a record from all of the representative muscles of the body at once, thus producing records representing the entire motor cortex and a display of its general pattern. Such a record, so far as the motor cortex of the brain is concerned, would in some respects yield more information than the conventional electroencephalogram which, in reality, is a representation of the average of the electrical oscillations produced by all of the cellular activity between two electrodes applied to the scalp.

The gross histology of the motor unit, that group of muscle fibres innervated by a single axon originating from a single anterior horn cell, has not yet been clearly delineated.[1,22] It is often assumed that the fibres of each motor unit lie together in a single cluster. There is evidence that this group of fibres is divided into smaller bundles which interdigitate with fibres of adjacent motor units.[22] If such is the case, exclusive recording from the single motor unit is understandably most difficult. The difficulty in understanding the functional aspect and histology of this unit is due to the fact that anatomical isolation of the unit is not possible by any techniques yet known. Investigators therefore have resorted to indirect means of determining the position of these fibres comprising the motor unit. For example, Buchtal and associates[1,22] have recorded simultaneously from several small adjacent areas of muscle using multichannel recording equipment. It has also been well established by Feinstein and his associates[65,66] that each muscle in the body has its own characteristics regarding the number of fibres innervated by each axon. For example, the lateral head of the gastrocnemius muscle is thought to contain approximately 2,000 muscle fibres per motor unit. The anterior tibial muscle, on the other hand, is thought to have only about 600 fibres per motor unit. These figures, called the muscle fibre to nerve fibre ratio, have been determined by actual counts of portions of muscle; thereby an estimate of the total number of fibres in the muscle is obtained. The number of motor fibres innervating the muscle counted was arrived at by counting the motor fibres in the peripheral nerve as it penetrated the muscle.[65] But there is an inherent error in such counts, since a considerable number of gamma fibres are included.

In our experience, the most efficient probe for making motor unit action potential counts is the Buchtal (Adrian, Bronk) single coaxial electrode. However, it must be used as a single-ended mode. The internal electrode with a 1 mm diameter, in our practice, is led to a small single-ended preamplifier located very close to the muscle. The shield is grounded (Figs. 4 and 7).

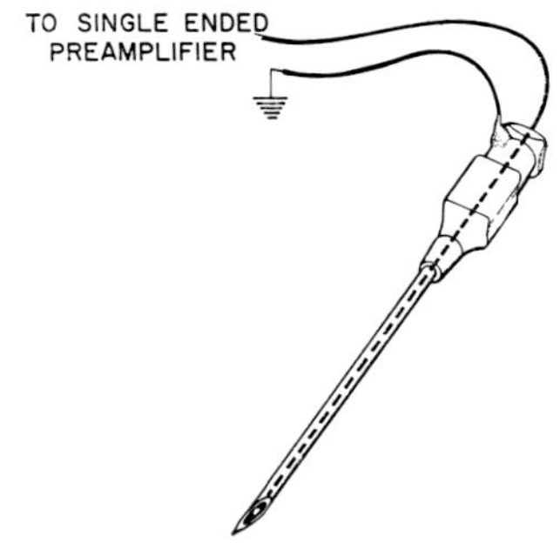

Figure 7. The Buchtal coaxial electrode adapted to single-ended preamplifier.

Coaxial shielding produces distortion due to differentiation, as discussed in Chapter II. In applications where high-fidelity recordings are required, such as electrodiagnosis, the circuitry demonstrated in Figure 4A is accepted as the method for producing the truest representation of the action potential waveform. Here, one takes advantage of the differential technique, and the absolute value of the action potential is demonstrated as well. This is achieved by the use of a single insulated, unshielded electrode and a nearby surface "indifferent" electrode placed over an isoelectric region. Thus, distortion is minimized, noise is virtually eliminated by use of the differential mode, and the absolute value of the action potential is approached, since the second connection to the differential amplifier is to an isoelectric region (Fig. 4A).

As one proceeds cephalad, the motor units are in general more finely innervated. That is to say, each axon innervates a smaller group of muscle fibres. Perhaps this is best exemplified by the case of the extraocular muscles, which perform extremely fine movements. In these muscles we find motor units which contain as few as four or five muscle fibres.[104] In contrast, the number of muscle fibres in the motor units of the lateral head of the gastrocnemius muscle is approximately 2,000, according to Feinstein *et al.*[65]

These considerations suggest several possibilities in instrumentation. One can "integrate" or totalize the electrical activity produced by a volume of muscle during contraction. Since one can count the resulting individual action potentials, using the electronic counter, a relationship between these two values is expected. Therefore, for a single muscular contraction:

Q/C is proportional to the size of the motor unit.
Q = total electrical activity for a given interval.
C = total action potential count for the same interval.

Using the absolute-value integrator (which measures the total electrical activity of all rectified forms of the electromyogram, giving a total in microcoulombs for a contraction) and the highly sensitive modern electronic counter which counts the action potentials of the same contraction, one can easily and simultaneously obtain these two values for a single muscular contraction (see Chap. IV). We have found it impossible to accurately record the real tension developed by any of the muscles studied without employing bone-pin techniques.[41] One can, however, standardize such a procedure of integrating and counting without the insertion of pins by putting the same requirements on the muscle to be tested each time (see Chap. IV). Thus the quadriceps is studied by requiring the muscle to extend the knee at a constant rate, lifting a known weight attached to the ankle. Such a procedure is satisfactory for the study of isotonic contractions. For isometric contractions, one can often obviate bone-pin techniques by attaching a force gauge[170] and by requiring the muscle to maintain a certain tension during the interval of recording. The same sort of arrangement has been used for analyses of the dynamics of the biceps brachii. One can require the arm to hold a known weight or exert a certain isometric tension as measured by force gauges. Muscle length, an important consideration in evaluation of the action potential, is easily controlled by attention to the position of the elbow and shoulder.

The foregoing thus indicates that electronic analyses of muscle represent excellent methods of study of the action of muscles which the student of the future will be required to employ. It is apparent that he must have a basic understanding of these investigative techniques even to keep abreast of the advance of neurophysiology in the future. It is therefore the purpose of this monograph to describe some modern medical electronic techniques as well as their physiological application, particularly with regard to the electronic approach to the study of neuromuscular physiology that will be of importance in orthopaedic surgery. Some of the units and techniques employed have been developed in this laboratory.

However, we do not go so far as to say that biological systems such as the brain, spinal cord, peripheral nerves or muscles should necessarily lend themselves directly to quantitative recording by mechanical or electronic devices. Most of the available units were designed for measurement of physical or electronic events not comparable to those found in biological systems. We have found it necessary sometimes to construct special units to meet the requirements for certain applications found in biology. As will be described in Chapter VII, our analyses have shown that the event to be recorded—the passage of the action potential down the membrane of the muscle—is quite rapid and should not necessarily be subjected to the wide bandpass of the ordinary preamplifier. The wide bandpass increases the volume of active muscle included in the recording. These large portions of muscle surrounding the recording electrode combine their components, producing averaged potentials of lower frequency. We believe, therefore, that we must consider the use of a frequency band in which the lower frequencies are restricted in certain applications or confuse the record with distant large volumes of muscles whose action potentials combine to produce a nonspecific large waveform (Chap. II).

Though the "rate" of firing of the single motor unit of the typical skeletal muscle may not exceed 40 or 50 per second,[21] intervals between each action potential—longer in duration than the action potential—are found. They represent the refractory period that follows each motor unit action potential. These intervals are series of only approximately equal periods.

The "frequency" of the action potential depends upon individual waveform duration. This is very short. The action potential of the single motor unit has a duration of a little over 2 msec and therefore a high frequency. It is the frequency of the action potential rather than the slower "rate" of action potentials which clearly indicates the requirements of relatively high frequency recording equipment if one wishes to make a true record of the electrical activity produced by the firing of single units or groups of units. The requirements of the recording system concerning frequency range are considered in Chapters II and VII.

As will be discussed in Chapter VII, single motor axon stimulation makes possible recordings restricted to the single motor unit excited. The characteristics of the resulting waveform will dictate the preamplifier specifications best suited to record from the single motor unit.

Large numbers of motor units may influence the recording by firing synchronously, combining their action potentials to produce large waveforms. This is particularly true when dual internal or surface electrodes are used. A broader frequency bandpass is indicated under these circumstances (Fig. 8).

The primary use of recordings of the electrical activity of muscles with which we are concerned herein is the study of muscle function. Recordings of the series of action potentials produced during muscular contraction make it possible to observe the precise time of onset and the termination of muscular activity, as well as the extent to which the muscle participated. These recordings, made synchronously with cinematic recordings of the moving body parts, serve well in the analysis of the muscles of the lower extremity during locomotion or of the muscles of the hand and forearm during voluntary activity.

In order to better interpret these synchronized records of muscle electrical activity and body motion, additional knowledge of the nature of the articulations concerned is necessary. For example, the axis of the lower extremity extends from the center of the femoral head to the point of weight bearing or from the center of the femoral head to the center of gravity of the lower extremity during non–weight-bearing conditions. The changing position of this axis explains the variable actions of the iliopsoas muscle and the adductor muscles (Figs. 37, 38).

An analysis of the axis of the subtalar joint, which determines the orientation of the rotation of the calcaneus about the talus, was made. The position of this axis in the foot determines the functions of muscles whose tendons pass in relation to it. This applies equally to muscles whose tendons are transferred in such a way as to pass the subtalar joint. Using metallic pin targets, the motion of the subtalar joint during the walking cycle was calculated in eight subjects. The total range of motion at this joint was also recorded in each case (Figs. 33, 48, 59; Tables I, II).

Most of the work previously reported concerned the more or less involuntary muscles of the lower extremity. A newly developed method was used for recording in a single process the electrical activity of as many as six muscles of the forearm and hand with a synchronized cinematic recording of the body part. This method depended upon a dual-lens, rotary prism-shutter camera, which simultaneously recorded the traces from the oscilloscope screen and the motion of the forearm and hand (Chap. VI). Extensive studies were thus made of the muscles of this region in normal subjects. A series of studies of muscles before and after tendon transfer in the upper extremity was carried out. Although the rhythmic (phasic) character of the behavior of the muscles of the lower extremity during walking (Chap. V) was not seen, certain patterns of activity were often observed. Differences between the virtually

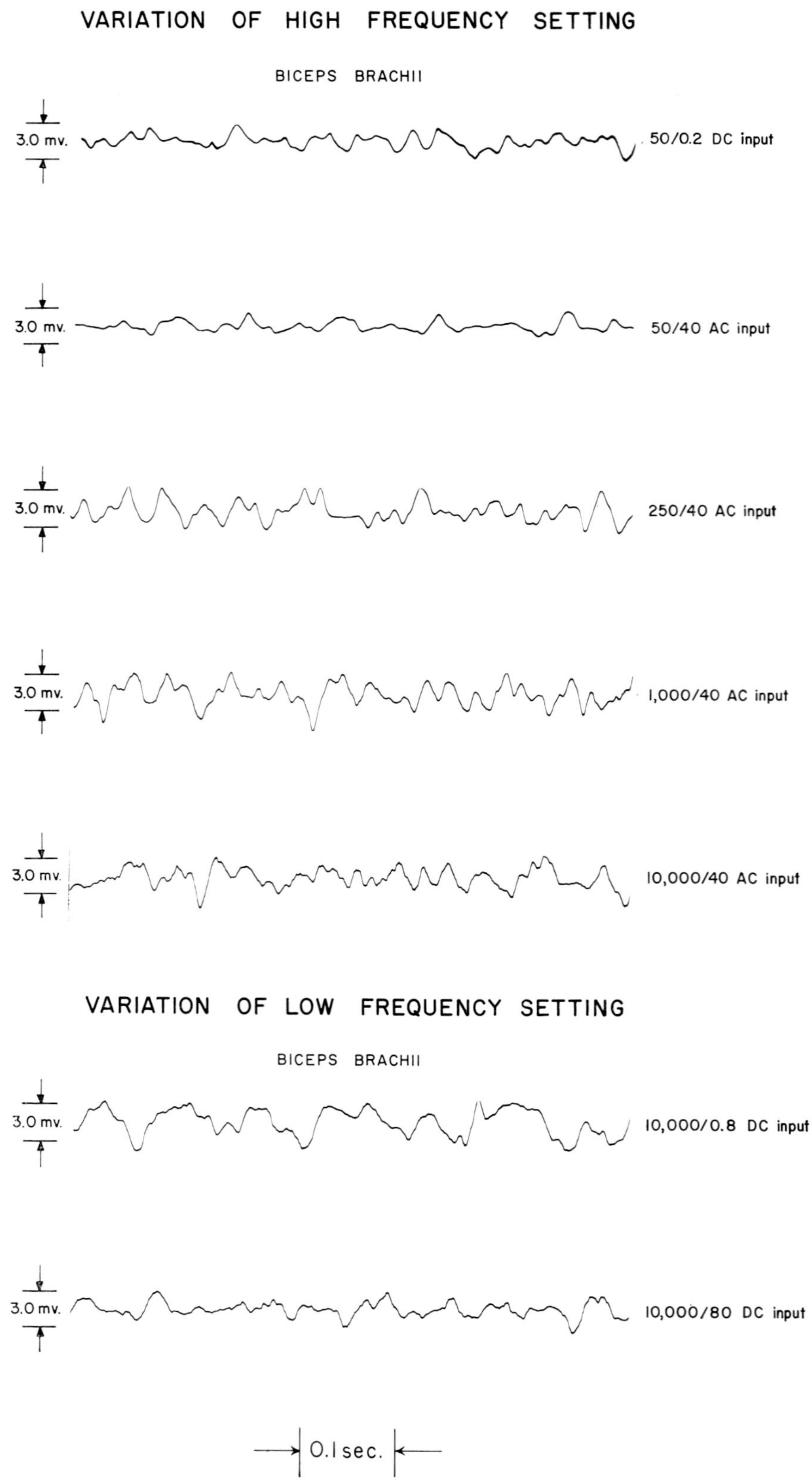

Figure 8. Muscle action potentials recorded at various frequency ranges.

totally voluntary, less consistent activity of the muscles of the upper extremity and the more involuntary phasic activity of the muscles of the lower extremity are shown in the illustrations in Chapters V and VI.

We are thus involved in studies of the means by which motion takes place: the segments of the body, the nature of their articulations, and their centers of gravity, centers of rotation, and axes of rotation. With a knowledge of the geometry of this framework, we are better equipped to assign the functions of various muscles that produce motion of the body and its parts and make it possible to resist external forces such as gravity. Thus the study of the living and moving form is a dynamic and interesting approach to the study of anatomy, in contrast to the study of motionless parts—gross anatomy—to which our observations were more confined in the past.

However, having acquired some knowledge of body parts, their articulations and their motors, we are equipped to study the essential communications system or the many composite signals by which and through which the many muscles, combinations of muscles, and parts of muscles are directed. This includes a feedback (sensory) proprioceptive system which keeps the appropriate centers informed of the position and state of all parts and muscles in their various states of contraction.

This study has been directed toward the motor system, and this particular study may properly be called a neuromuscular approach, for we are studying the peripheral parts: the body segments, their articulations, their muscles, and that part of the nervous system called the "final common path" through which the ultimate signals for neuromuscular activity are ultimately delivered. Of particular interest to us at this time is the nature and importance of the basic means for propagating impulses or messages—the action potential.

Cardiac and skeletal musculature in many ways are very much alike. Both produce their forces through an actin-myosin system arranged longitudinally in fibres. Both depend upon conduction of an impulse, an electrochemical event, to produce action potentials which will activate them. But there is a great difference in the terminals of their control systems. Human skeletal muscle has a myoneural junction for virtually every muscle fibre. Hunt and Kuffler demonstrated multiple innervation of muscle fibres in the cat and frog.[96] Thus the strength of the contraction can be finely regulated and directed. The muscle fibres, grouped as the terminals of motor units, may be fired in various degrees of both frequency (rate) and number; in contrast, cardiac muscle, innervated by a syncytium, must fire all of its fibres in synchrony to be effective as a pump. Thus the impulses that precede each systole are known to have a single and simultaneous point of origin, and cardiac fibres are capable of spreading the impulse from one fibre to another.

There have been many vigorous and prolonged studies of the membranes of both nerve and muscle fibre. These fine membranes, each possessing differential permeability to intracellular and extracellular fluids, are capable of rapid changes of their delicate state, permitting action potentials to pass under some conditions and not others. The action potential depends upon changes in this delicate, differential membrane permeability. During periods of inactivity, the outer surface of the membrane is polarized to a precise degree, while the potential inside the cell is approximately −90 millivolts. The state of equilibrium can be disturbed by impulses which are said to depolarize the structure as they pass along it at velocities in the order of several meters per second. Thus the concentration of sodium and potassium ions on either side of the membrane becomes of the greatest importance, as does the size of each ion, which is also a factor in differential membrane permeability. Many studies have been done in order to gain insight into the nature of the action potential: the nature of the membrane; the concentration of sodium, potassium, chloride and calcium ions; and the velocity of the conduction of the action potential which begins when all of the necessary conditions are arranged. However, the great body of information thus gained has been acquired from studies on lower forms where the nature of the thin membranes of nerve and muscle fibres differs from that of man, and the nerve fibres are far less likely to be myelinated. Thus it would be of great interest to gain similar information from higher animals, e.g. primates more comparable to man. Many studies of neuromuscular activity such as muscle tension and the action potential of the single motor unit can be undertaken, using such laboratory animals as the cat.[129,130] It is our opinion that such studies of higher animals would be far more likely to be translatable to similar concepts in the case of man.

In the studies reported herein, electrical silence has been considered coincident with absence of muscular activity. In the resting muscle, complete absence of electrical activity is uniformly the case. However, records from intracellular recording electrodes, taken during periods of inactivity[107] have demonstrated "miniature end plate potentials," representing spontaneous activity at the junctional regions.

Chapter II

MUSCLE—SOME BASIC CONSIDERATIONS

Series of recorded action potentials are expressions of neuromuscular activity in terms of peripheral motor behavior. Single action potentials are usually the subject of study of clinical electromyography. However, where the minute characteristics of the waveform of the single motor unit action potential are not the subject of study—where one is interested in the onset, termination and extent of muscular activity—series of muscle action potentials are of great value in the study of functional anatomy. To appreciate these electrical recordings from muscle, one should be aware of basic neuroanatomy, the organization of muscle and the composition of its motor units. The aggregate action potentials of the fibres of the motor unit appear virtually in unison and make possible the recorded action potential commonly seen in recordings from muscle.[24]

When one studies muscular behavior such as the pattern of phasic activity, he is actually examining the behavioral patterns of the central nervous system through observation of peripheral neuromuscular activity. The groups of muscle fibres of the motor unit represent the motor terminals of the central nervous system. Their aggregate behavior is the final expression of complex motor pathways.

The main portion of the action potential of the single motor unit consists of a single spike followed by a minor secondary oscillation of opposite sign.[42] A rough representation of the form of the single motor unit action potential is observed when one records from a very small, restricted volume of muscle, using a single recording probe with the other recording electrode placed in an isoelectric region (Fig. 4A). When two electrodes are inserted into muscle and the signal recorded by the differential mode, as in the use of dual internal-wire electrodes in the study of functional anatomy, waveforms of motor units tend to combine, producing irregular oscillations.[26,27] Thus normal series of action potentials recorded by the differential technique are somewhat diphasic and irregular, their shapes depending upon the recording method employed.

Motor unit action potentials are made possible by the accumulated transmembrane potential changes occurring along the many muscle fibres in a region near the tips of the recording electrodes.[69,149,164] This change of electrical potential coincides very closely with the process of innervation. Where the single motor unit action potential can be observed, under normal conditions it is found that tension begins just before the termination of the primary spike (Fig. 176), which has a duration of slightly over 2 msec.[42] The above time durations are found when the muscle is at resting length or under slight passive stretch. The onset of tension is influenced considerably by change in the initial passive stretch (see Chap. VII). These observations are made only under experimental conditions. The clinical or percutaneous technique for the recording of muscle action potentials gives no clear identification of the structure from which the recording is made. Thus the tip of the recording electrode may be influenced not only by one motor unit but by adjacent motor units or subunits of other interdigitating motor units. An indefinite number of these structures influence the recording due to the effect of tissue as a volume conductor. A thorough study of the effect of the volume conductor is found in the work of de Nó.[122]

Each grossly diphasic waveform of continuous recordings is ordinarily thought to represent the firing of one or more motor units. Since virtual synchrony can frequently be demonstrated using pairs of internal electrodes in striated muscle (Figs. 1, 2), it is evident that many of the larger spikes which have a fast rise time represent the simultaneous contraction of a number of motor units. Since we are unable to visualize histologically the position of electrode tips, it is difficult to identify the origin of action potentials. However, when two tiny coaxial electrodes are independently recorded from, it is our observation that almost identical pulses are frequently seen. The duration between the pulses from each electrode is the same when the electrodes by chance are in a position to record from a predominating motor unit. This indicates that the motor unit of most skeletal muscle has a refractory period and a maximum rate of firing of approximately 40 to 50/sec.[142] This supports the contention that the motor unit behaves according to the all-or-none law. There is, however, evidence that leads us to question the all-or-none law (Ralston; Chap. VII, Fig. 160).

The multiple innervation of skeletal muscle by many axons is in contrast to that of the syncytial innervation of cardiac muscle. Normally the entire ventricular mass of cardiac muscle behaves as a unit according to the all-or-none law. In this type of muscle, excitation spreads from fibre to fibre. When two external electrodes are used, a diphasic type of oscillation is recorded. It is well known that the character of this diphasic oscillation varies with each position at which the electrodes are placed. Using standard positions of electrode placement, deviations from the expected wave form have been found to indicate certain types of pathology. This system of analysis of the electrocardiogram has been established to a great extent by empiricism.

The motor unit has been defined as the group of muscle fibres innervated by a single axon (and the anterior horn cell of that axon). As such, each motor unit represents a single anterior horn cell. During contraction, the resulting diphasic, somewhat complex, wave varies in the normal, presumably because of slight asynchrony of fibres within the motor unit (Fig. 53). Recruitment can thus be interpreted as the excitation by the central nervous system of additional motor units to take part in the activity. Actually, the innervation of the motor unit has been found to be less simple. Overlapping of motor units by neighboring axons has been reported.[1,22] Each motor unit in the human anterior tibial muscle is thought to contain approximately 600 fibres and to have a diameter of about 1.5 mm.[66] The axon innervating the motor unit, however, may in some instances give off branches to a few fibres of adjacent motor units. Identical action potentials have been obtained from a single fibre by stimulating separate axons. Thus the innervation of the motor unit is not always necessarily confined entirely to one axon. The division of the motor unit into subunits[1] which interdigitate with those of other motor units explains the difficulty in obtaining records restricted to a single motor unit. In pathological conditions where the density of surviving motor units is reduced, the opportunity to record from the solitary motor unit is greater.

As the axon penetrates the motor unit, it branches to supply many fibres. It has been shown in certain animals, such as the hamster,[82] that there are more muscle fibres than motor end plates. In some of the skeletal muscles of the hamster, there are three times as many muscle fibres as axons.

The fibrillary potential is of much shorter duration[75] and usually of lower amplitude than the action potential of the normal motor unit.[76] Amplitude depends to some extent upon the proximity of the recording electrode tip to the fibres or fibre concerned. However, the duration of both the fibrillary potential and the motor unit depends to a considerable degree upon the type of electrodes and recording media used. The fibrillary potential, representing the denervated but living muscle fibre or possibly a small group of such fibres,[1] can usually be recorded 14 to 21 days after interruption of the peripheral nerve.[168] After approximately one year, the denervated fibre no longer produces the fibrillary potential.[2] It has a characteristic appearance on the recording media as well as a high-frequency clicking sound in the loudspeaker. It is a very useful diagnostic sign.[126,172] The disadvantage of this sign as a diagnostic aid is that one may have to explore systematically a large portion of a muscle to rule out the presence of fibrillary potentials. Very fine probes are used for this type of examination. The differential technique, using a single monopolar electrode (Fig. 4A), has a number of advantages. With one electrode in an isoelectric region, it is possible to record the absolute value of the action potential—taking advantage of the differential mode with its ability to cancel artifact (noise).

The double coaxial electrode may lead to confusion in interpretation of records. It consists of two tiny filaments inside a No. 23 gauge needle. These filaments, 75μ in diameter and separated by less than 200μ, are led to a differential preamplifier. High-frequency spikes are expected, even from normal muscle. These spikes are easily mistaken for fibrillary potentials. In Figures 9, 27 and 160, records from the double coaxial electrode are compared to records made simultaneously from the same muscle, using the single coaxial electrode (Fig. 7). Disadvantages of the shielded single coaxial electrode are discussed in another part of this chapter.

The fibrillary potential is very valuable in distinguishing the primary myopathies such as muscular dystrophy from secondary myopathies such as poliomyelitis or lesions of peripheral nerve. In the primary myopathies, no fibrillary potentials are ordinarily found and the action potentials are of low amplitude but high rate. In progressive diseases such as amyotrophic lateral sclerosis or the acute stage of poliomyelitis, one would expect to find fibrillary potentials. They indicate involuntary contractions of single or small groups of temporarily or permanently denervated muscle fibres. Their continuation indicates the presence of a progressive process of denervation. If denervation persists, atrophy of the fibre follows, and it may ultimately cease to produce electrical activity. Reinnervation may result in the "nascent"[94] motor unit action potential or the large action potential if axon branching occurs. The nascent motor unit action potential thus represents a group of fibres that have been reinnervated to form a functional, reorganized motor unit. Axon branching, which may take place in the recovery stage of poliomyelitis or in lateral column disease of diabetes mellitus, represents the rescue of denervated fibres by remaining axons which send out new branches to them.

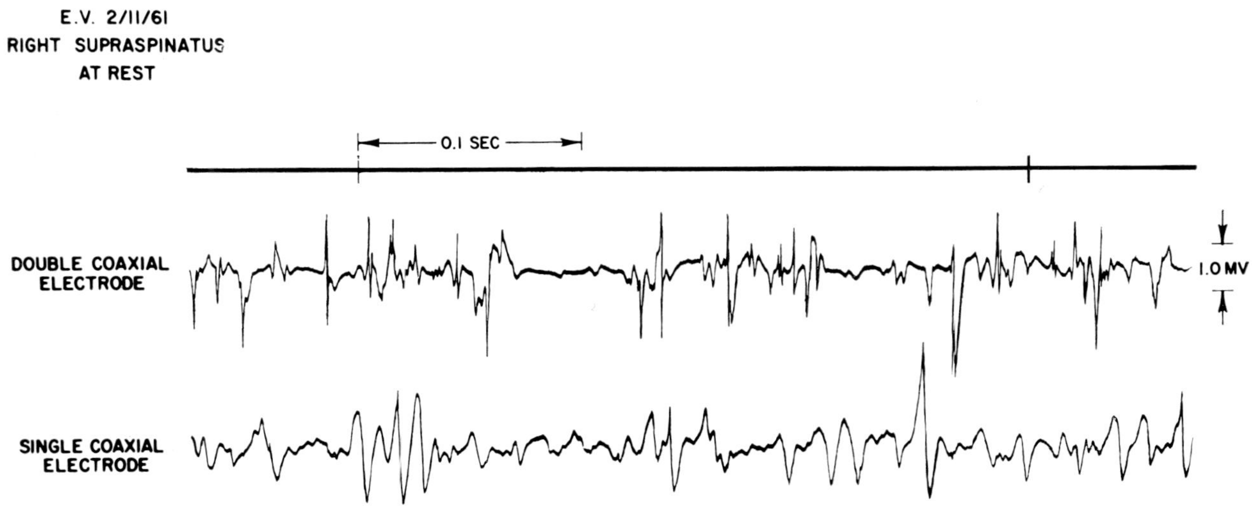

Figure 9. Records from Visicorder made from the supraspinatus muscle at rest. The difference between the records from the double and single coaxial electrodes is shown.

Since this axon continues to supply its own original fibres and has acquired new fibres, a large motor unit action potential may result.

Fasciculation is quite different from fibrillation. Fasciculation represents involuntary activity of entire motor units in contrast to involuntary activity of single fibres or small groups of fibres as found in fibrillation. Fasciculation is often found in amyotrophic lateral sclerosis.

There are a variety of useful electrodes or applicators. Depending usually upon the type of electrode or probe used, different types of preamplifiers are employed (Fig. 4).

Surface electrodes may be easily applied, using electrode jelly. This type of applicator is excellent with large superficial muscles or groups of muscles. But surface electrodes lack the specificity of internal electrodes such as the internal coaxial, the monopolar or the internal-wire electrodes as described below.

With the coaxial or the monopolar electrode, one can make recordings from a few or sometimes even a single motor unit. These are the applicators most often used to study the pathology of the motor unit and to ascertain the presence or absence of fibrillary potentials. We have found the Buchthal type of coaxial electrode (Fig. 7) very useful when the small internal recording probe (diameter 0.1 mm) is considered the live electrode and the surrounding shield is grounded. These two connections can then be led to a very small battery-operated preamplifier.* This makes it possible to conduct an amplified signal ($\times$ 100) from the source of the action potential to the recording oscilloscope. Thus the relatively weak signal from the coaxial unipolar electrode is much increased and amplified at the ideal place—its source—before it is conducted to the oscilloscope preamplifier. This procedure usually makes it possible to eliminate the shielded room for the recording of small potentials. The resulting larger signal conducted by long cables to the oscilloscope preamplifier is influenced far less by the ever-present outside electrical interference than the original much weaker signal would be.

When employing the No. 502 dual-beam Tektronix oscilloscope, we have eliminated the instability of the base line of the electromyographic trace by replacing the 0.1 microfarad alternating current input coupling capacitors. They are replaced with 0.004μf capacitors. The low-frequency response limit is thus raised from two to approximately 40 cycles per second (Fig. 28) when the alternating current coupled input is used.

Prior to making this change in the preamplifier alternating current input coupling capacitors, an electromyographic signal was subjected to the following analysis. A signal was led to the standard alternating current coupled input (Fig. 10, upper trace). The signal was also led to the second alternating current coupled channel after passing through 0.004μf capacitors (Fig. 10, lower trace). As the two traces of Figure 10 demonstrate, the addition of the 0.004μf capacitors in series results in no serious distortion of the signal or loss of amplitude.

For variation of the upper-frequency response limit, the No. 532 Tektronix oscilloscope was used. For multichannel variable frequency range applications, Type FM122 Tektronix low-level preamplifiers were used. The signal of each of the preamplifiers was simultaneously displayed on the No. 502 dual-beam Tektronix oscilloscopes.

Dual internal electrodes (Fig. 6) have many advantages. They transmit a large signal that does not require excessively high amplification. This results in minimal distortion and interference. Differential amplifiers are used. The electrodes are specific and permit stimulation of the muscle for purposes of identification. This is very important in orthopaedic functional studies where the observer wishes to record individual muscle action. Such electrodes are painless and remain in place for long periods. Although they do not sample the entire muscle, they sample a large volume of muscle—enough to represent the muscle's activity. These electrodes are made from flexible, insulated, fine copper wires of diameter 0.003 in. (75μ) and

* Tektronix No. 123, preamplifier.

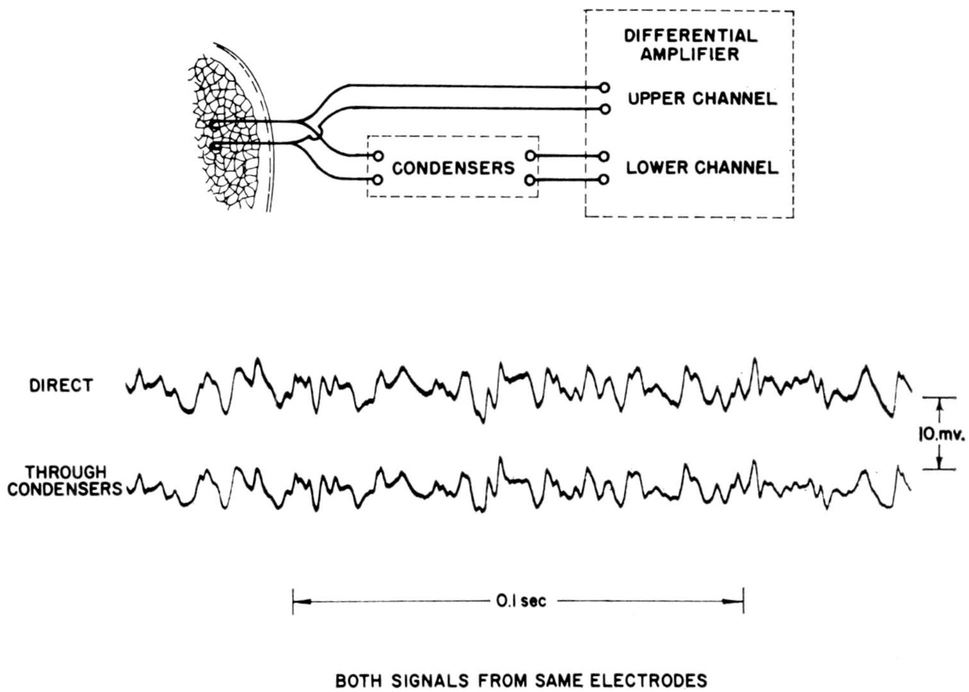

Figure 10. The signal is hardly discernibly altered by the addition, in series, of $0.004 \mu f$ capacitors to the inputs of the differential preamplifier. The recordings were made on the Visicorder.

are inserted with No. 24, 25 or 26 hypodermic needles. With the electrodes in place (Fig. 6) the subject may be observed during walking or other activities. The signal produced is ideal for recording on the sound track of the sound motion picture camera (Figs. 11 and 12). This results in a synchronized display of muscular activity and the accompanying audio representation of the resulting electromyogram.

The magnitude of the signal obtained varies with several conditions:

1. Types of electrodes (Fig. 4).
2. Size of exposed tips of electrodes.
3. Distance between and orientation of electrodes (if two are used as in differential method).
4. Mode of amplification used.
   a. Differential (Figs. 11 and 12).
   b. Single-ended (Figs. 11 and 12).
5. Resistance between electrodes.
6. Input impedence of preamplifier; this must be high, several times the resistance between the electrodes.
7. Choice of muscle—upper extremity, lower extremity, etc.
8. Pathological state of the muscle.

For example, in primary myopathy the signal is small, while in anterior horn cell disease, such as poliomyelitis, after recovery the signal may be of greater amplitude, longer duration and lower rate.

The action potentials as demonstrated reflect the activity of the motor units in the area surrounding the two electrodes. Although individual motor unit discharges can often be seen, muscles during vigorous contraction produce electrical activity consisting of superimposition of many motor unit action potentials one upon the other. The electrical content of the area examined, however, is represented by the test, until the contraction becomes so great that simultaneous signals of opposite sign tend to cancel out one another.

The single coaxial electrode and the monopolar electrode are highly selective and may be used to display action potentials from small

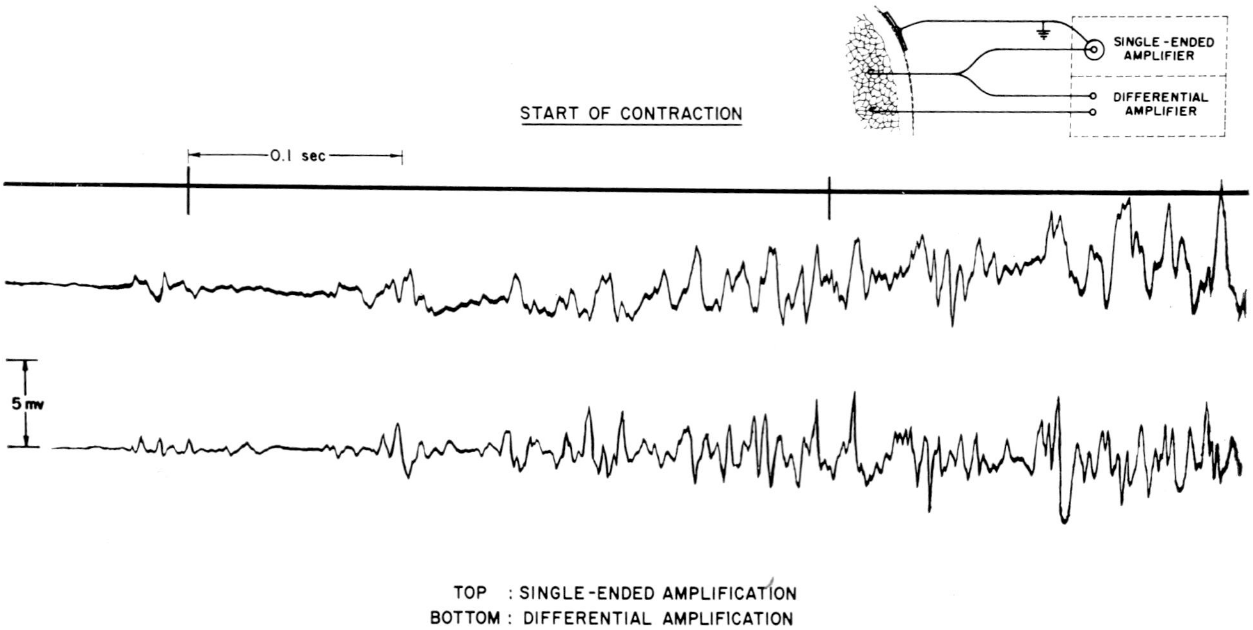

Figure 11. Comparison of single-ended and differential methods of recording.

portions of the muscle. With patience, one can at times apparently demonstrate the activity of the single motor unit. Interdigitation of motor units (see Chapter VII) with adjacent motor units obviously makes exclusive single motor unit recording exceedingly difficult, even when minute probes are used. Some consider that upon inserting the coaxial electrode the "insertion potential" is helpful in some diagnostic considerations.

Another type of small electrode is the double coaxial electrode. This provides two leads from two tiny probes very close together at the needle tip (Fig. 160), and differential ampli-

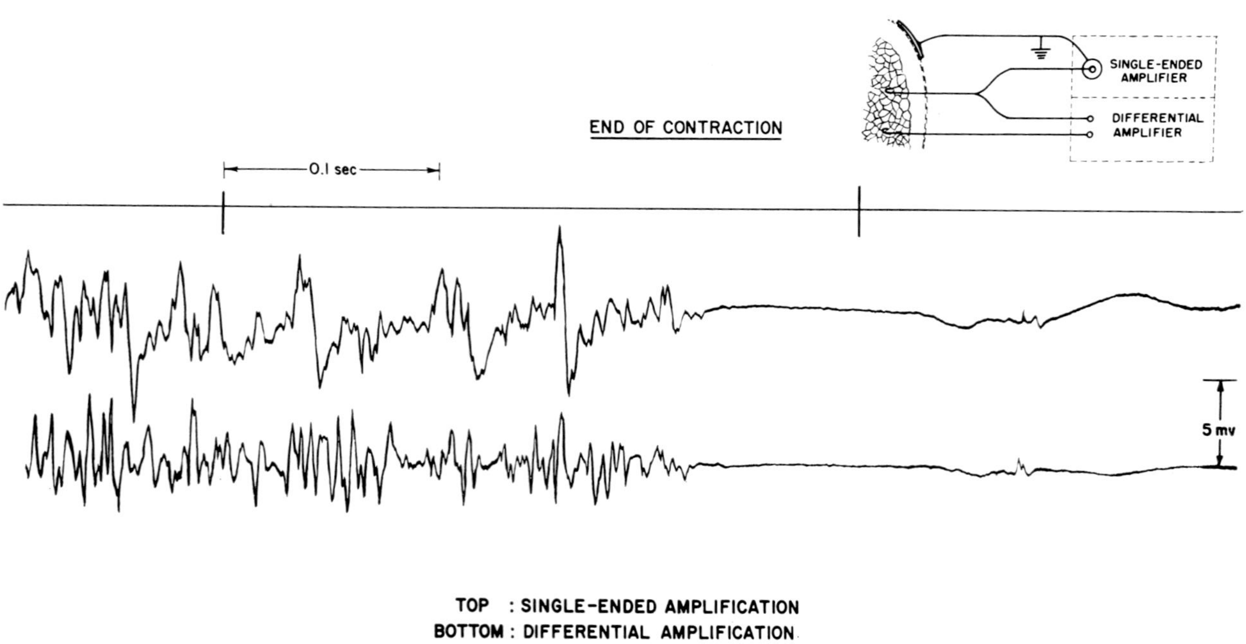

Figure 12. The records are made simultaneously.

fication is used. Since this results in a comparison of potential changes in some regions from tissue very close to each electrode tip, the resulting spikes are of very high frequency. This resulting high frequency may cause confusion. Some normally fast rise time potentials (as recorded by such a method) are often indistinguishable from the fibrillary potential (Fig. 9).

Dual internal wire electrodes simultaneously sampling several motor units usually fail to record the shape of individual motor unit potentials, since these, to a great extent, are superimposed as groups (Fig. 13). For purposes of locomotion studies and muscle-action analyses, these flexible wire electrodes are usually placed 1 cm apart (Fig. 6). With these electrodes in place, one may record from deep and otherwise totally inaccessible muscles such as the psoas major or posterior tibial muscles. Furthermore, using these electrodes, one can easily stimulate the muscle electrically, thus identifying beyond any question the anatomical position of the electrodes. If one is not primarily concerned with the precise nature of the exact wave form but rather with timing and general assessment of the state of muscles or groups of muscles, then the internal dual copper wire electrodes are ideal. They employ differential amplification, and their leads are conducted to the oscilloscope amplifier via shielded cable. The differential amplifier consists of two matched and balanced amplifiers. Each electrode is connected to a single amplifier which is referred to ground. The difference between the outputs of the two identical amplifiers is the resulting signal. This method thus accurately represents the difference in potential between the two points at which the electrode tips are positioned.

An alternative is the single-ended input employing the single probe and ground plate over the muscle or ground consisting of a second electrode in an inactive portion of the muscle (Figs. 11 and 12). This method produces a continuous record of the potential fluctuation at the site of the recording electrode. In reality, for rapidly contracting muscle, it does not have any particular advantage over the easily applied differential method. Since in either case a large number of fibres are recorded from, the potential changes during a strong contraction

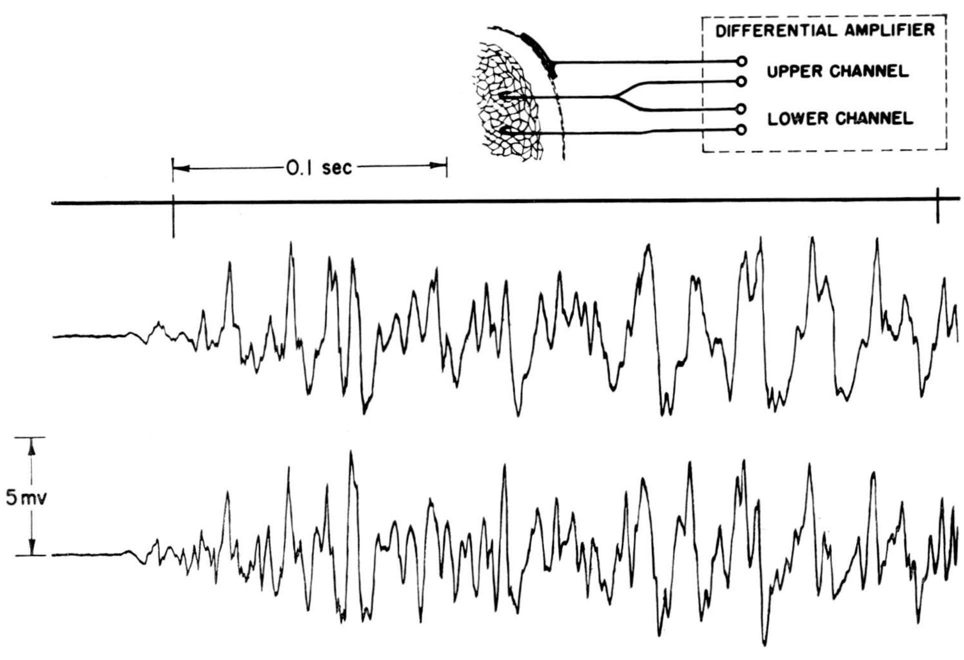

Figure 13. Differential amplification. Recorded at 10 in./sec (Visicorder). *Upper channel,* single internal electrode and surface electrode. *Lower channel,* dual internal electrodes.

actually represent averages. Both methods then fail to record individual identifiable motor unit action potentials. When the single-ended electrode is used with a ground plate over the muscle, an excellent record is obtained during relative inactivity when the antagonist contracts. However, if the ground plate is placed over another muscle (such as the muscle's antagonist), contraction of the antagonist may produce considerable electrical activity (Fig. 83). This represents the undesirable activity between the recording electrode and the ground plate—the intervening tissue. Where the ground plate can be confined to the muscle tested, it appears that a satisfactory recording results—a reproduction of the potential change at one area. It is possible to use differential amplification (Fig. 13) or the single-ended mode (Figs. 11 and 12).

In our studies we have used electrical stimulation in order to identify the position of the recording electrodes. Muscle or nerve stimulation might also be used to assess the state of peripheral motor nerve damage; for example, that immediately following axillary nerve injury. Here stimulation could be quite misleading, since for a variable period direct muscle stimulation would be possible, although considerable injury to the nerve may have taken place. Often surgical decision must be made more promptly. For example, in the case of the radial nerve, fractures of the mid-shaft of the humerus often make it quite difficult to decide from clinical examination whether or not the nerve has been completely interrupted. If the nerve has been completely interrupted, there will be electrical silence in the muscle with fibrillary potentials appearing in two or three weeks. If, upon voluntary contraction of the muscle whose nerve has been injured, action potentials can be obtained, one may conclude that the nerve has at least not been completely interrupted. At this crucial time, the electromyographic test is of great value. Without it, one must make a decision based on purely clinical examination or wait for a considerable period to see if recovery takes place. At that time it may be too late for surgical treatment of the nerve.

One might summarize the methods discussed according to types of electrodes or applicators and mode of amplifier input. The muscle action potential is readily recorded by either differential or single-ended amplifiers (Figs. 11 and 12). As stated, the alternating current coupled method is more suitable than the direct current coupled method. One can use the following arrangements:

Type of Applicator	*Mode of Amplification Input**
1. Dual internal electrodes. a. Two active electrodes b. One active, one inactive electrode.	Differential input (Fig. 13).
2. Dual internal electrodes with one electrode grounded.	Single-ended input.
3. Single internal electrode with plate applied to skin (plate grounded).	Single-ended input (Figs 11 and 12).
4. Single internal wire electrode and ground (internal or skin with intervening preamplifier).	Single-ended input (Figs. 11 and 12).
5. Coaxial electrode.	Single-ended input (Figs 4 and 9).
6. Coaxial electrode with preamplifier at source.	
7. Double coaxial electrode.	Differential input (Fig. 9).
8. Surface electrodes.	Differential input (Fig. 4).
9. One internal, one surface electrode.	Differential input (Fig. 13).
10. One internal, one surface (indifferent) electrode for electrodiagnosis.	Differential input (Fig. 4A).

Differential amplification using dual internal electrodes has yet to be improved upon for studies of locomotion or for muscle function analyses about joints. In our work, we have systematically explored all of the above combinations. When one uses the single internal electrode with the ground plate earthed with the rest of the system, the resulting amplitude compares favorably in frequency and amplitude to that of the dual-wire differential application (Figs. 11 and 12). Counts of each have been recorded and compared. In the differential application, one compares the potential existing in the region of one electrode to that existing at the location of the other electrode. Hence the farther apart the electrodes are placed, the more identifiable and significant the resulting information. A customary laboratory

* See Figs. 4 and 4A.

method is to place one of the differential electrodes in an inactive area and the other into a relatively active area (Chap. VII). The inactive electrode may be placed in tendon. The difference in potential between the active electrode referred to ground and the inactive electrode referred to ground is the resulting wave form displayed on the oscilloscope. This is true of all differential methods. The method of differential recording is excellent for testing single isolated muscles as described in Chapter VII, since it confines the recording to the volume of muscle surrounding the two electrodes.

In Figure 13, tracings from two differential inputs—one consisting of one live electrode and ground, the other consisting of two internal electrodes (both led to the oscilloscope preamplifier in differential fashion), are shown.

There is danger in "integrating artefact." This hazard is anticipated when one employs low-frequency equipment which fails to reproduce a true representation of the action potential. It is recommended that investigators employ apparatus of sufficient quality to record the motor unit action potential to the exclusion of artefact. Such artefact is usually of low frequency. A frequency bandpass of approximately 40 to 10,000 Hz is ideal for most applications (Figs. 28, 161).

As recording instruments, one may use the oscilloscope or, for multichannel recording, an assembly of high-frequency galvanometers.[86] It is demonstrated (Fig. 14) that recordings from each of these two instruments, oscilloscope and multichannel Visicorder made at high speeds when accurately calibrated, are quite comparable, if not identical. The Visicorder* is a multichannel recorder which employs galvanometers. Miniature galvanometers with recording ranges of 0 to 8,000 Hz (flat to 5,000 Hz) (Fig. 15) are available. This instrument requires a multichannel galvanometer amplifier which must be placed between the oscilloscope vertical signal output and the recorder (Fig. 16). Preamplifiers may be employed in place of the oscilloscope. A four-channel study made simultaneously on four selected muscles of the shoulder is shown in Figures 17 and 18.

It can be shown that by sharply cutting down the upper frequency limits of the input of a good oscilloscope preamplifier with variable frequency settings, one distorts, misrepresents and reduces the amplitude of the orginal signal (Figs. 8, 20, 21). The resulting "integrated" curves made from such records have little meaning. One must also take into consideration the time constant of common "integrators" in use. Since this time constant is often as long as one tenth of a second, it may mislead us in judging the beginning and end of the muscular performance. There is an artificial time shift in the pattern of activity—an asynchronous recording. The so-called R-C integrator in common use might better be termed an "averager" (see Chap. IV). The averager (RC integrator) distorts the input signal (Fig. 84). As an alternative, we have employed the "absolute-value

* Minneapolis-Honeywell.

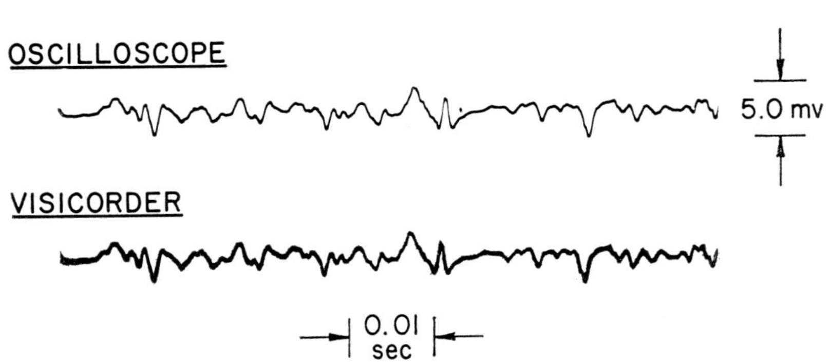

Figure 14. The same signal is simultaneously displayed on the oscilloscope and Visicorder. Recorded at 50 in./sec.

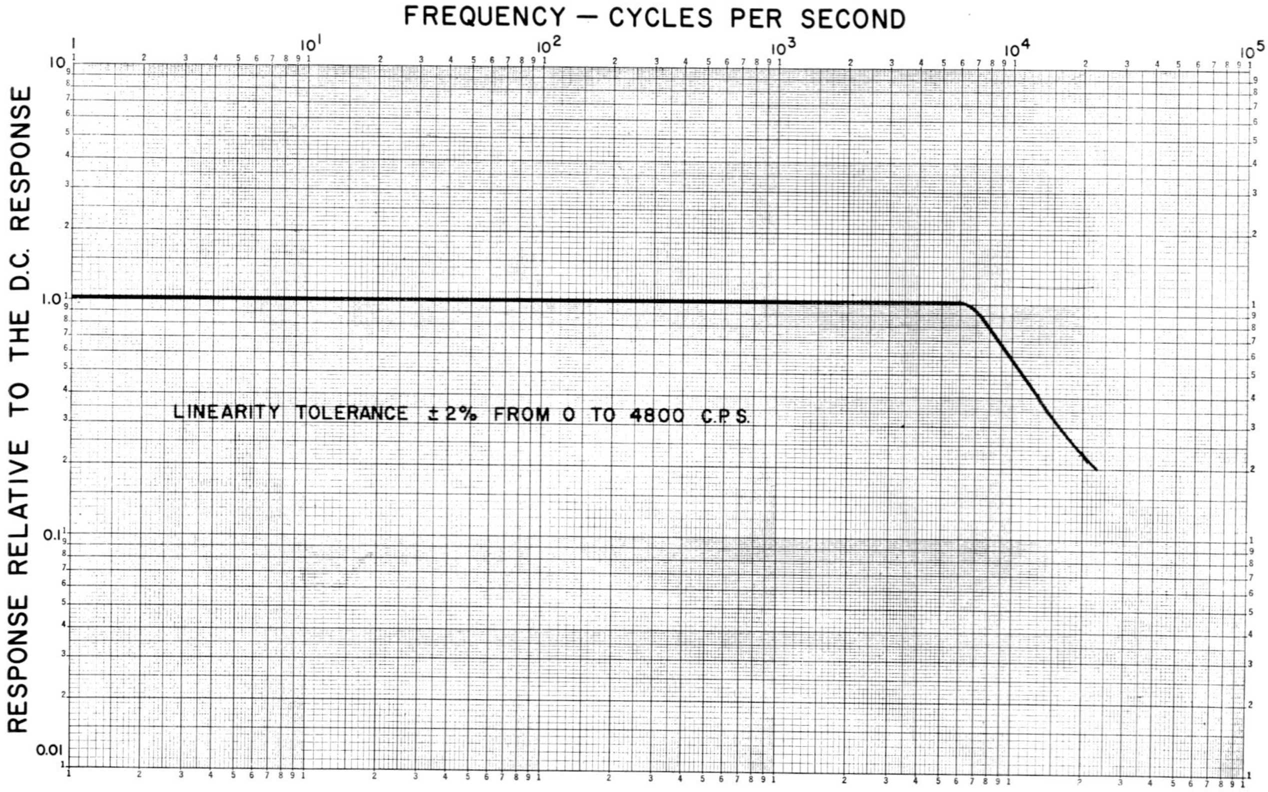

Figure 15. Frequency response of Heiland M8000 galvanometer used in multichannel recordings.

integrator" which rectifies all negative forms and gives a total value expressed in microcoulombs for each contraction. With this instrument, we are able to ascertain the total amount of electricity produced in the region of the electrodes during contraction. We have found considerable use for this instrument, a function not provided by any other known instrument, as described in Chapter IV.

While one does not depend upon the electromyogram to make the diagnosis in itself, it is often a valuable adjunct to a neurological study. Sensitive equipment readily records the fibrillary potential, thus enabling one to appreciate peripheral nerve injury, such as damage to the radial nerve in fractures of the shaft of the humerus. The decision to explore the radial nerve may depend upon the findings gained from the electromyographic test. Here, in searching the involved muscle for fibrillary potentials, one must remember that it may be 14 to 21 days before these potentials appear in denervated muscle and that the ultimate fate of fibrillating fibres will be atrophy with loss of action potential of any kind, or ultimate reinnervation, usually with a somewhat abnormal result. The resulting motor unit action potential is due to regeneration of the axon or reinnervation of the fibres by branching of uninjured axons, producing the large motor unit action potential.

In distinguishing primary myopathy from myopathy secondary to nervous system pathology, it is apparent that primary myopathy does not produce fibrillary potentials so characteristic of the muscle with a lower motor neuron lesion at certain stages. Since our clinical and pathological descriptions of some muscular and neuromuscular entities overlap, their boundaries being somewhat vague, the value of the electromyogram may be appreciated when it is pathognomonic, as in myotonia congenita. Here the electromyogram produces a characteristic sound in the loudspeaker and may be

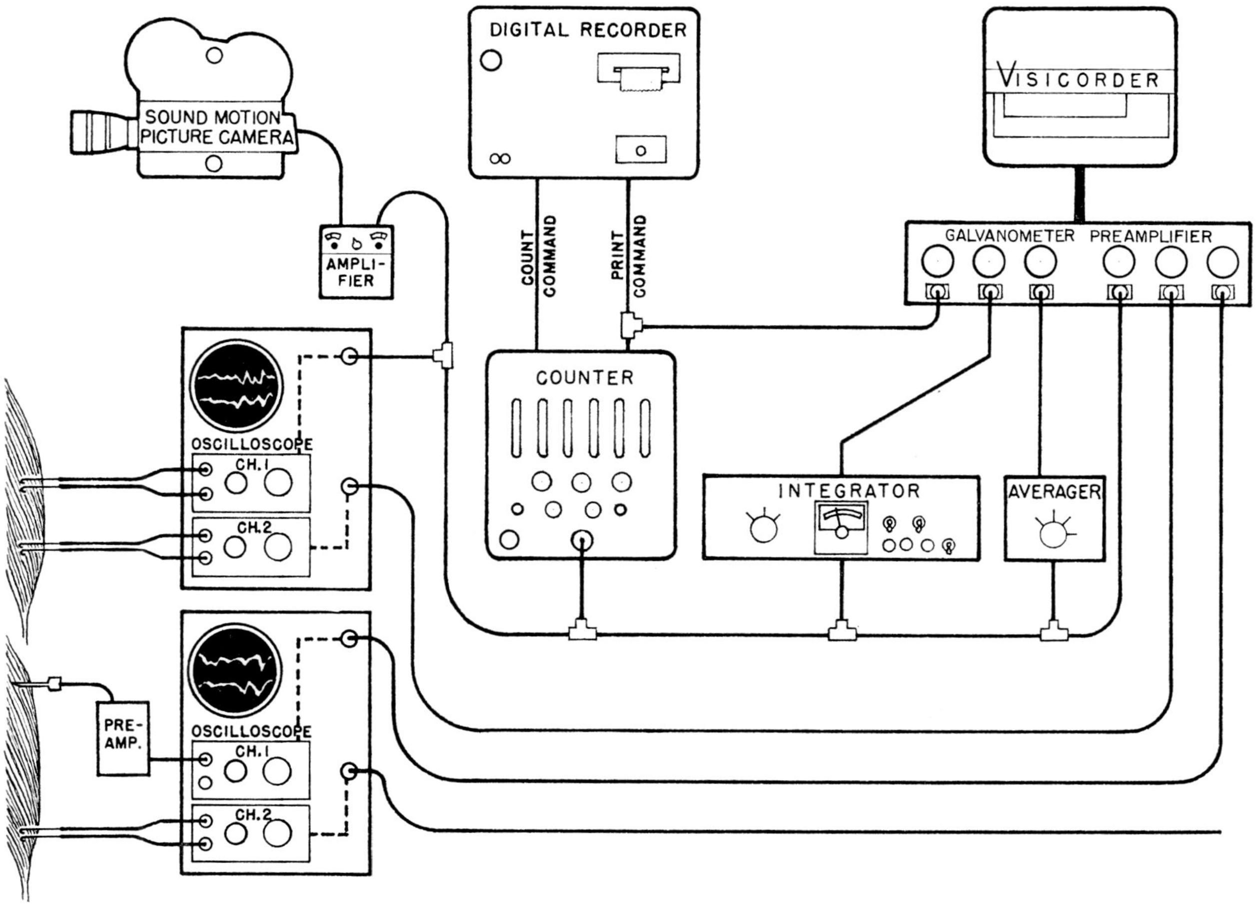

Figure 16. A plan for simultaneous recording from four muscle sites—three by differential, one by single-ended amplification. Other useful information, such as the analog output of the absolute-value integrator or the print command signal produced as the counter issues a count, may be added as separate Visicorder channels. The third channel shows the internal single coaxial electrode in place, connected to a single-ended preamplifier with adjustable voltage gain to ×100. Its frequency bandpass is 3 to 25,000 Hz. Its input impedance is 10 megohms.

the most definite means by which a conclusive diagnosis can be made. Here, a frequency, not a rate, is produced—at least during a portion of the recording. The characteristic "dive bomber" sound of myotonia congenita gives the disease picture definite identification that one is not likely to forget. This may also be sometimes observed in dermatomyositis.[118] The primary myopathies thus may be contrasted with diseases due to nervous system lesions, such as progressive muscular atrophy, amyotrophic lateral sclerosis, myotonia congenita and syringomyelia.[116]

In chronic poliomyelitis, motor unit recordings are characteristically of longer duration and higher amplitude. The phenomenon of synchronization has been pointed out by Hertz and co-workers.[89] This is a finding of identical electrical behavior in various parts of involved muscle which has been attributed to fusion of anterior horn cell activity, due to disease. A more reasonable explanation is axon branching,[23,111] the rescue of denervated fibres by remaining intact axons. Such a process may easily explain the large action potential of the "giant motor unit," in which there is far less interdigitation of adjacent subunits. This simplifies the individual action potential. These muscles,

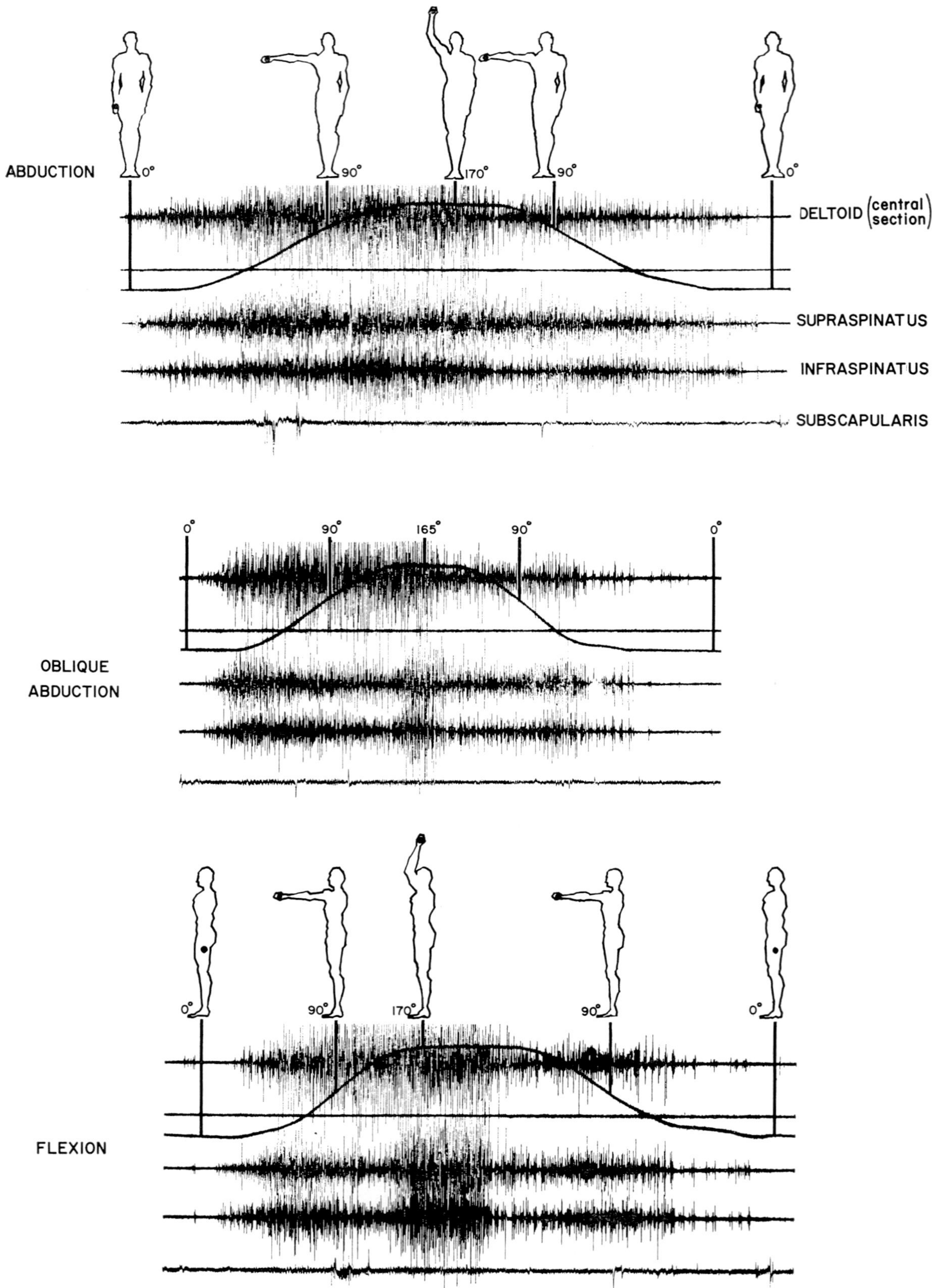

Figure 17. A multichannel study using two dual-beam oscilloscopes and the Visicorder as shown in Figure 16.

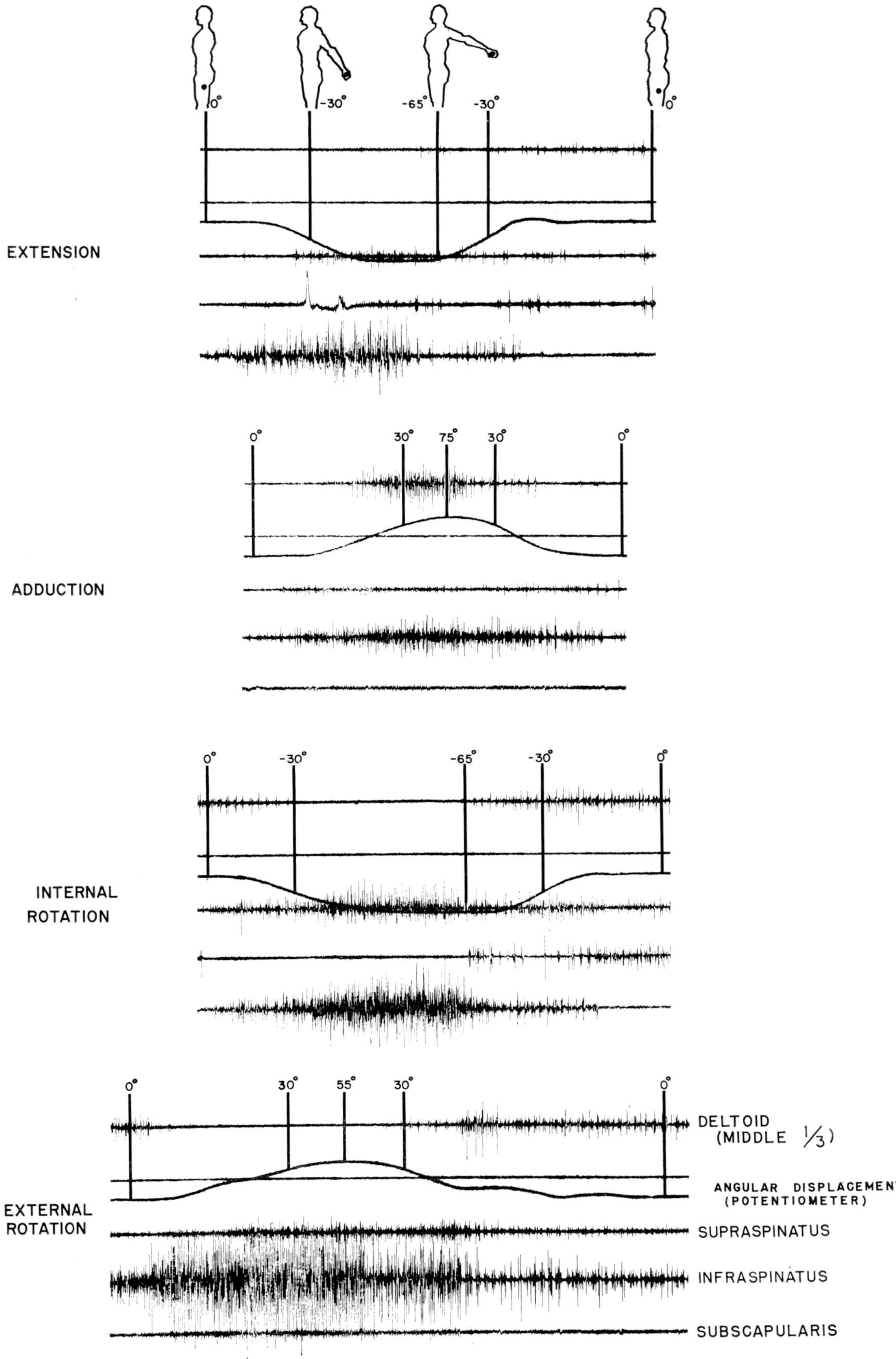

Figure 18. Continuation Figure 17.

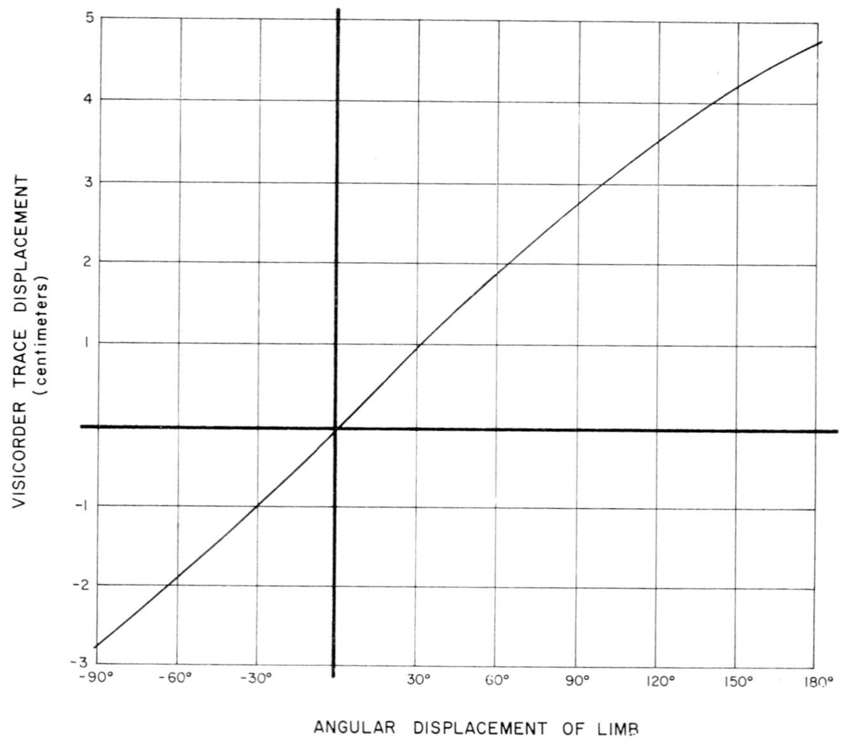

Figure 19. Angular displacement recorded by use of potentiometer. (*Second trace, Figs. 17 and 18.*)

though demonstrating higher amplitude, have a much lower motor unit action potential frequency and rate.

In 1941 to 1944, Schwartz and associates[156,157] demonstrated by electromyographic means evidence of spasm in acute poliomyelitis. Early spasm was observed in the antagonist as well as the protagonist. This spasm at times exceeded voluntary muscular contraction. During this stage of the disease, marked electrical activity was observed on passive stretch. Spasm was often observed electrically in apparently unaffected muscles, but here the electrical activity of voluntary contraction exceeded that of spasm. There was no spasm if voluntary contraction was impossible, and spasticity was considered a generalized phenomenon in early poliomyelitis. No correlation was observed between initial spasm and final residual weakness. The "spine sign" may thus be a reflex of uninvolved muscles in polio. Muscle, of course, is incapable of spasm unless the lower motor neuron is preserved.

In determining the state of innervation of muscle, one can often observe partial denervation. Progressive regeneration can be observed. For example, fibrillary potentials are usually first observed in the denervated muscle 14 to 21 days after nerve injury. If denervation is complete, fibrillary potentials gradually disappear as the fibres themselves degenerate. The electromyograph thus records the progressive changes in Wallerian degeneration. If denervation is incomplete and regeneration follows, groups of fibrillary potentials may gradually be transformed into a composite, somewhat irregular wave form. This wave form may be low in amplitude and short in duration. The recovered fibres are "out of phase" with each other—the "nascent motor unit."[94] These potentials finally resolve themselves into the somewhat diphasic form of the normal motor unit.

Clinically, it is often difficult to differentiate certain myopathies and atrophies of central origin. In primary disease of muscle without known nervous system involvement, fibrillary potentials are almost never observed. An example of such myopathy is pseudohypertrophic muscular dystrophy—without known nerve lesion and without fibrillary potentials but with a family history of disease and disturbance of creatine-creatinine metabolism. In central atrophies, the distal portions of the extremities are often first involved. An example of this type of involvement is the Charcot-Marie-Tooth syndrome.[116] Here fibrillary potentials are usually found during progressive stages of the disease, since anterior horn cell as well as posterior column and Clarke's column degeneration may occur. Spasticity is often observed, in contrast to the primary myopathies, where it is absent. Other examples of central atrophies are progressive muscular atrophy, amyotrophic lateral sclerosis, and myotonia congenita. Peroneal muscular atrophy is another example of central atrophy where there is involvement of peripheral nerves or anterior horn cells as well as the pyramidal tracts. Muscular involvement is secondary and there are usually fibrillary potentials. These would be expected during the process of degeneration of anterior horn cells or peripheral nerves.

The electromyographic method has been useful in establishing the level of cord lesions in such conditions as cervical or lumbar herniated intervertebral discs. The test may be of value if myelography fails to localize the lesion. In disc lesions, posterior muscles innervated by posterior primary rami may demonstrate electromyographic evidence of root compression, though no abnormalities are found in the muscles innervated by the anterior rami. If one is considering a diagnosis of herniated cervical intervertebral disc or brachial plexus injury, the differentiation can often be easily established electrically on the basis of involvement of the posterior spinal musculature innervated by posterior primary rami.[50] In plexus lesions, the spinal muscles are spared and produce normal action potentials. In lumbar intervertebral disc lesions, progressive root compression is very unlikely if one cannot demonstrate evidence of fibrillary potentials in the posterior musculature of the lumbar spine. The disadvantage of this test is that multiple needle insertions may be necessary if the test is to be conclusive. The clinical value of the electromyogram depends upon thorough searching for abnormal action potentials. The extent of the search may approach making of insertions to represent the activity of all parts of the muscle, if definite conclusions are to be made.

Extensive use has been made of the electromyogram, using multichannel equipment during cortical stimulation in order to gain further appreciation of the extent of cortical representation.[73,74] If a practical surface applicator could be developed, multiple simultaneous records of representative muscle throughout the body would be a most useful index of the pathways from the motor cortex. Due to variation in the amount of subcutaneous tissue, action potentials recorded by surface electrodes are not comparable regarding amplitude or frequency. We have yet to improve upon the technique of internal wires and multichannel recording equipment of appropriate frequency range (Fig. 16).

The normal electromyogram consists of a series of recorded action potentials produced by motor units firing at various rates and frequencies. At high frequencies or during strong contractions, motor unit action potentials are superimposed upon one another. This produces the familiar normal interference pattern. Through recruitment, the central nervous system is thought to be able to summon the participation of supposedly previously idle motor units to increase the strength of the contraction. At the same time, active motor units may increase their rate of firing.[21] Since the individual motor unit is thought to behave for the most part according to the all-or-none law, the resulting tension of contracting muscle may depend upon the total number of action potentials produced. Total or analog counts of action potentials should represent values proportional to the muscle's tension, if all action potentials represent equal contributions to the tension developed by the muscle. As tension increases, it is not known whether increase of action potentials by recruitment of additional motor units or by increase of the frequency (rate) of those already in action will make the greater contribution to tension. Many of the motor units functioning are perhaps in, or near, a state of tetanus. Are others idle or contracting at a comparatively low rate? There is increasing evidence that masses of motor units do not fire in this haphazard or irregular manner but rather tend to fire synchronously. This is strongly suggested by the findings of synchrony in adjacent masses of muscle, even though large probes are used (Figs. 1 and 2).

The axons penetrating the muscle vary in diameter.[58] This suggests that their motor units are of varying size and are composed of different numbers of muscle fibres. The action potentials of such motor units should, accordingly, vary in magnitude. It is likely that the fibres of adjacent motor units interdigitate with the one producing the principal action potential. This would influence its form and confuse the interpretation of the single action potential even if recorded from a very small probe. Thus, each action potential does not bear the same significance regarding its contribution to the total tension developed by the muscle. However, if it were possible to obtain action potential counts from significantly large volumes of muscle, the rate recorded would be representative of that found throughout the muscle. This concept is reinforced by the knowledge that groups of motor units tend to fire synchronously, particularly as the strength of the contraction increases. Electronic counting of recorded potentials is thus a most promising index of intensity of the action of muscle.

Two questions inevitably arise: How typically is the muscle actually represented by records from the two electrodes from which the count is taken? Would electrodes placed in another part of the muscle convey a comparable record? On occasions, using normal subjects, several pairs of electrodes were inserted. The tracings of Figure 1 were taken from the upper and lower portions of the anterior tibial muscle and recorded simultaneously with a dual-beam oscilloscope. The tracings were recorded at a camera speed of 20 in./sec. No sweep is employed with this method, the constant motion of the film providing the base line or horizontal (time) dimension. It is apparent that during most of the recording, the traces from the upper and lower portions of the anterior tibial muscles are remarkably similar. This phenomenon of synchronization at times appears to be almost complete for a considerable portion of the muscle and leads one to question the importance of the mechanism of individual recruitment of inactive motor units as one of the exclusive means of increasing or grading muscular tension. Recruitment would have to be a completely uniform progression of increasing frequency (rate) throughout the muscle, since rate of contraction increases synchronously in different portions of the muscle. It appeared that contractions even at moderate magnitude were likely to involve the entire muscle with all of its motor units participating at approximately the same rate. It seems likely that nature would make some use of all available motor units, increasing their rate of contraction to increase tension. A muscle during a given contraction thus might have uniformity of rate throughout its mass. Identical concurrent pulses are often found at any portion of the muscle during a contraction of moderate to large magnitude. In the presence of such complete synchronization, increase in rate appears to be a more important mechanism than recruitment for increasing the strength of contraction. If recruitment were the essential mechanism, its effect could hardly result in identical changes (synchrony) in the oscillograms of volumes chosen at random from different portions of the muscle (Figs. 1 and 2). Synchrony was more complete as the intensity of muscular contraction increased.

Thus increase in rate often involves synchronization and is to be distinguished from integration. Integration represents the total electrical activity of only the amount of muscle exposed to the electrodes. If rate is uniform throughout the muscle, it represents the general level of activity throughout the muscle. Total integration is therefore not so closely identified with whole muscle behavior as is action potential rate which may represent the predominating rate to be found throughout the muscle. Integration, however, has been found in our studies to vary in almost linear fashion with increase in rate. Mathematical differentiation of the analog curve of the integration process might produce a curve very closely resembling the analog curve of count rate which we have been able to record.

Synchronization was described by Buchthal and Madsen in 1950[23] and previously by Buchthal and co-workers in several earlier reports. They recorded similar action potentials from leads from various parts of normal and pathological muscle. Single and double coaxial electrodes 200μ apart* were used. Synchronization was observed particularly on maximal effort when "recruitment" was thought to be complete. It has been suggested that these synchronous records originated from the same motor unit.[111] The present technique for the investigation of synchrony (Figs. 1 and 2) employed pairs of internal electrodes. The exposed tips were one-half inch in length, thus recording activity from many motor units. This technique would not ordinarily produce records confined to the single motor unit in normal muscle. This meant that groups of motor units fired simultaneously in synchrony. Synchronization was observed throughout almost the entire recorded material in one third of the examinations made on the normal anterior tibial muscle. Where complete synchronization was not found in various parts of the anterior tibial muscle, uniformity of total motor unit action potential count and total integrated electrical activity were found. Count rate is thus an excellent numerical index to the intensity of muscular activity.

It appears that ideal frequency recording ranges have yet to be established. The optimum frequency bandpass is considered to be in the range of 40 hertz to 10,000 Hz. Under certain conditions, we have found it undesirable to extend the frequency recording range into the lower limits (Figs. 28, 29, 31). Figures 8, 20 and 21 demonstrate quite clearly what happens

* This measurement is made on the bevelled end of the needle. The electrodes are actually closer together, (Fig. 160).

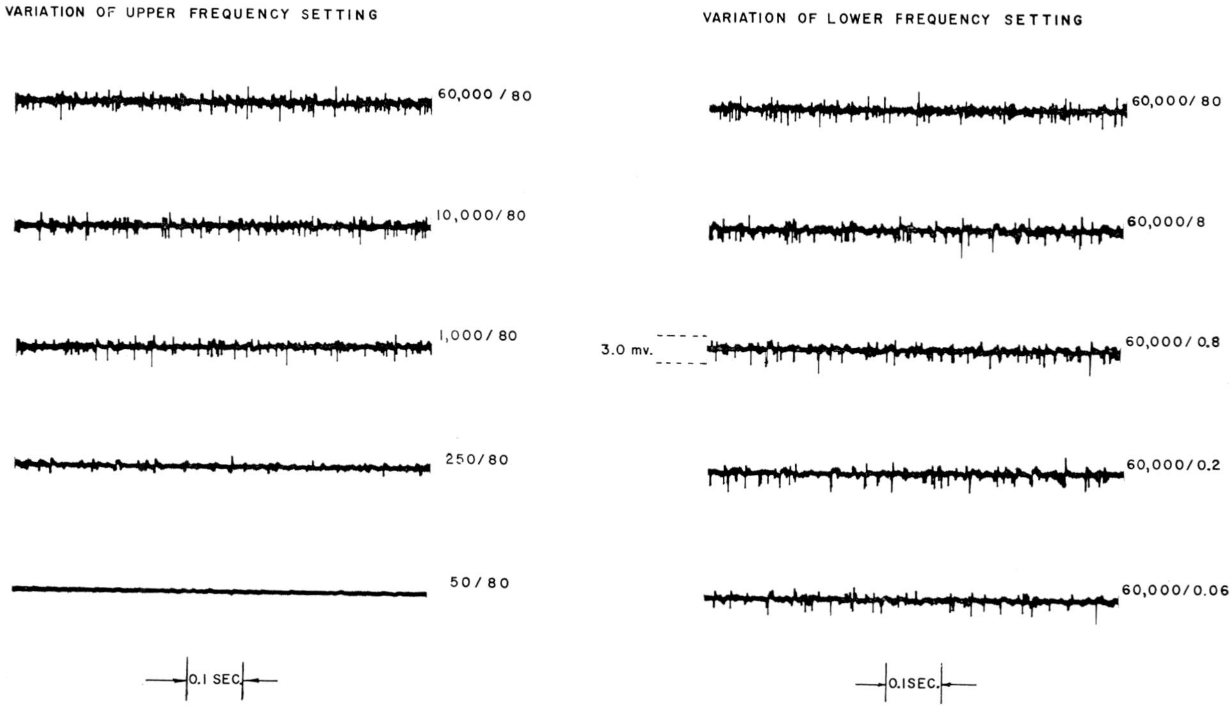

Figure 20. Effect of variation of frequency recording range on oscillogram of vastus-medialis during extension of knee; 1 cm equals 0.5 mv. Recordings made from skeletal muscle, under controlled conditions, using various frequency recording ranges. (From Instructional Course Lectures, American Academy of Orthopaedic Surgeons, 1959.)

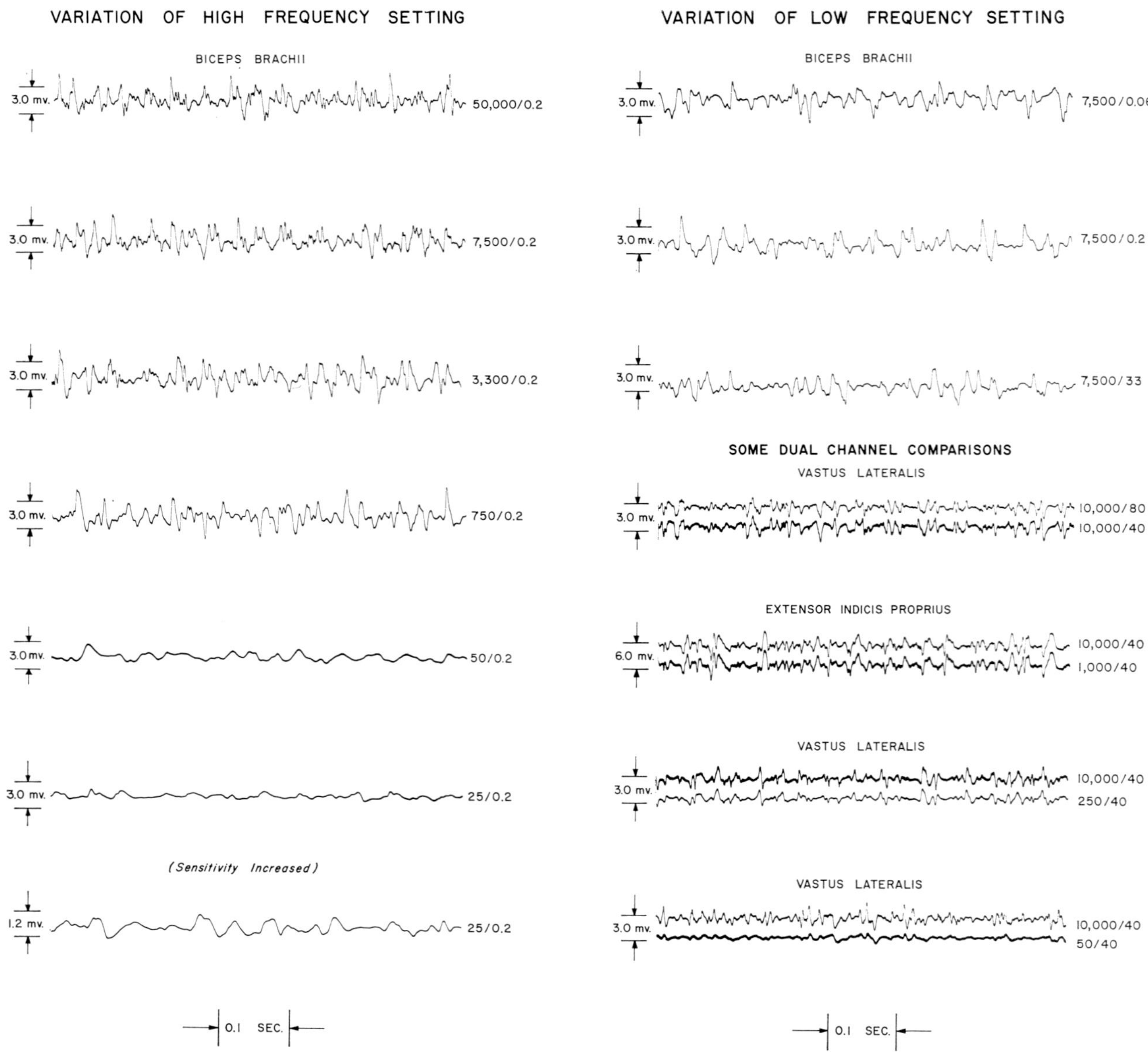

Figure 21. Muscle action potentials recorded at various frequency ranges. Recordings from biceps brachii with elbow at 90° holding 10-lb weight (isometric).

when one uses a preamplifier with variable recording ranges. As the lower frequency limit remains constant, the upper frequency limit may be changed during the contraction. The column on the left of Figure 20 demonstrates the effect on the electromyogram of changing the upper frequency setting. It is quite clear that there is considerable difference in records made using an upper frequency setting of 10,000 Hz and 1,000 Hz. The column on the right shows that there is far less effect when one varies the low frequency. In Figure 21, the deficiency of the 250 Hz upper limit is demonstrated.

By alteration of the preamplifier inputs, the option of setting the recording range at the upper limit of 10 kilohertz or 1.5 kHz is available. A lower limit of 1.2 kHz has been employed. As shown in Figure 32, this produces a much more uniform series of wave forms which are more suitable for counting and integrating. It is felt that the distance of the electrodes from the active area reduces the frequency of the wave form, thus lowering the apparent frequency requirements. Restriction of the frequency bandpass to the desired limits makes it possible more definitely to confine the recording to the desired region under examination.

Actually the ideal frequency recording band has not yet been decided. It is considered highly desirable, however, to complete wave form analyses on records conventionally obtained in order to ascertain the predominating rates. In Chapter VII, a method for ascertaining the upper and lower frequency limits of the motor unit action potential of the anterior tibial muscle is described.

In recent years, equipment has become available that will record simultaneously from several muscles without compromise regarding the ideal frequency recording range. Unfortunately, the apparatus often employed formerly was the "electromyographic amplifier" and pen-writing electroencephalograph multichannel recorder. This recorder, however, has a totally unsuitable upper frequency recording limit often not exceeding 75 Hz. The linearity of the response of pen writers to the signal from the amplifier is questionable. Its low-frequency response with resulting low amplitude necessitates higher gain. There is greater amplification of artefact as well as signal.

A convenient alternative is the new type of miniature galvanometer.* Such galvanometers can be arranged in sets of at least one dozen, enabling one to record directly and simultaneously from as many as twelve muscles. The recording frequency range of such a galvanometer is quite suitable, 0 to 8,000 Hz flat to 5,000 Hz (Fig. 15). Records made from the Visicorder are compared to those made simultaneously from the same electrodes and displayed on the oscilloscope (Fig. 14).

The ideal equipment for display of the action potential from muscle is the oscilloscope. With this instrument, the signal can be critically analyzed on the oscilloscope screen before or during recording. Many oscilloscopes have a number of controls such as calibrated sensitivity and variable upper and lower limits of frequency. One can thus easily appreciate the magnitude and character of the signal recorded. The oscillation may be swept across the screen for analysis by the horizontal (sweep) amplifier. The horizontal dimension may be introduced by continuous recording equipment such as the 35 mm Fairchild camera. Some standard oscilloscopes can be easily adapted for multichannel use by employing multi-trace plug-in amplifiers. In one instrument, the Tektronix #565 oscilloscope, two four-trace amplifier plug-in units may be used. Each plug-in unit contains four separate, indedependent amplifier channels (Fig. 131).

The signal produced by many of these instruments can be "integrated." It is erroneous to consider integration a continuous analog display of the level of electrical activity at the site of the electrodes in the muscle under study. The curve produced by the RC "Integrator" is, in reality, a continuous average curve. Integration results only if the area under the averaged curve is measured, using such an instrument as the planimeter. The so-called "integrators" which rectify the signal producing a neat-appearing analog curve depend upon resistors and condensors. They therefore introduce some distortion into the recording of the "integrated" or averaged curve. R multiplied by C is called the "time constant" of the instrument (Fig. 84). Here, R is in megohms, C is in microfarads.

The loudspeaker may also be used to demonstrate the signal. The audio representation is valuable and can be conveniently used while the visible electromyogram is being recorded as another means of monitoring the phenomenon. The audio signal can be recorded by tape recorders or on the sound track of motion picture film. The latter results in a self-synchronized recording of the subject's activity and the electromyogram in audiovisual form.[34,39] The 16-mm sound film is an excellent medium for display of the electrical activity of individual muscles as they contribute to various body motions.

There are artefacts which cause confusion in interpretation of the electromyogram. Some of

* Minneapolis-Honeywell.

these result from 60-cycle current, diathermy, excessively high recording frequencies with the introduction of radio frequencies, and excessive amplification. Some of the better oscilloscopes have high signal-to-noise ratios which minimize the recording of artefact. Differentially amplified signals are less affected by artefact than those taken from the single electrode (Figs. 11 and 12). Preamplifiers placed near the source are very effective in increasing the value of the signal, markedly reducing the effect of "noise" as the signal is conducted to the main amplifier. This arrangement has made the single-ended mode a very practical method for diagnostic work using the single coaxial electrode. This relatively simple set-up makes the shielded room unnecessary.

The resistance measured between two internal recording electrodes or between a single electrode and the grounded subject is high, usually 35,000 to 45,000 ohms. A high impedance at the amplifier input is therefore desirable. Preamplifiers, having comparatively low input and output impedance, may therefore be unsatisfactory.

THE DIFFERENTIAL EFFECT AND DISTORTION CAUSED BY COAXIAL SHIELDING IN ELECTROMYOGRAPHIC ELECTRODES

When an electrode is inserted into skeletal muscle, the electromyographic wave appearing on it is essentially a time-dependent phenomenon. However, one might imagine a multitude of electrodes inserted in a line parallel to the path of the ionic impulse passing along the muscle fibre membranes. In this case, the potential on each wire may be thought of as a function of both space and time. If we suppose that the wave is not dispersed as it travels, it may be represented as follows:

$$f\ (x - vt)$$

where x is the distance the impulse passes along the muscle, v is the velocity of conduction, and t is the time parameter. This may be differentiated on both x and t, and the term $(x - vt)$ may be called z for the time being, for in common mathematical usage,

$$f'(z) = \frac{df(z)}{dz}$$

Hence $$\frac{\partial f}{\partial x} = \frac{df(z)}{dz}\frac{\partial z}{\partial x} = f'(z)$$

and $$\frac{\partial f}{\partial t} = \frac{df(z)}{dz}\frac{\partial z}{\partial t} = -vf'(z)$$

Since $-v$ may be taken to be constant,

$$\frac{\partial f}{\partial t} \backsim \frac{\partial f}{\partial x} = f'$$

and in a study where shape and relative amplitude, but not necessarily absolute amplitude is important, one may be substituted for the other.

It is immediately clear, then, that if two electrodes are inserted in the muscle at x and $x + h$ (where h is very small) the voltage appearing across them at any time will be

$$f\ (x + h - vt) - f\ (x - vt)$$

but

$$= \frac{f\ (x + h - vt) - f\ (x - vt)}{h}\ h$$

and

$$\lim_{h \to 0} \frac{f\ (x + h - vt) - f\ (x - vt)}{h} = \frac{\partial f\ (x - vt)}{\partial x}$$

Then if $\frac{h}{v}$ (the time required for the impulse to travel the distance between the electrodes) is much shorter than the time necessary for any of the features of the impulse to occur, the limit will be closely approached and

$$f\ (x + h - vt) - f\ (x - vt) \approx h\,f'\ (x - vt)$$

Strictly speaking, this is the case only when the line connecting the tips of the electrodes is parallel to the direction of propagation of the impulse; if the angle between them is θ,

$$f\ (x + h\cos\theta - vt) - f\ (x - vt) = h\cos\theta\ f'\ (x - vt)$$

However, it is likely that this picture cannot

be held in general to be exact. There is no reason to suppose that all impulses are isolinear or, more damaging to the concept, that all of the effects contained in an impulse are isolinear. With small electrode separations, the distance becomes comparable to the size of the electrodes themselves, and this fact influences the interelectrode field. Yet, these qualifications do not destroy the differentiation effect completely; the human eye can make necessary allowances and thus detect definite differentiation.

In the past, double and single coaxial electrodes have been used in diagnostic procedures. In the latter case, the impulse is taken from the single coaxial electrode between the central electrode and the grounded outer shield. When using the double coaxial electrode, the output is taken from between the two central electrodes, while the shield is either grounded or left floating. For the purposes of electromyography, it makes little difference whether the shield is grounded, since the electrical capacity of the body is great enough for it to seem to be "ground" for the small impulses.*

The shield has a profound effect on the nature of the electromyogram obtained from these electrodes. The speed and degree of conduction is much greater in metals than in ionic solutions. The result is that in the case of a single coaxial electrode, the electromyogram is the potential in the center of the mouth of a long tube which has the same potential over its entire surface. Consequently, a wave of the form f (x — vt) passes down the muscle. The impulse g (x — vt) seen by the single coaxial electrode is by no means identical to f (x — vt). Likewise, the voltage appearing across the double coaxial electrode is not equal to f′ (x — vt).

Considerable similarity is found between g (x — vt) and f (x — vt), although output as recorded using the coaxial electrode is much lower. This is due to a change of the voltage of the isopotential lines about the tip of the electrode. Since the electrode is hollow (neglecting for the moment the wires inside), there are no isopotentials inside. One can visualize the isopotentials as "breathing" in and out as an impulse passes. The insertion of a small wire down the center of the tube essentially preserves this effect. When this is done, the resulting voltage corresponds fairly closely to f (x — vt), although it is obviously of lower amplitude. The shape is distorted due to the nonlinearity of the effect, but the actual degree of distortion is virtually incalculable.

The effect of the shield on the double coaxial electrode is even more marked. The distance between the electrodes is on the same order as the diameter of the shield. The anticipated

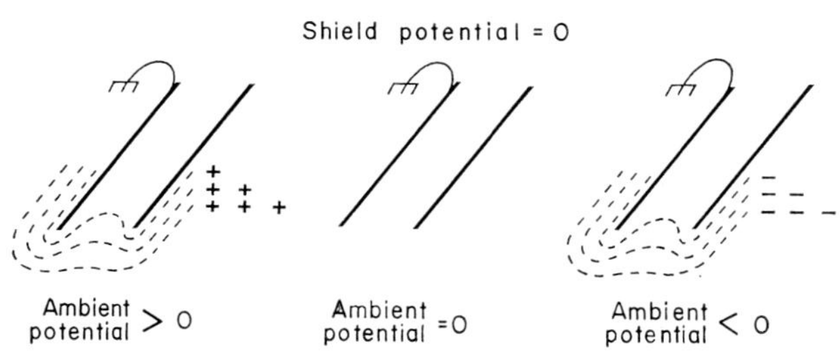

Figure 22.

distortion results. The fact that the shield is grounded weakens the supposition that the output of the double coaxial electrode is the derivative of anything physically important.

Differentiation, as discussed above, occurs when the same wave appears on two electrodes but with a phase difference between them due to their spatial separation. What is actually the case, however, is that the same wave does not appear on both electrodes, since their relations to the shield are different. For example, when the ambient field is positive, the potential field may be diagrammed as in Figure 23. It is clear that the B electrode is physically more separated from, and electrically more independent of, the shield than is the A electrode. Consequently, the excursions of the isopotential lines are greater in the region of the B electrode. The shape and amplitude will be slightly different on the two electrodes (the amplitude will be

* A "ground" is a source-sink from which or into which electrons may be moved with little effect. The reason that one can pump only so much electricity into a body is that it soon "charges up" to the charging potential; the electromyographic impulses are too small and too short for this effect to occur.

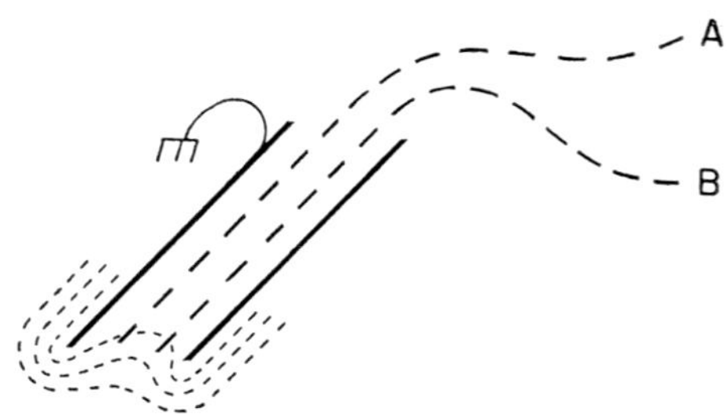

Figure 23.

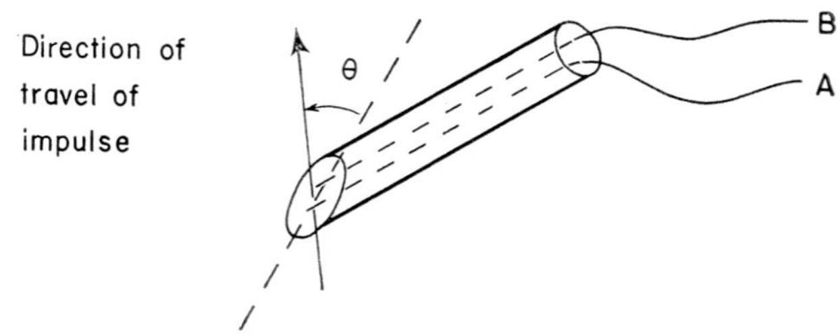

Figure 24. The angle θ of direction of passage of impulse with the orientation of electrodes.

greater on B). The resultant voltage will therefore be somewhat different than

$$f(x + h - vt) - f(x - vt)$$

Rather, ignoring the differences in shape and considering the amplitudes, it is a good approximation that the voltage appearing may be expressed:

$$V_{AB} = (1 + \epsilon)\, g(x + h - vt) - g(x - vt)$$

where, as before, $g(x - vt)$ is the wave $f(x - vt)$ which has been modified by the presence of the shield and ϵ is a constant which is close but not equal to zero.

It is pertinent to distinguish between impulses travelling in different directions. For the impulse in Figure 24, the effective separation of the electrodes is $h \cos \theta$, and

$$V_{AB} = (1 + \epsilon)\, g(x + h \cos \theta - vt) - g(x - vt)$$

When $\theta = 90^\circ$, ideally there should be no output at all, since the two electrodes should be at the same potential at all times. For the reason mentioned above, they are not. In this case:

$$V_{AB} = \epsilon\, g(x - vt)$$

there is no differentiation effect. When $\theta = 0^\circ$,

$$\begin{aligned} V_{AB} &= (1 + \epsilon)\, g(x + h - vt) - g(x - vt) \\ &= \epsilon\, g(x + h - vt) + g(x + h - vt) \\ &\quad - g(x - vt) \\ &\approx \epsilon\, g(x - vt) + h\, g'(x - vt) \end{aligned}$$

Here, it is not immediately clear whether the $h\, g'(x - vt)$ term overpowers the $\epsilon\, g(x - vt)$ term, nor has it been established whether ϵ and h are of the same polarity. The answer to the latter supposition depends on how the particular electrode is fitted in to the analysis: if the A electrode is at x and the B electrode is at $x + h$, ϵ and h are of the same polarity. The answer to the former depends on the construction of the coaxial electrode: the more bevel there is on the electrode, the greater the asymmetry and the more significant the value ϵ.

These effects are best assessed experimentally. The following recordings were obtained by making a single double coaxial electrode act simultaneously as a single coaxial and a double coaxial electrode by the following method (see Fig. 25).

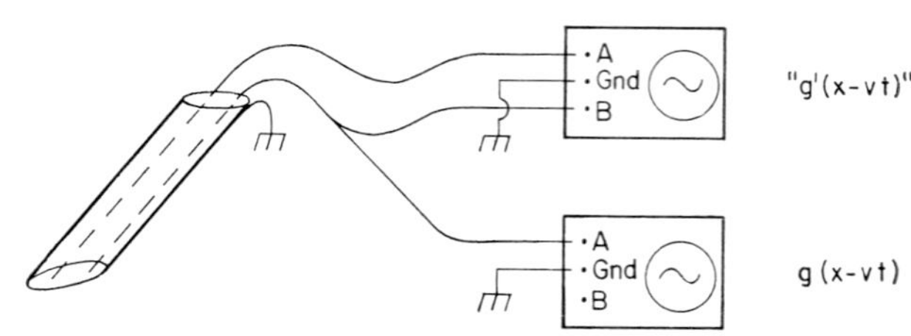

Figure 25. A method for simultaneous display of single-ended and differential records from one double coaxial electrode.

The data were taken at various electrode positions from 0° to 90°. The angle θ is formed by a line drawn through the electrode tips and the long dimension of the muscle (Fig. 24). There are instances where a strong differentiation effect is evident. In others, the effect is completely absent, showing only $g(x - vt)$. The instances of differentiation predominate in the 0° group; the 90° group contains very few. Isolated impulses were selected because of the difficulty of interpreting cascades of electromyographic impulses.

In the lower tracings (Fig. 26), the data were recorded at $\theta = 90^\circ$. The traces are quite similar. However, in this case, the gain for the differential channel was increased. Had the gains been the same, as in the above records of

Figure 26, the differential record would have been of an amplitude too low to permit analysis.

There is no reason for discrimination against fast impulses when they exist in the muscle tested, using either coaxial electrode. This is demonstrated by the records (Fig. 26). Fast rise pulses at the output of the single coaxial electrode appear approximately differentiated in the double coaxial recording. Sharp spikes appear which correspond temporally to the fast

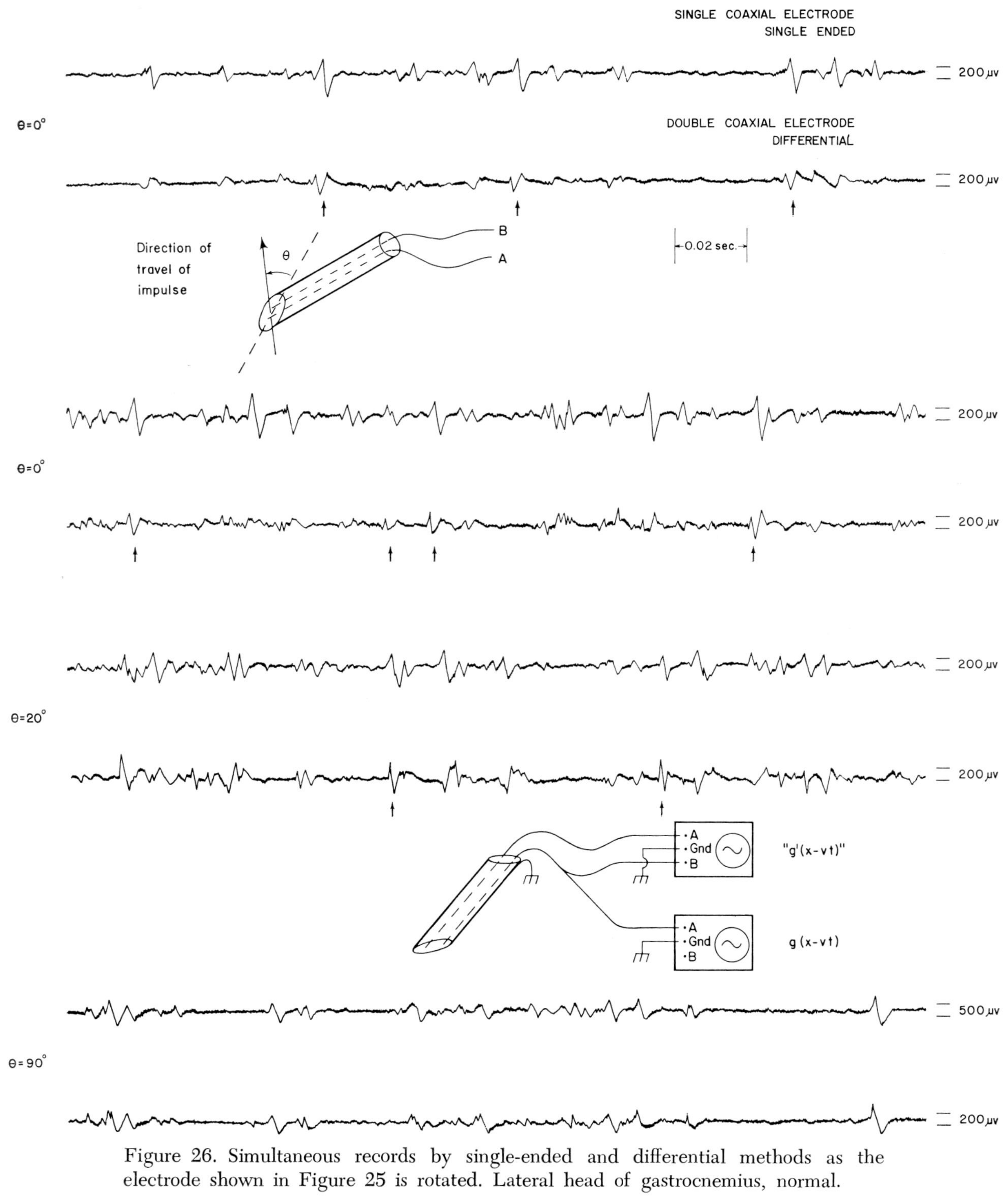

Figure 26. Simultaneous records by single-ended and differential methods as the electrode shown in Figure 25 is rotated. Lateral head of gastrocnemius, normal.

rise. This is not particularly important in itself if one understands the basis of the method employed. However, sharp spikes are often called fibrillation potentials, commonly considered indicative of progressive denervation in the muscle. Since the double coaxial electrode is often used for the detection of such phenomena, serious errors in interpretation result if these two types of electrodes are used indiscriminately. As the recordings indicate, the difference is of sufficient magnitude to obviate completely the worth of the double coaxial electrode in this capacity.

Figure 27 is a series of records showing the various comparisons possible using the double coaxial electrode as in Figure 25. The records made from the A electrode, single-ended, appear almost identical to those similarly made from B. At times, the differential mode produces pulses of very low amplitude, while the single-ended pulses are of considerable magnitude.

THE USE OF SELECTED FREQUENCY BANDS IN RECORDING FROM MUSCLE

The characteristics of the single motor unit action potential of the rabbit anterior muscle is described in Chapter VII. The single motor unit action potential is found to be a relatively simple, reproducible wave form. Analysis of this wave form indicates the general level of the frequency recording range needed in recording instruments. Since this rather simple wave form truly represents the action potential of the single motor unit, are not accessory oscillations accompanying the principal wave form often reported in clinical work and attributed to the single motor unit, actually due to oscillations produced by an indeterminable number of other motor units in the field? Due to the lower frequency of the potential of these combined, more distant units, can their recording, with subsequent confusion of the record, be eliminated by raising the low-frequency limit of the recording apparatus? The calculations (based on the Fourier transform) in Chapter VII indicate an optimum low-frequency recording level of approximately 30 Hz.

If one could restrict the recording to a small volume near the end of the probe—to the exclusion of the larger, less specific, averaged volumes of surrounding muscle, a more realistic record of the action potential of the motor unit would result. This might be accomplished by appropriate setting of frequency limits as dictated by the data of Chapter VII.

Using full–frequency-range recording instruments (Fig. 28), one can record the action potential of the single motor unit as described in Chapter VII. In this case, where the single alpha axon was stimulated, the confusion of surrounding active motor units was eliminated. In order to eliminate these factors in favor of the discrete action potentials, the "restricted frequency inputs" of Figures 29 and 30 were designed. The effects of these attenuating circuits are shown in Figures 31 and 32. In each figure, the action potentials obtained using the attenuated circuits are compared to those from the conventional broad-frequency band. Although we have undoubtedly not yet arrived at the optimum recording range for human muscle, the 1.2 kHz to 10 kHz frequence band (Fig. 32) appears to yield more uniform wave forms. Further adjustment of these frequency limits must be made. It is, of course, not held that human skeletal muscle may have the same frequency characteristics as the rabbit muscle studied. It is presumed that the frequency recording range desirable for human skeletal muscle will be on the same order as that found in the rabbit.

When the broad frequency range is employed, one not only includes the rather indiscriminate periphery but the normal interference pattern is soon reached. This completely confuses the recording, due to the combining of many rapidly occurring potentials. No realistic count or integration can be made. The restricted frequency band, to a great extent, will exclude the averaged potentials of the

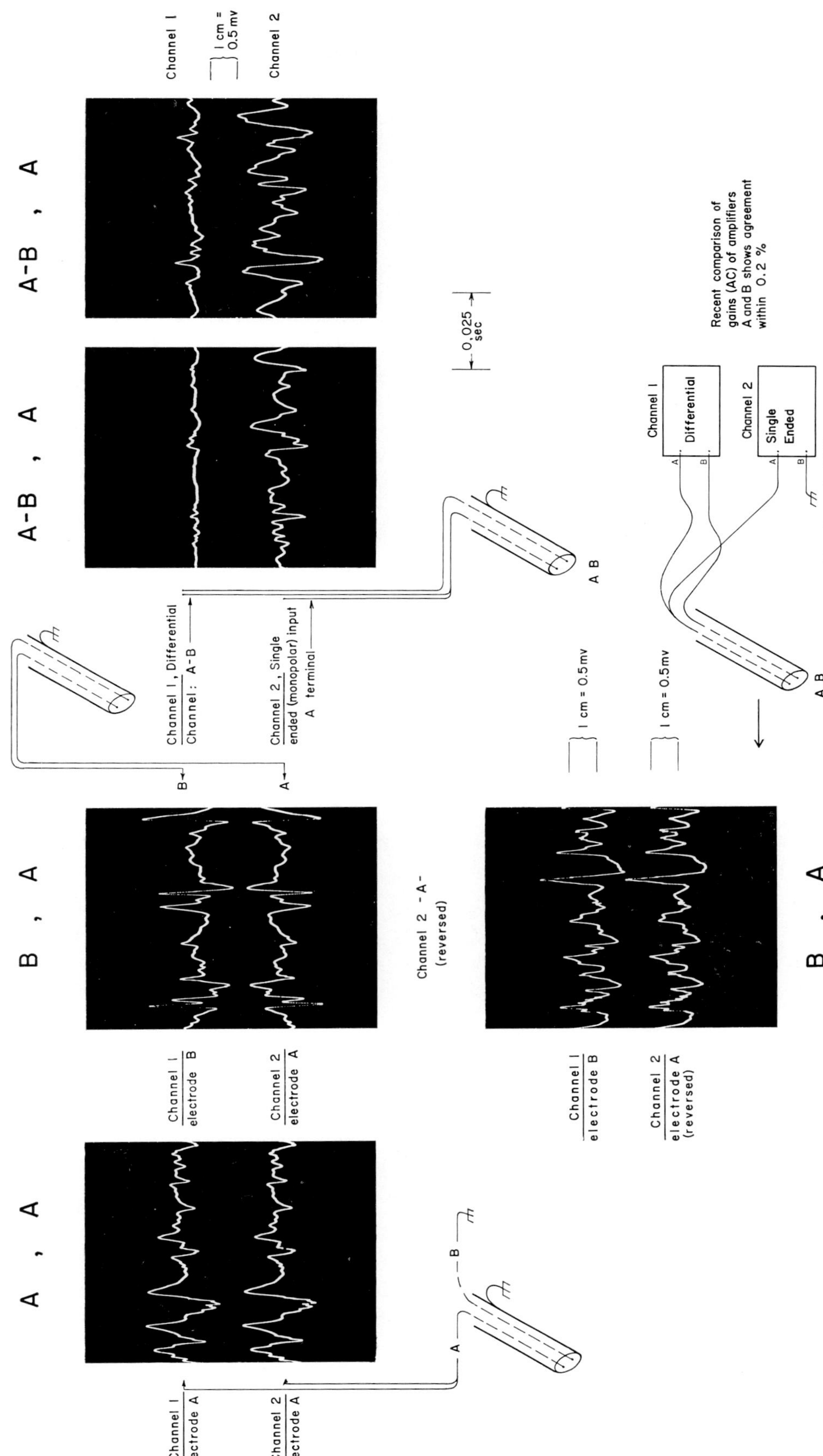

Figure 27. Action potentials obtained from normal anterior tibial muscle using method shown in Figure 25.

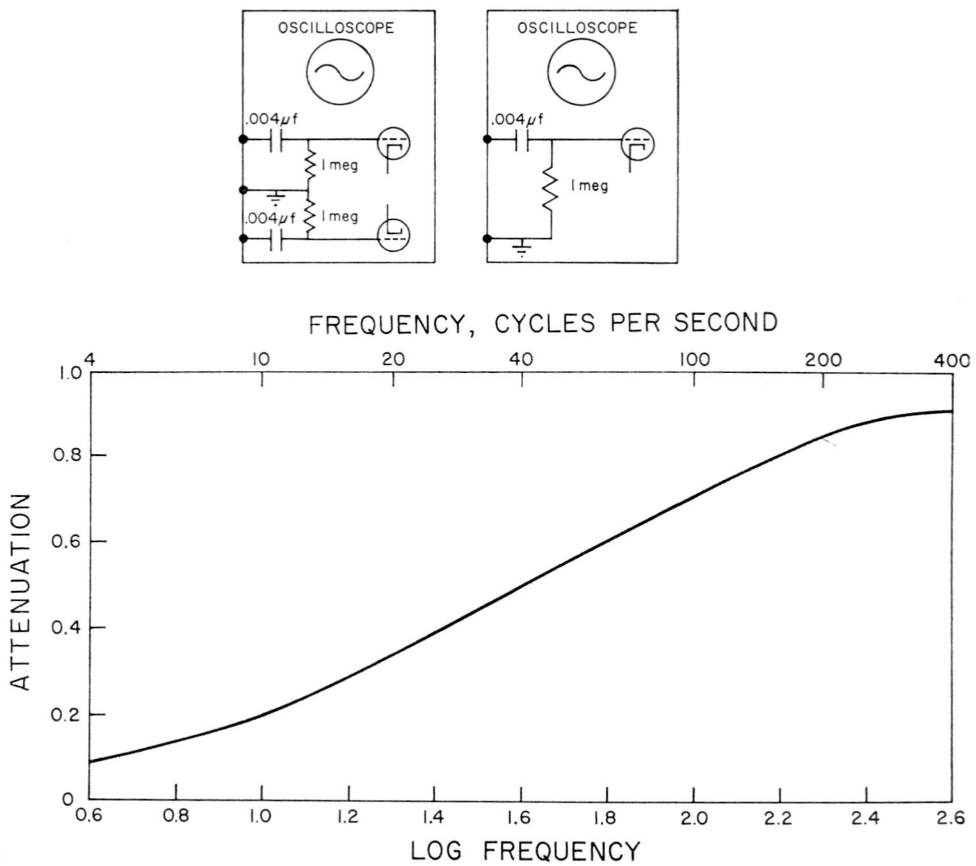

Figure 28. The frequency bandpass of approximately 40 Hz to 250 kHz for differential or single-ended mode using the same oscilloscope.

large imponderable volumes of active muscle in the vicinity of the electrodes. The elimination of these large portions of the muscle will result in the necessity for greater amplifier sensitivity, since the recording is restricted to a smaller but more specific volume of active muscle (Figs. 31 and 32).

Normally, it is observed that recordings attributed to the single motor unit are produced by abnormal muscle where the density of surviving motor units is greatly reduced. Using small probes, one can imagine a field in which only one motor unit influences the recording. Or, in normal muscle, very weak contractions may produce only a few, discrete action potentials. However, the failure of these two sources to produce the very simple motor unit action potentials similar to those recorded upon stimulation of the single alpha axon (Chap. VII) indicates that one solitary motor unit alone does not produce the action potential recorded. Proper frequency-band adjustment promises to bring us closer to the recording of a small volume, in some cases possibly the single motor unit, at the electrode tip. We may ultimately record the simple wave form of Chapter VII in clinical work to the exclusion of the numerous accessory oscillations which often precede and follow the main action potential spike of the motor unit. Thus, restriction of the frequency band at the proper limits lessens the confusion of volume conduction of impulses not of primary concern in diagnostic applications.

Recordings confined to the single motor unit action potential would facilitate analysis. Time duration and amplitude would be readily apparent (Fig. 32). A direct comparison of solitary action potentials in various peripheral neuromuscular and primary muscular diseases may be of great clinical diagnostic value.

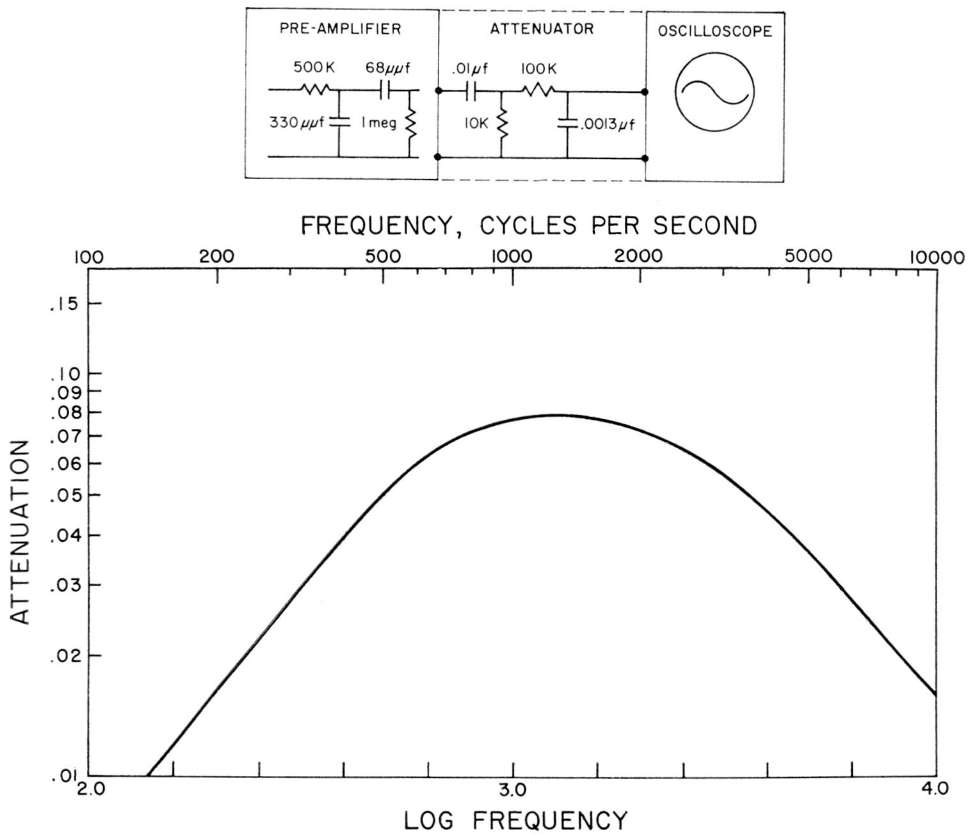

Figure. 29. Attenuation of input frequency bandpass to approximate the 1.2 kHz to 1.5 kHz range (employed to produce lower record of Figure 31).

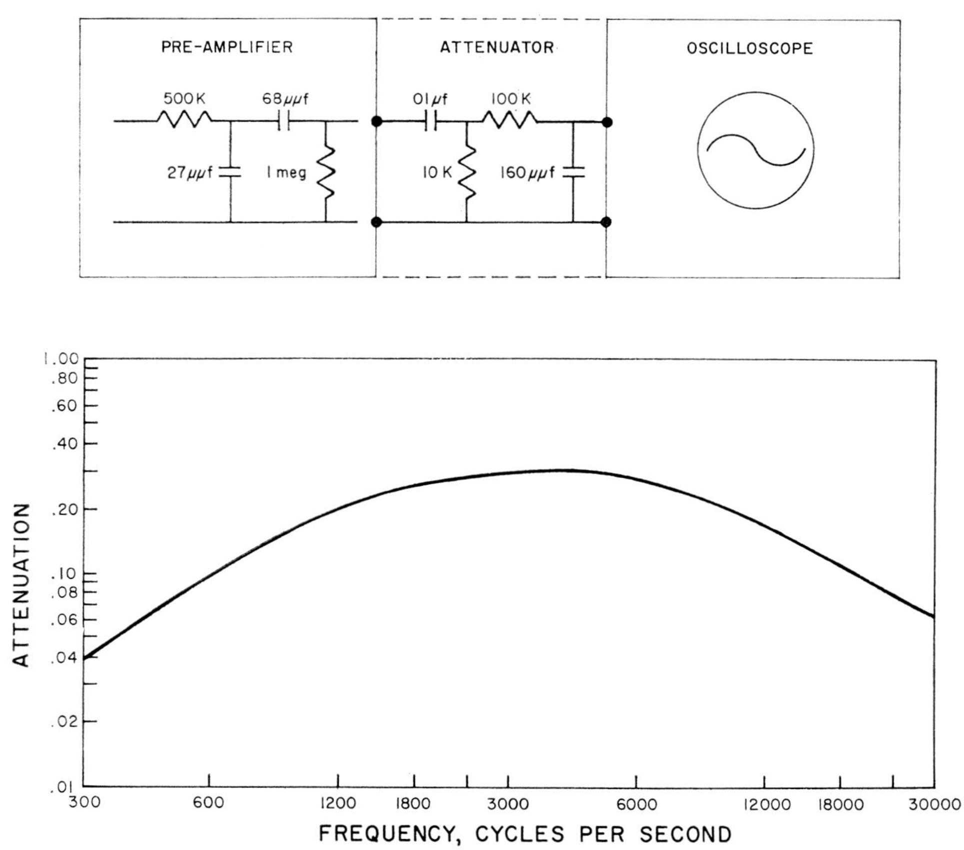

Figure 30. Circuit used to restrict input frequency bandpass to the 1.2 kHz to 10kHz range—circuit used to produce lower tracing of Figure 32.

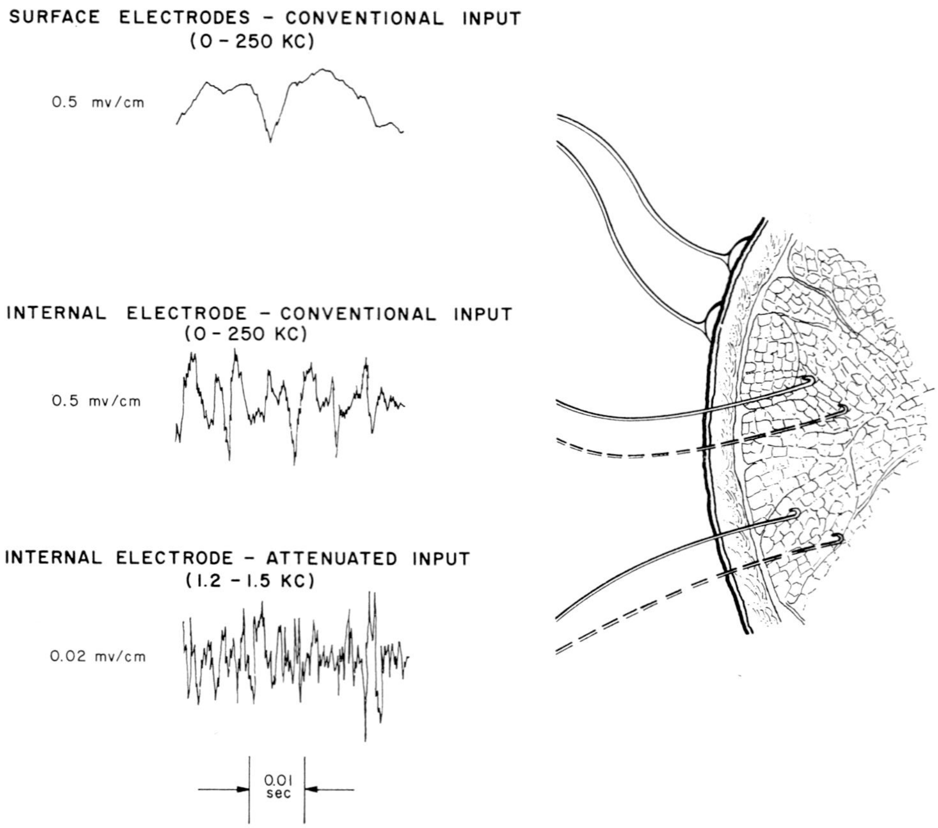

Figure 31. Normal muscle. Comparison of traces made at the same speed from (a) surface electrodes, *broad-frequency band,* 0 to 250 kHz, (b) internal electrodes by differential or single-ended mode, *broad-frequency band,* 0 to 250 kHz, (c) internal electrodes by differential or single-ended mode, *restricted-frequency band,* 1.2 kHz, to 1.5kHz.

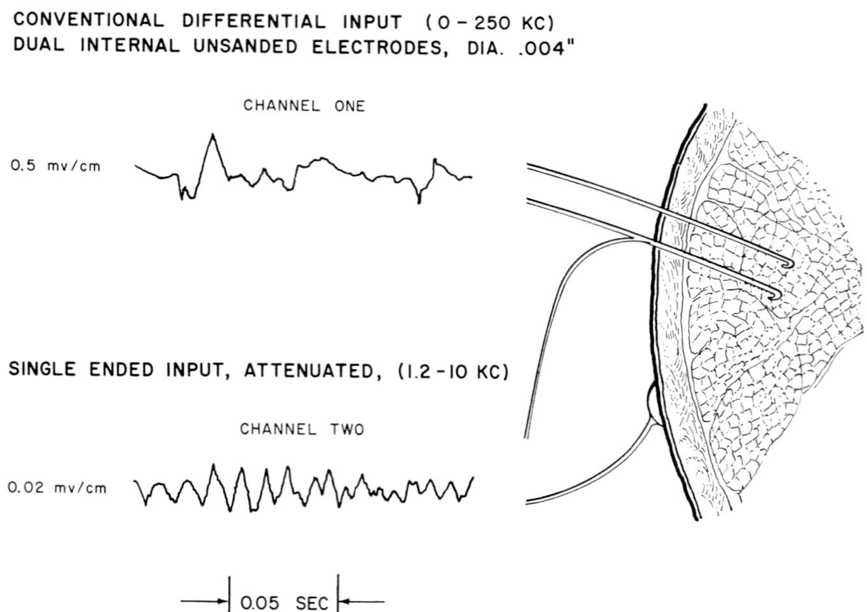

Figure 32. Muscular dystrophy. Comparison of recordings made using *broad-frequency band,* 0 to 250 kHz, differential mode (*upper trace*) and *restricted-frequency band,* 1.2 kHz to 10kHz, single-ended mode (*lower trace*).

It is observed that excessive elevation of low-frequency recording levels is not the approach for recording from the individual fibre. Since the motor unit is thought to behave ordinarily according to the all-or-none law, the synchrony within the motor unit makes individual fibre recording unlikely, using probes of this type. Recordings from single muscle fibres using a percutaneous technique have been demonstrated by Pierce and Wagman.[141] A tungsten electrode with a 1μ tip was introduced by a micromanipulator.

We now know, at least for rabbit and cat muscle, the precise levels required for the upper- and lower-frequency settings (Chap. VII). We must provide high-frequency equipment that is capable of recording over the levels arrived at, based upon the data of Chapter VII. These data can be extrapolated to indicate the probable range required for human muscle. The lower-frequency level, as discussed, must be set to avoid the surrounding volumes of unknown extent discussed above. These volumes, combinations of varying numbers of motor units, contribute low-frequency oscillations to the recording. The optimum low-frequency level, which would virtually eliminate these confusing forms, has not yet been arrived at.

Methods employed to raise the lower-frequency limit are shown in Figures 28, 29 and 30. In general, the method consists of alteration of the R-C preamplifier input. For example (Fig. 28), the original bandpass of the instrument (1–2 Hz to 250 kHz) was changed to approximately 40 Hz to 250 kHz by inserting 0.004μf alternating current input coupling capacitors in place of the standard 0.1μf capacitors. Further attenuation of the band width was accomplished by the addition of separate units in series (Figs. 29, 30).

Spectrum analysis and wave form analyzers have been used by Walton[166] and Becker.[17] Spectrum analysis was applied to the conventional electromyogram by Taverner[118] (see Chap. XII, Licht[118]). The waveform analyzer was employed by Becker to identify action potentials in the somewhat higher-frequency range produced by passive stretch.

If we are able to produce recordings from the single motor unit in the human subject, direct wave form analyses will be possible. The procedure described in Chapter VII is safe and could easily be adapted for recording from the human subject in the operating room. It is hoped that this diagnostic procedure of the future may lead to further insight and provide valuable data in neuromuscular diseases and primary myopathy. The procedure, correlated with conventional methods of biopsy, is harmless to the patient. It might be called the "live biopsy."

Chapter III

SOME AXES AND CENTERS OF ROTATION OF THE SKELETAL SYSTEM

A consideration of the axes or centers of rotation about which the muscle may be presumed to act is helpful in understanding the actions of many muscles. The electromyogram is of great value in such studies to demonstrate the time and extent of the activity. With flexible wire electrodes in the muscle belly, its role is revealed by the electromyograph. If the single muscle can be considered for the sake of convenience to possess individual action, then one can usually construct an imaginary axis or center of rotation about which the muscle may produce motion of skeletel parts. For example, the axis of the subtalar joint (Figs. 33, 34, 35, 36) decisively influences the actions of the muscles whose tendons pass it on the way to their insertions in the bones of the foot. Only by considering the center of rotation of the hip joint can one appreciate the action of the iliopsoas muscle which changes its position (or line of action) with respect to the center of the femoral head in various positions (Fig. 37). The complex actions of the adductor longus and adductor magnus muscles are a mystery with regard to the rotary effect of these muscles, unless one studies the axis about which they rotate the lower extremity (Figs. 38, 39, 40). Since this axis changes with various positions of the lower extremity, the study of its associated muscles is complex (Fig. 38).

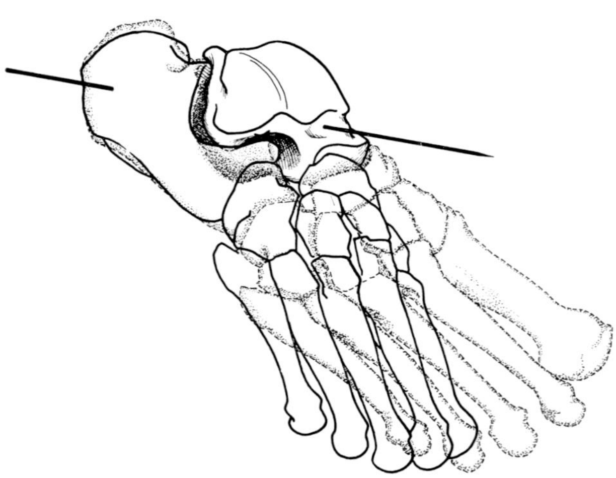

Figure 33. The action of the foot about the compromise axis of the subtalar joint.

One can assess the action of the adductor musculature in various positions of the extremity if one realizes the position of the axis of the lower extremity about which this complex muscle group functions. In Figure 38, the axis is demonstrated in one extremity as it passes from the center of the head of the femur to the point of weight bearing of the lower extremity. It thus becomes obvious that the adductor muscles are very powerful internal rotators of the hip, particularly in certain positions. Knowledge of this changing axis of rotation is of great value not only to explain the action of the several adductor muscles but in providing the surgeon with essential information when adductor tenotomy is contemplated.[36]

When employing the electromyograph for the study of muscle function, it is not sufficient merely to demonstrate whether electrical activity is present at certain times. The equipment should be at least capable of distinguishing between real action potential and artefact and should preferably be of sufficient quality to demonstrate the differences between normal and pathological muscle. The resulting record then reveals not only whether there is muscular

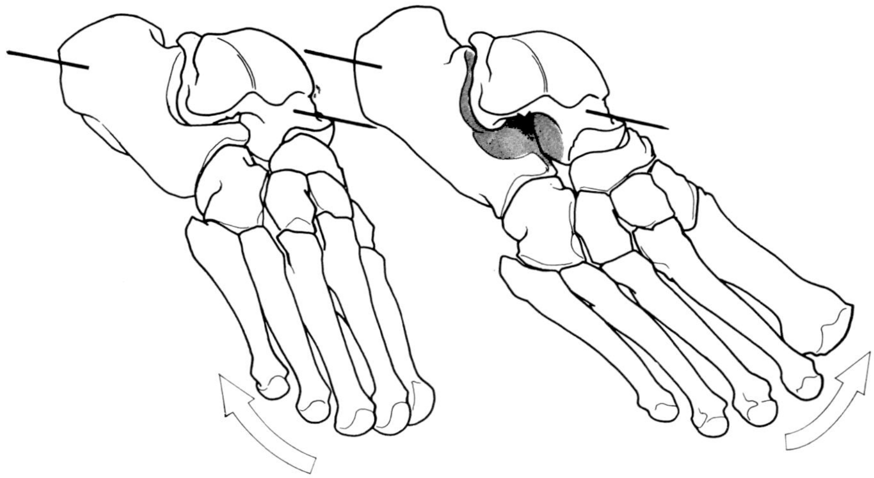

Figure 34. External and internal rotation of the foot about the compromise axis of the subtalar joint.

activity during a certain period but to some degree the quantity, quality and amplitude of the action potentials. Results are of little value if the method employs units of inappropriate frequency ranges or without high signal-to-noise recording ratios. A precise means of absolute identification by electrical stimulation of the muscle to be studied is essential. The differential amplifier, using electrodes placed close together in the muscle, insures restriction of the recording to the muscle under consideration.

One of the most profitable applications of the electromyograph method is in the study of locomotion. With the use of flexible internal dual electrodes, the activity of various muscles, including the deep muscles of the lower extremity, the hip, and also the iliacus and psoas muscles, can be recorded as the subject walks or performs other acts. Where tendon transfer has been made, muscles are readily evaluated postoperatively.[40] The time of contraction and the magnitude of electrical activity can be appreciated. Preoperative study of the muscle to be transferred may be of great value. Using the instrument postoperatively, one may readily ascertain the degree of recovery of the muscle and thus decide when sufficient training has been given. This can best be appraised if a preoperative study has been done.

Such an analysis of the muscle before and after surgery greatly enhances the orthopaedic surgeon's insight regarding the value and application of the various known transfers. For instance, experience has repeatedly shown that it is folly to expect the stance-phase peroneus longus muscle to adopt the swing-phase character of the anterior tibial muscle when transferred anteriorly. The anterior tibial muscle, however, when transferred posteriorly into the

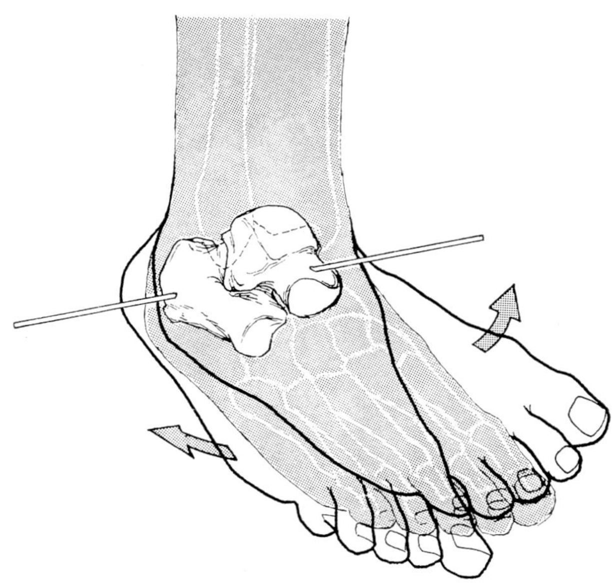

Figure 35. External and internal rotation of the foot at the subtalar joint (pronation and supination). (From Instructional Course Lectures, American Academy of Orthopaedic Surgeons, 1959.)

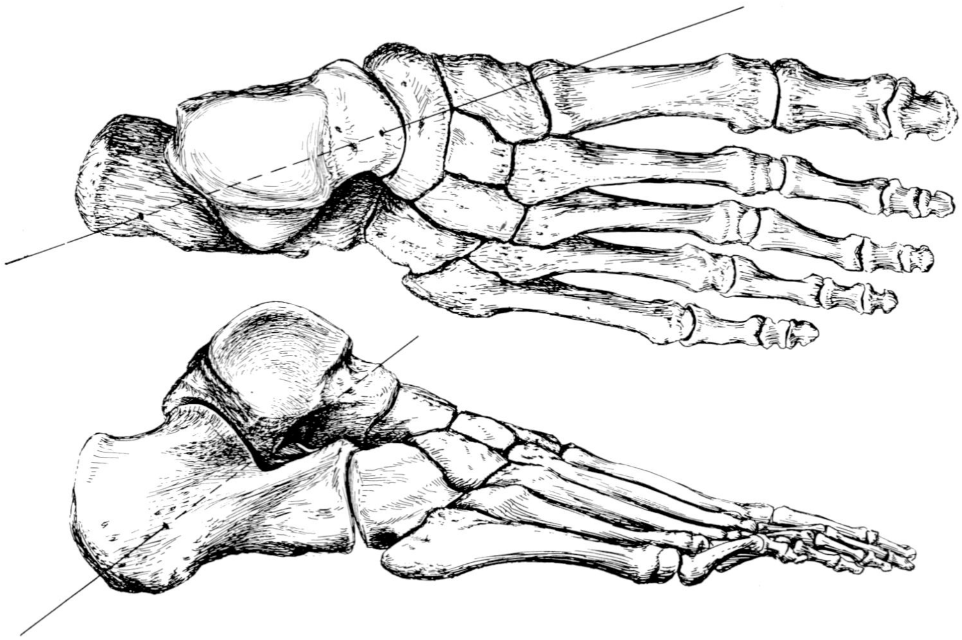

Figure 36. The position of the axis of the subtalar joint in the normal foot.

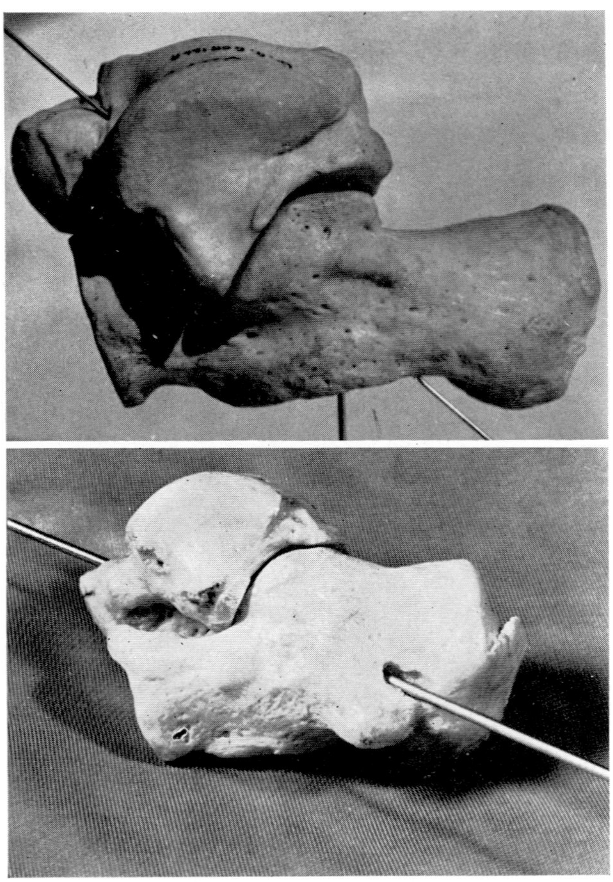

Figure 36A. The talocalcaneal joint of the gorilla (*top*) and the human foot (*bottom*) showing development of the anterolateral process of the calcaneus in man and its marked suppression in the gorilla.

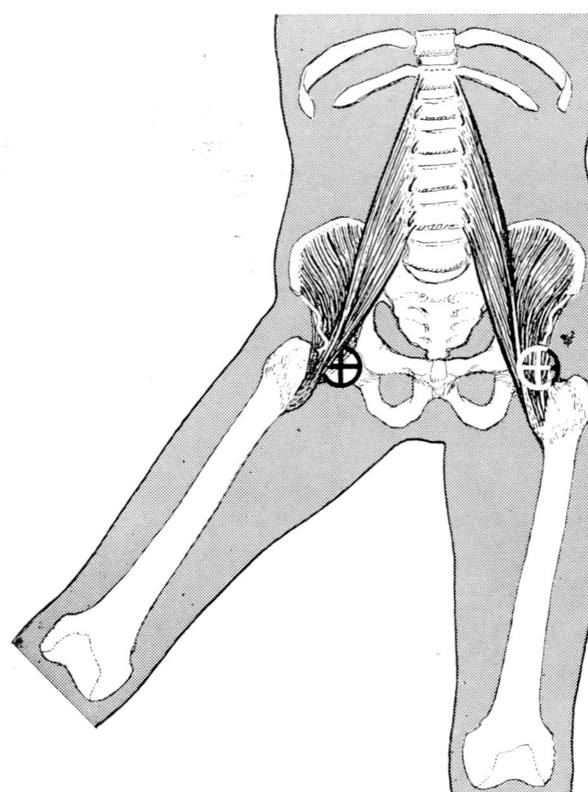

Figure 37. The position of the distal portion of the iliopsoas muscle with respect to the center of rotation of the hip joint in abduction (*left*); normal position (*right*).

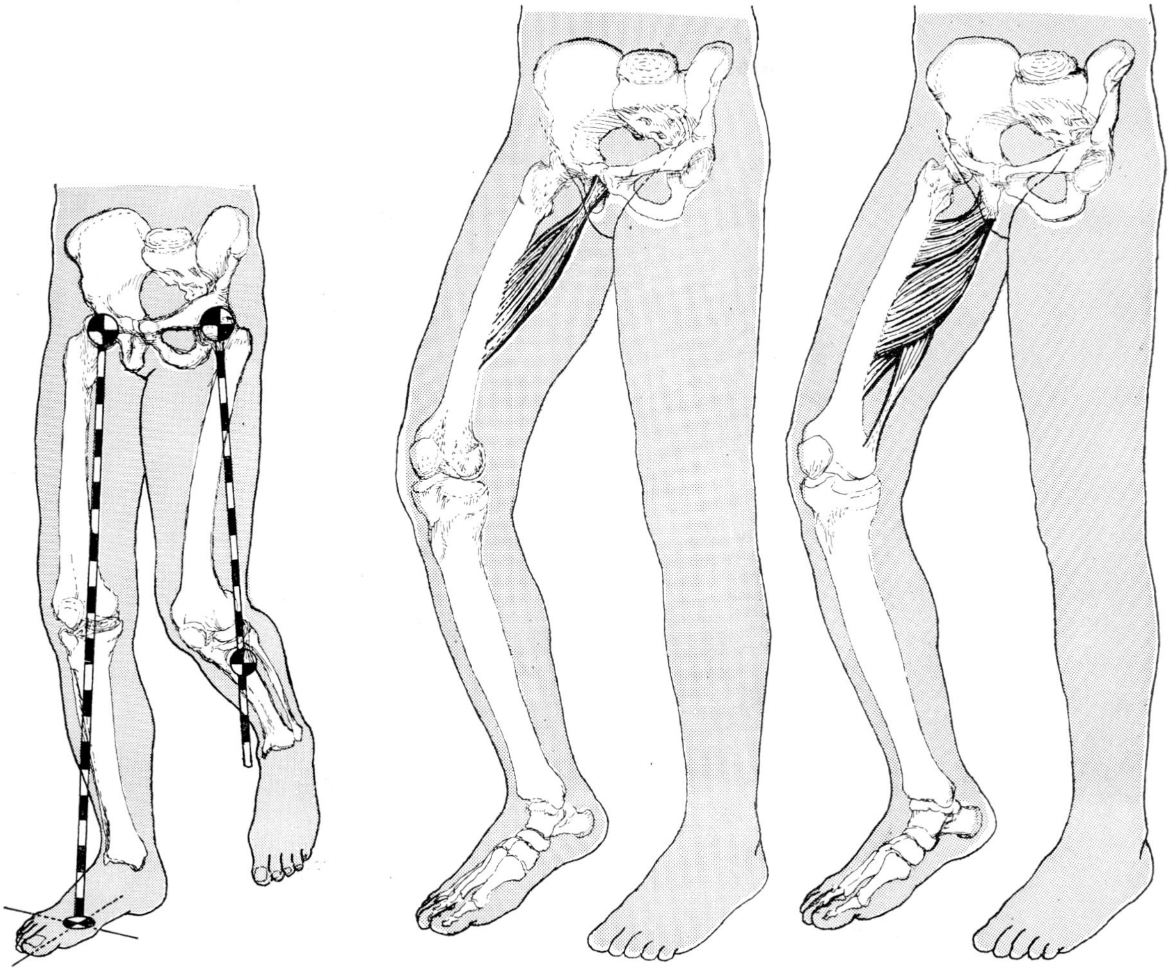

Figure 38, *left*. The axis of the lower extremity with extremity in contact with floor; with extremity in swing (*right*). (From Instructional Course Lectures, American Academy of Orthopaedic Surgeons, 1959.)

Figure 39, *center*. The adductor longus muscle.

Figure 40, *right*. The adductor magnus muscle, showing the anterior, middle, and posterior heads.

calcaneus has an excellent chance to become a stance-phase muscle (Chap. V).

Traditional methods of determining muscle function are

1. By study of the origin and insertion of the muscle.
2. By observation of the effect of electrical stimulation.
3. By observation of paralytics in whom certain muscles are paralyzed or weakened.

A more comprehensive study results from observation of actual participation of the muscle during specific movements. This is possible with the use of equipment which makes possible a synchronous display of the electrical activity of the muscle and the moving body parts concerned (see Chaps. V, VI). Records of electrical stimulation, voluntary movements and the use of the muscle during normal walking are easily obtained. Considerable insight into the function of muscles can be gained by combining this information with a knowledge of the axes of the joints of the lower extremity about which the muscles act.[34,35]

The well-known axis of the subtalar joint,[32,125] which enters the lateral aspect of the tuber of the calcaneus and emerges from the dorsal aspect of the head of the talus, passes through the center of the spherical portion of the head of the talus (Figs. 33, 34, 35). The position of this axis varies in each individual. A method for determining the position of the axis in the living subject has been developed by Wright, Desai and Henderson.[171] By designing a functional brace with ankle joint and adjustable subtalar joint, these investigators have been able to arrive at the position of the axis for a given individual.

To a great extent, the axis of the talocalcaneal joint influences the action of the anterior tibial (Fig. 41), peroneal (Fig. 42), posterior tibial (Fig. 43), and the long extensor muscles (Fig. 44). When tendon transpositions are made about such axes, the position of the axis and the time of contraction of the muscle are

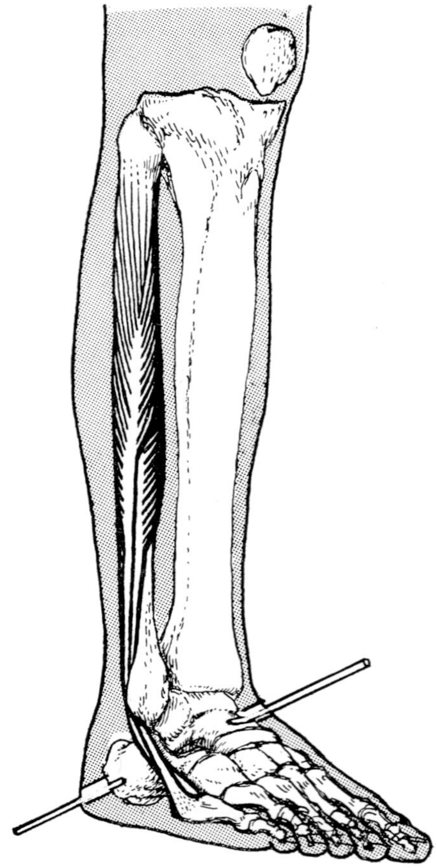

Figure 42. The position of the tendons of the peroneus longus and brevis with respect to the subtalar axis. (From Instructional Course Lectures, American Academy of Orthopaedic Surgeons, 1959).

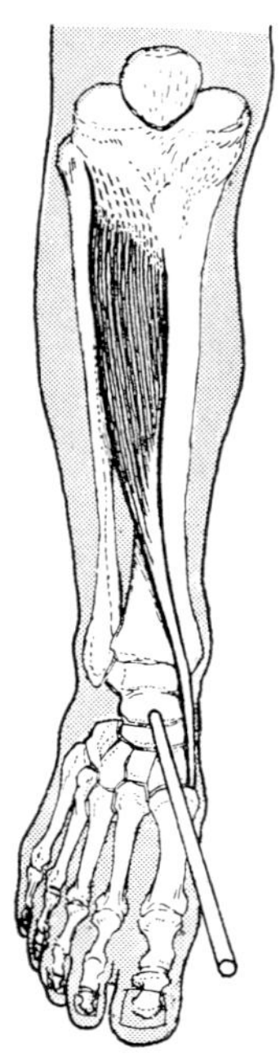

Figure 41. The position of the tendon of the anterior tibial muscle with respect to the axis of the subtalar joint. (From Instructional Course Lectures. American Academy of Orthopaedic Surgeons, 1959.)

paramount considerations. For example, with paralysis of the anterior tibial muscle (Fig. 46), one may wish to transfer the extensor digitorum longus into the dorsum of the foot. The surgeon must then determine the ideal position for insertion of the transferred tendon with respect to the axis of the subtalar joint (Fig. 47). Also in the varus foot with paralysis of the extensors and perhaps the peroneal muscles, one may elect to transfer the tendon of the remaining anterior tibial muscle laterally with the hope of overcoming varus while preserving efficient dorsiflexion. Here, knowledge of the anatomical position of the subtalar axis is essential in order to arrive at the ideal position on the dorsum of the foot for the new insertion of the anterior tibial muscle.

The movement of the foot about the subtalar

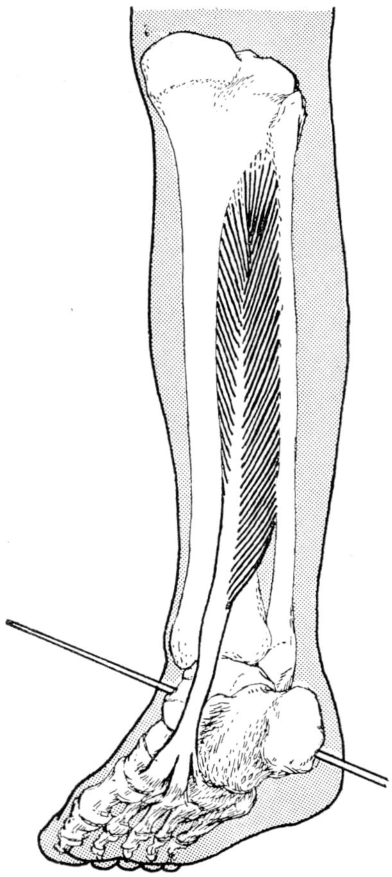

Figure 43. The action of the posterior tibial muscle about the axis of the subtalar joint. (From Instructional Course Lectures, American Academy of Orthopaedic Surgeons, 1959.)

joint is important in analyses of motion in feet of different types.[61] Using metallic pins as targets, the angular motion of the calcaneus with respect to the talus during the stance phase of locomotion was calculated.[38] The pins were inserted into the lateral aspect of the calcaneus and the lateral aspect of the neck of the talus in order to form an angle in a plane perpendicular to the calcaneotalar axis. The angle was continuously recorded by 35-mm film projection in two perpendicular planes. Several determinations were made on eight individuals. The first three subjects were selected as follows:

1. A subject with a normal foot.
2. A subject with a moderately pronated foot.
3. A third subject having a very flexible flatfoot.

X-ray films were made on all of the individual feet examined in order to rule out any possibility of bone or joint abnormality. The motion of the calcaneus on the talus in each of the three subjects is shown in Figure 48, with each curve representing several walking cycles which were calculated and averaged. The curves were made from averages of individual walking cycles where each point represents a calculation for consecutive projections and the camera was running at a constant speed of 50 frames/sec.

These calculations of the motion of the subtalar joint were thus made from simultaneous photographic projection in two planes using (a) the direct side or lateral view and (b) the view from directly below. This view was made possible by a 45° inclined mirror (the subject walking on a glass walkway).[38] The direction cosine[55] of each target was measured by veniered protractor in three planes for each 35-

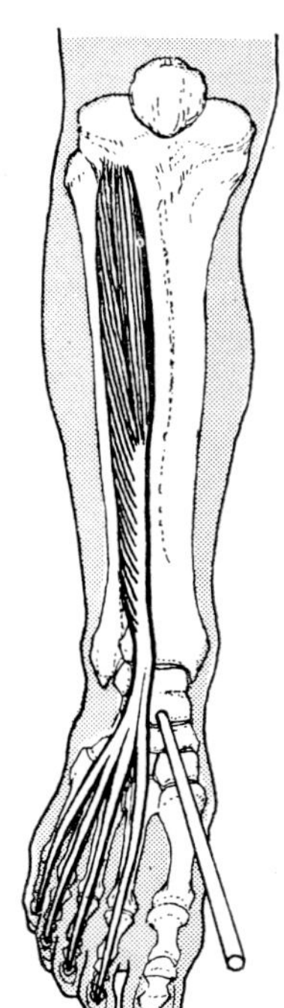

Figure 44. The relationship of the extensor digitorum longus tendons to the subtalar axis. (From Instructional Course Lectures, American Academy of Orthopaedic Surgeons, 1959.)

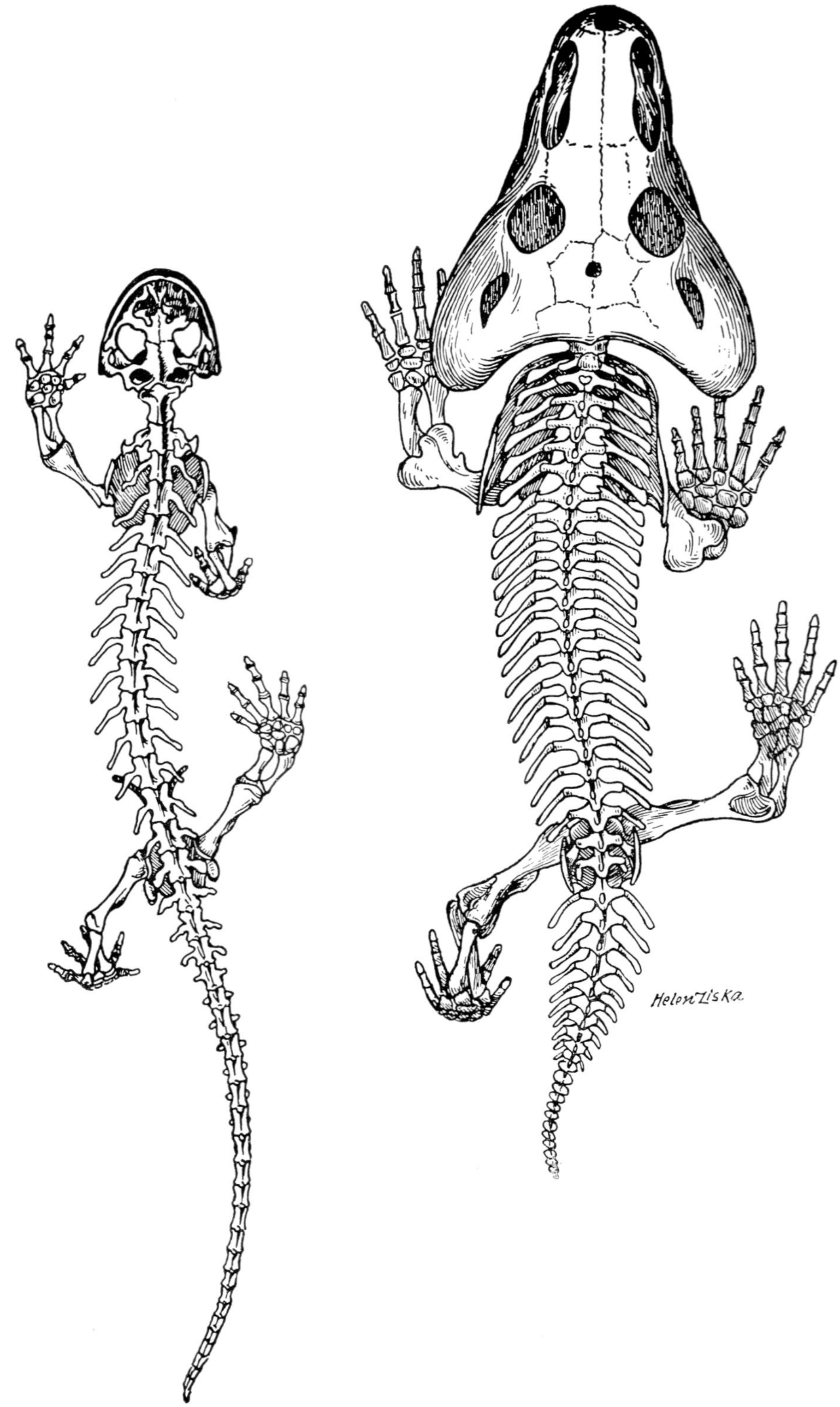

Figure 45. A caudate and a labyrinthodont demonstrating horizontal rotation in quadruped.[154] (Courtesy of American Museum of Natural History.)

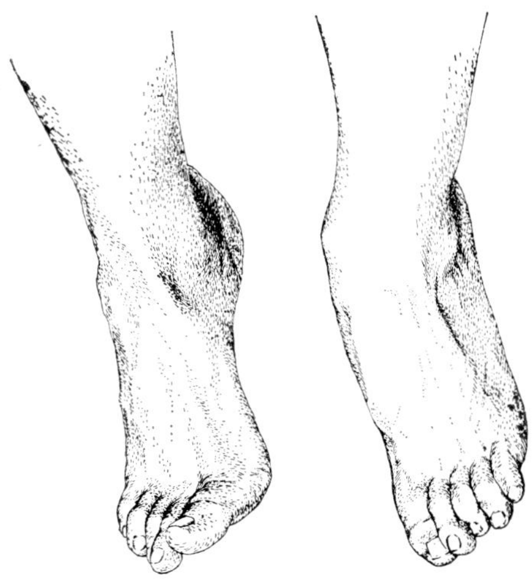

Figure 46. In the absence of a useful anterior tibial muscle (*right*), the foot externally rotates at the subtalar joint on dorsiflexion. This demonstrates the importance of the long extensors as external rotators of the foot. (From Instructional Course Lectures. American Academy of Orthopaedic Surgeons, 1959.)

mm frame. Thus the angle formed by the two pin targets could be calculated for each frame as the subject walked.

The curves of Figure 48 readily show the differences in subtalar motion in the various types of individuals.

1. The normal (somewhat cavus) foot demonstrated very little motion during walking.
2. The moderately pronated foot showed 5 or 6° of rotation about the subtalar axis during the stance phase of walking.
3. The individual with the very flexible flatfoot repeatedly demonstrated 11 or 12° of motion of the calcaneus about the talus.

TABLE I
CLASSIFICATION OF THE HUMAN FOOT ACCORDING TO POSITION OF THE SUBTALAR AXIS*

	Motion During Stance Phase of Walking	*Total Range of Voluntary Motion*	*Medial Deviation of Axis*
Cavus foot	Slight 3 to 4°	Least 11°	Least <16°
Normal, somewhat pronated foot	Moderate 5 to 6°	Moderate 24°	Average 16°
Flatfoot	Pronounced 16°	Greatest 28°	Greatest >16°

* Based on data from Subjects 1, 2, 3.[38]

With the targets in place, the motion of the calcaneus about the subtalar axis was also demonstrated during active pronation and supination of the foot with the subject sitting. This motion also might be called external and internal rotation at the subtalar joint. The measurements, of course, were actually made at the talocalcaneal joint. Passive motion demonstrated the same relationships: there was far more motion in the flatfoot than in the moderately pronated foot. The somewhat cavus foot had the least active voluntary motion of all three.

These curves representing the motion of the talocalcaneal joint during the stance phase of walking lend themselves to a considerable amount of analysis. Where the joint is active, i.e. where there is a considerable amount of motion of the calcaneus about the axis of the subtalar joint—for example, in the lower curve of Figure 48—there is considerable supination or internal rotation at the beginning of stance and as full weight bearing approaches. This motion is reversed, as is demonstrated in the

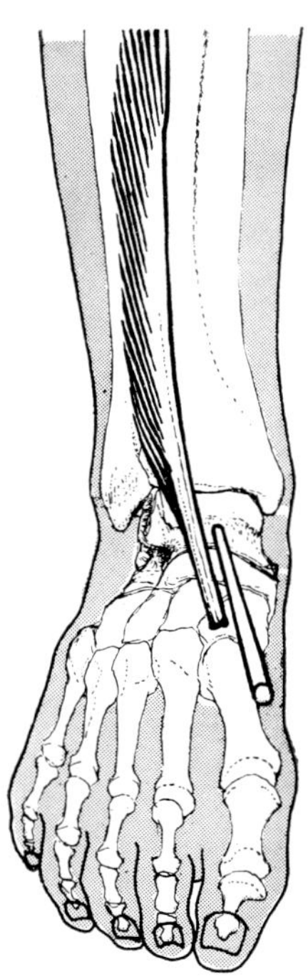

Figure 47. Neutral placement of the extensor transfer into the dorsum of the foot. (From Instructional Course Lectures. American Academy of Orthopaedic Surgeons, 1959.)

diagram of Figure 48, as the stance phase of walking terminates and the extremity is being relieved of weight-bearing. The motion is far less pronounced in the normal type of foot (see upper curve, Fig. 48). Internal rotation of the calcaneus about the subtalar joint actually represents supination.

The application of the study provides an understanding and further insight regarding such deformities as the severely pronated foot, the valgus foot of polio and the proper mechanical approach toward insertion of the subtalar bone block of Grice.[81] Before attempting the bone block, it is assumed that one has balanced musculature acting about the axis of the subtalar joint. In our experience, a small single graft placed perpendicular to the subtalar axis, avoiding all articular surfaces and extending well into the joint itself, has been most successful (Fig. 49). The understanding of the above concepts regarding the subtalar joint—its axis, and its motion in various types of individuals—

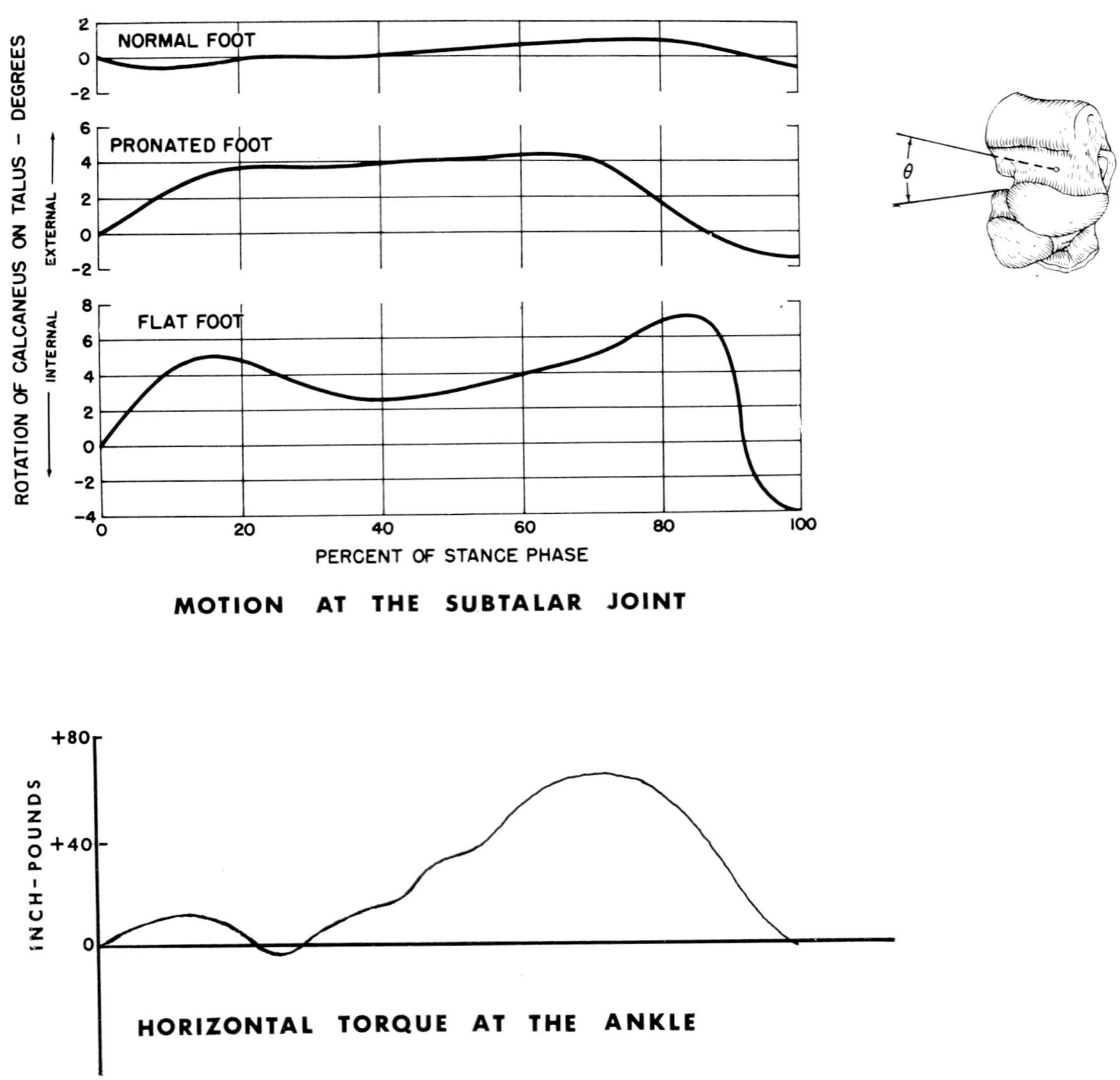

Figure 48. The motion at the talocalcaneal joint during the stance phase of walking in the normal foot, the pronated foot and the flatfoot. (Averaged curves taken from a report to the Advisory Committee on Artificial Limbs, Series 11, Issue 24, May, 1953.[38]) *Below*, A curve representing the horizontal torque at the ankle in the normal individual.

has been most helpful in developing a satisfactory technique for the procedure.

Similarly, the peroneus longus, if transfer is resorted to, must be inserted into a certain site in the dorsum of the foot. It must oppose any remaining internal rotators of the subtalar joint. Its proximity to the axis is thus a matter of balance, with other remaining muscles acting about the same axis (Fig. 50).

In the consideration of axes, one must not neglect the ability of the ankle joint to move in more than one plane.[12,33,37] The trochlea of

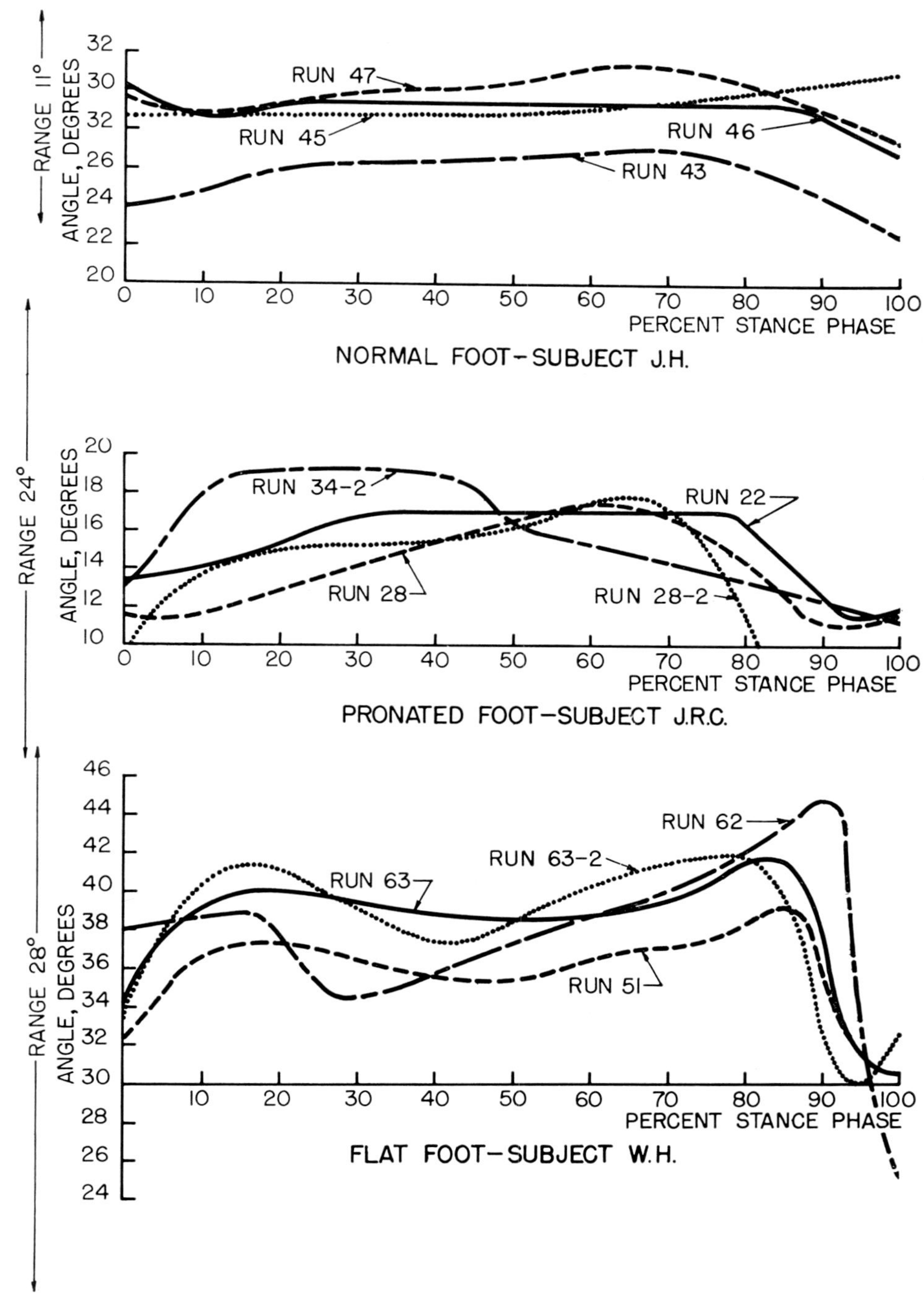

Figure 48A. Rotation of the calcaneus on the talus during normal walking by three subjects. The curves from which the averaged curves of Figure 48 were calculated.

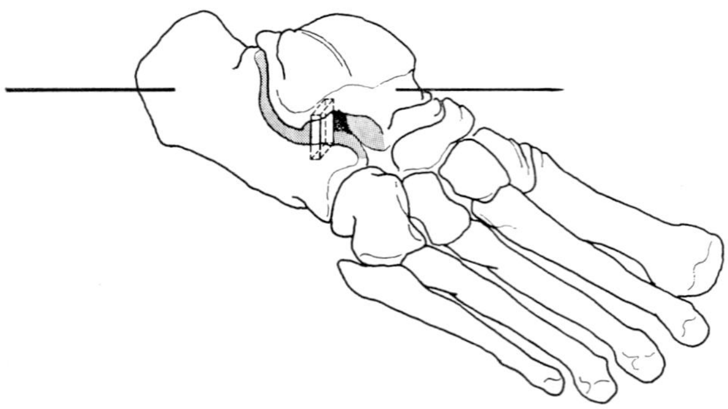

Figure 49. The subtalar bone block[81] placed perpendicular to the axis of the subtalar joint.

the talus is actually a portion of a truncated cone, with its apex located medial to the ankle (Fig. 51). (A line drawn between the malleoli, commonly referred to in clinical practice, does not represent the functional axis of the ankle joint.)

It has been demonstrated that normal subjects develop considerable horizontal torque between the leg and the foot during walking. This torque is hardly changed in the amputee or the patient with triple arthrodesis.[51] Transfer of weight bearing from one extremity to another in the quadruped (Fig. 45) makes obvious the necessity for horizontal rotation and torque in the lower extremities.

Figure 52 is a roentgenogram of the foot and ankle of a patient, age 36, who had a successful triple arthrodesis at age 13 for weakness and deformity of the lower extremity following poliomyelitis. The abnormal changes at the ankle joint are evident. These arthritic changes, with near dislocation of the ankle, are undoubtedly the result of horizontal torque repeatedly applied during walking. This torque has placed additional requirements on the ankle joint following fusion of previously mobile segments distal to the ankle joint.[37]

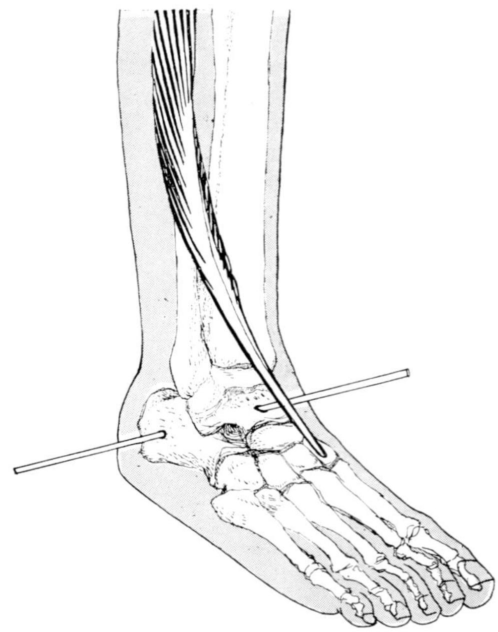

Figure 50. The new insertion of the transferred peroneus longus tendon showing its relationship to the axis of the subtalar joint. (From Instructional Course Lectures. American Academy of Orthopaedic Surgeons, 1959.)

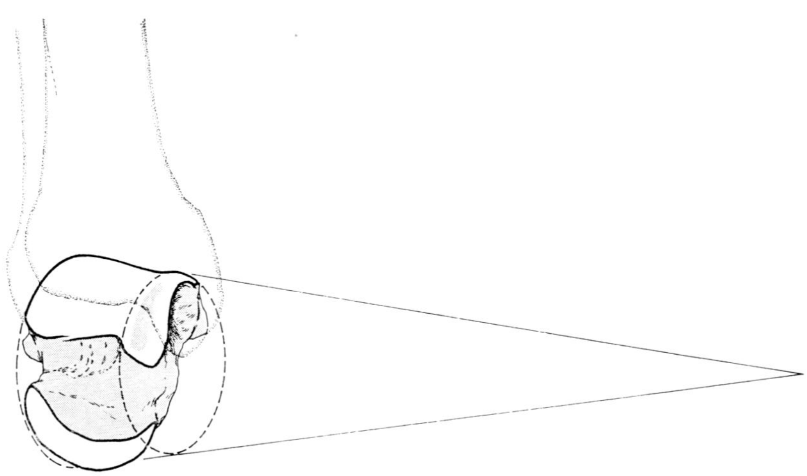

Figure 51. The action of the ankle joint. The trochlea of the talus is a portion of a truncated cone, its apex located medial to the ankle joint.

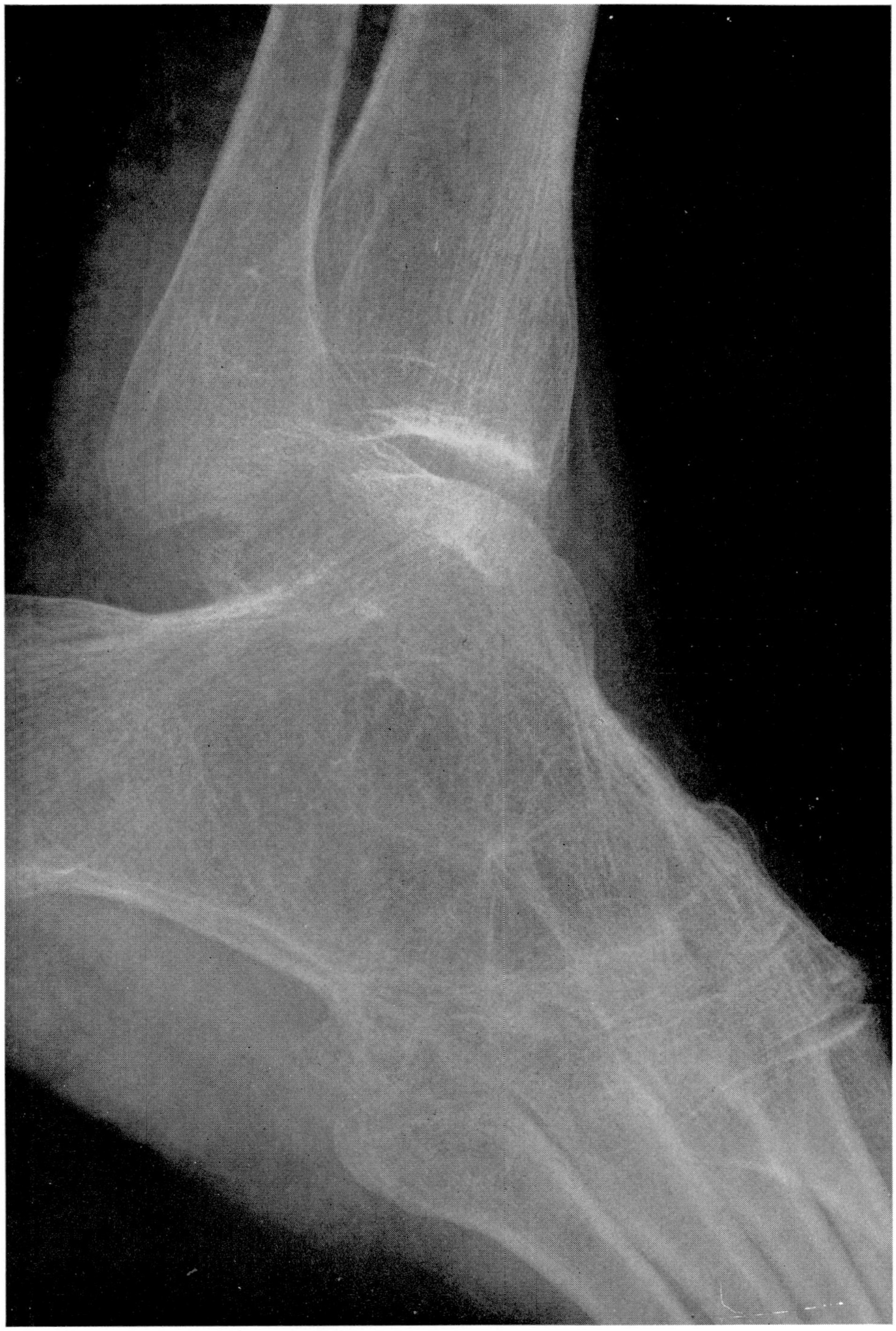

Figure 52. Marked arthritic changes at the ankle 23 years following triple arthrodesis.

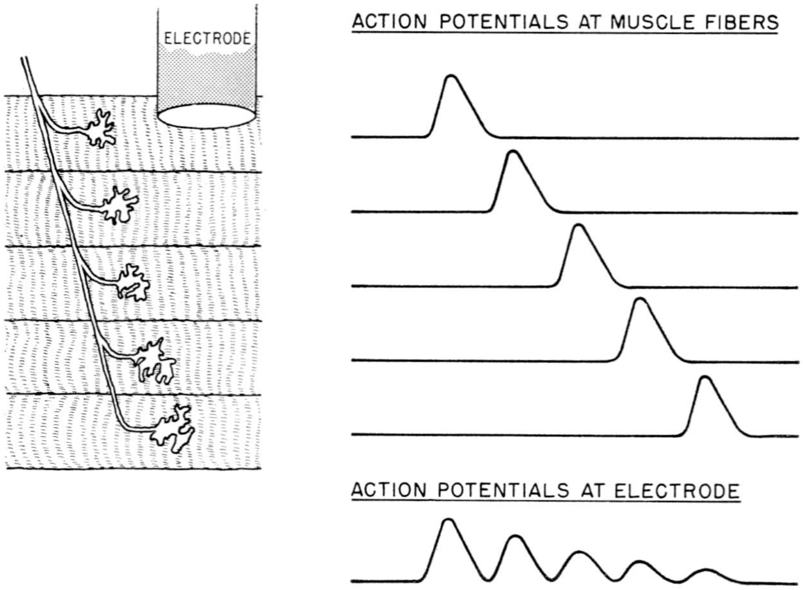

Figure 53. The action potentials of muscle fibres near the recording electrode vary in their contribution to the final recording. The differences in time and amplitude are exaggerated. The aggregate potentials of such spatially distributed muscle fibres produce irregularity in the action potentials of the motor unit.

The motion about the axis of the subtalar joint, crucial in most of the activities involving the lower extremity, was described in 1859 by Henke.[87] Further descriptions were contributed subsequently by Meyer,[131] Donitz[54] and Fick.[68] A method for establishing the precise position of the axis about which the calcaneus rotates with respect to the talus was described by Manter.[125] It was defined as a line deviated 42° from the walking surface and 16° medial to a line extending from the center of the calcaneus to a point between the first and second metatarsal heads. The position of the axis varies according to the type of foot. Its position also varies in each individual foot during pronation and supination. The position of the subtalar axis determines the action of muscles inserting to either side of the axis regarding their capacity to produce *internal* or *external* rotation about the axis.

This important axis, inclined medially and upward, determines the crucial subtalar motion. *Internal* rotation of the calcaneus about the axis occurs only with *inversion* and some *plantar flexion*. The resulting motion defined in three dimensions is *supination*. *External* rotation of the calcaneus about the subtalar axis occurs only with *eversion* and some *dorsiflexion*. The resulting composite motion is termed *pronation*.

In addition to internal and external rotation at the subtalar joint, Manter[125] also observed a small amount of forward translation of the talus along the axis during *pronation*. This motion was thought to be reversed during *supination*. The translation, according to Manter, was approximately 1.5 mm/10° of rotation.

Observation of passively produced subtalar rotation in dissected fresh specimens clearly establishes the nature of this motion. Pronation and supination involve essentially the talocalcaneal and the talonavicular joints. Any contribution of the other intertarsal joints and the tarsometatarsal joints may, for the most part, be neglected in this consideration of the functional anatomy of the foot. Their participation may be shown to be minimal. Only two truly mobile joints, the talocalcaneal and the talonavicular, contribute significantly to subtalar motion. Thus, one may consider the subtalar motion to be one between the talus and a "fused" group of tarsals beneath, permitting simultaneous rotation at the talocalcaneal and the talonavicular articulations.

The orientation of the axis accounts for variations in the type of foot. In the pronated foot, the axis is deviated medally more than 16°. In the supinated foot, the angle of deviation is significantly less than the average of 16°.

The position of the axis determines also the amount of motion found at the subtalar joint. The *pronated* foot, with axis deviated more *medially* than normal, has a greater range of subtalar motion. The *supinated* foot, with axis less deviated, demonstrates less subtalar motion.[61] The *supinated* foot thus takes advantage of a more favorable position. The head of the talus is closer to a position superior to the calcaneocuboid joint, under which is the strong long plantar ligament. The tarsals at the transverse tarsal joint are aligned closer to the vertical than those of the pronated foot. Transverse tarsal joint motion becomes far easier to stabilize. Subtalar motion is virtually eliminated during the stance phase of walking (Fig. 48 and Subject b, Fig. 59). The supinated human foot then more closely resembles an efficient lever. The more directly the talonavicular joint is placed over the calcaneum, the more completely immobilized are the subtalar and the transverse tarsal joints and the more effective is the foot as a lever. In contrast, the chimpanzee foot, as described by Elftman and Manter,[61] actually "breaks" at the transverse tarsal joint during the stance phase of walking. The tarsals of the chimpanzee are arranged, for the most part, in the horizontal plane. This results in a series of joints very difficult to control. The chimp foot characteristically "breaks," the heel rising from the walking surface while the structures distal to the transverse tarsal joint are still flat on the ground. As in the gorilla foot (Fig. 54), the subtalar axis is deviated over 45° from the sagittal plane. The foot of the arboreal gorilla is used for a somewhat different purpose than that of the lowland gorilla. In the latter form, a longitudinal arch appears to be developing, and in man, where the foot serves as a lever, the axis is inclined far less medially (Fig. 54).

METHOD

Added to the three previously reported subjects,[38] five additional subjects ranging in age from 20 to 29 years were selected at random as far as type of foot was concerned. This report is thus based on 8 subjects. Using local anesthesia, 5/64 in. threaded stainless steel pins were fixed securely into the lateral aspect of the neck of the talus and the tuber of the calcaneus. To the extent possible, the subtalar axis was located, and the pins were placed perpendicular to it. Subtalar motion then demonstrated motion of the two pins in parallel planes.

The subject then walked on a level glass walkway, with the targets in view of a 35-mm motion picture camera 21 feet distant. The camera lens was at the same height as the glass walkway. Its optical axis was perpendicular to the border of the walkway at its center, where the foot met the walking surface for the walking cycle to be analyzed. Each subject walked without shoes.

Directly beneath the glass walkway, an optical mirror was placed at an angle of 45° to the horizontal. Each frame of the motion picture camera thus showed the subject's foot, leg and targets in two perpendicular views. The camera was run at a known constant rate of 50 frames/sec.

Tests were made of known angles. The error due to the proximity of the camera (21 feet) was calculated. The greatest error resulting from any known source was a fraction of 1 percent.

The subject was asked to walk in his usual manner. He was positioned in order to favor the chance that the pin-loaded foot would strike the center of the walkway. All subjects appeared to walk in their normal fashion. Only one had symptoms of any kind during the pro-

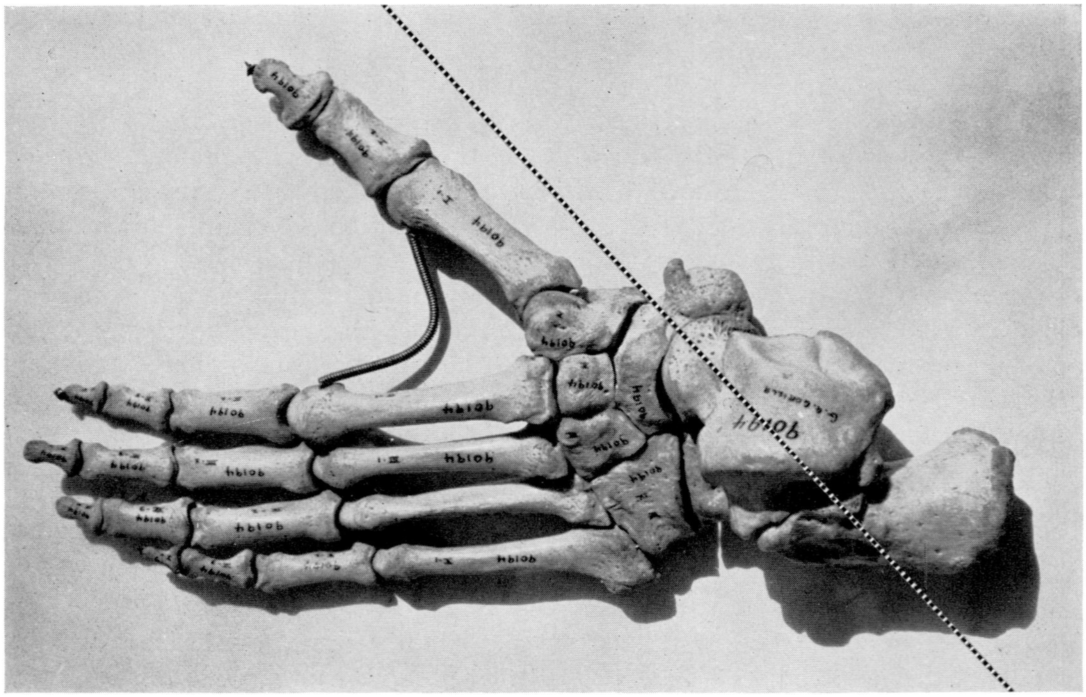

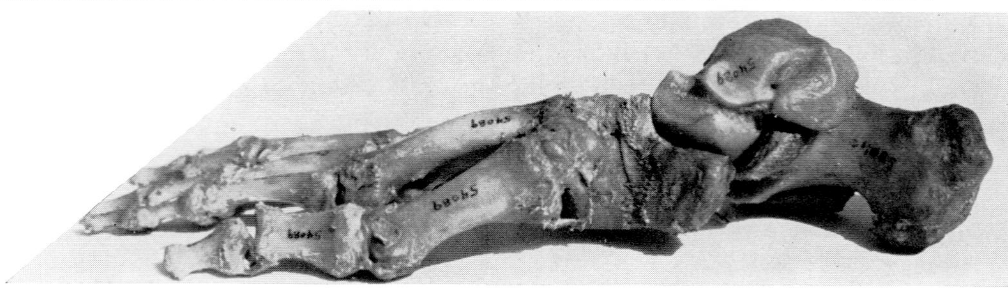

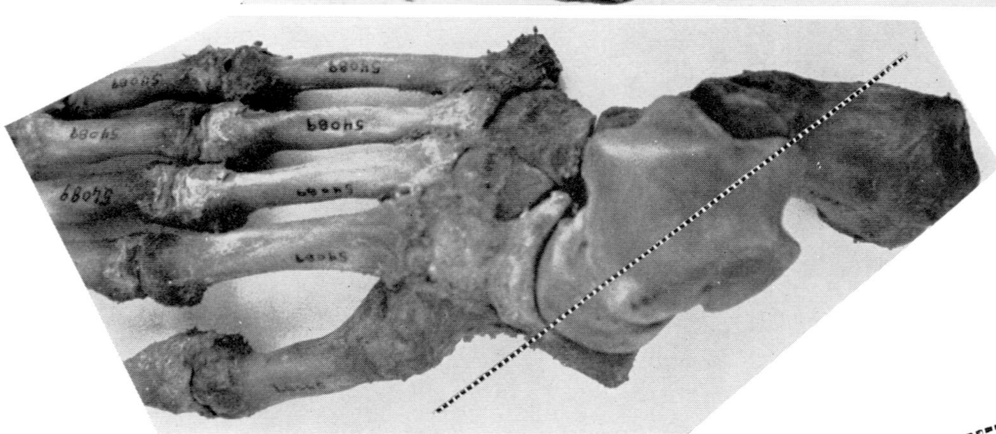

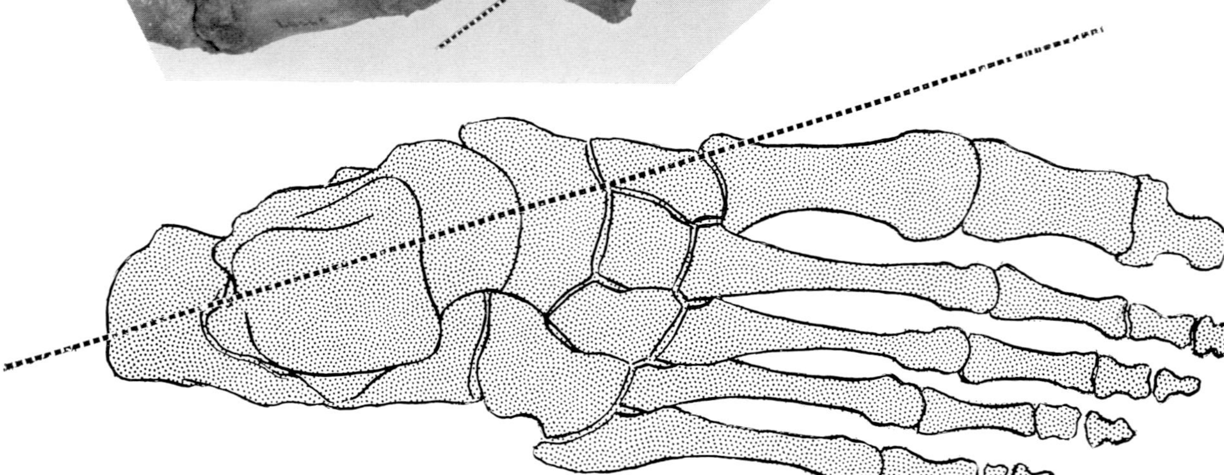

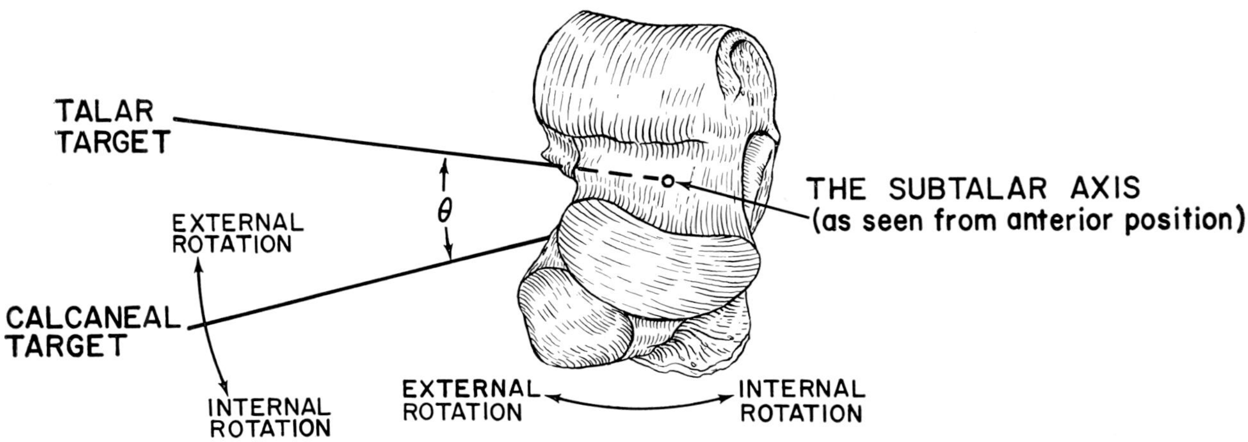

Figure 55. Orientation of talar and calcaneal targets for observation of the angle θ. Right talus and calcaneus with pin targets inserted to form the angle θ. The angle increases as the calcaneus rotates internally about the subtalar axis. The angle decreases as the calcaneus rotates externally about the axis. (From *Clinical Orthopaedics and Related Research, 50,* 1967.)

cedure. This subject complained of mild discomfort. His gait appeared to be undisturbed.

From the two perpendicular-plane series of photographs of the talar and the calcaneal targets, x, y and z or three-dimensional projections were measured. The x, y and z projections of the measured portion of the talar and the calcaneal pins were obtained by verniered protractor measurements of these projections in each 35-mm frame. From these values, the direction cosines of the targets, the angle θ between the pins may be calculated for each frame:

$$\text{Cos } \theta = \frac{T_xC_x + T_yC_y + T_zC_z}{\sqrt{T_x^2 + T_y^2 + T_z^2}\,\sqrt{C_x^2 + C_y^2 + C_z^2}}$$

After finding the angle θ for each consecutive frame during walking or voluntary pronation and supination of the foot, the resulting series of values is plotted. The resulting curves, with time as the horizontal dimension, represent the angular change or the rotation at the subtalar joint about the subtalar axis. Curves were obtained for normal walking and for active internal and external rotation as the subject demonstrated the total range of motion—the termination of supination and pronation.

The angle θ, formed by the skew lines (the talar and the calcaneal targets), was computed for each 35-mm frame according to the above equation. The data, consisting of the projections of the pins in three dimensions, was placed on the punchcards (Fig. 58). Then the punchcards were processed by the IBM No. 7040 computer, which was programed according to the equation for θ. The value of the angle θ was plotted for each frame. Heel strike was considered frame number 1. Successive frame numbers appear on the abscissa (Fig. 59).

The target pins placed perpendicular to the subtalar axis do not intersect; they are not parallel; they are called *skew lines.*[18,55,137] There is only one line in space that is perpendicular to each of two skew lines.[55,137] In order to make the calculation of the angle θ between the two targets, the assumption is made that the talar and the calcaneal targets are perpen-

---

≪ Figure 54. The position of the subtalar axis in the horizontal view: *top,* arboreal gorilla; *center,* lowland gorilla; *bottom,* human. (From Close, Inman, Poor, and Todd, *Clinical Orthopaedics and Related Research, 50:*159–179, 1967.)

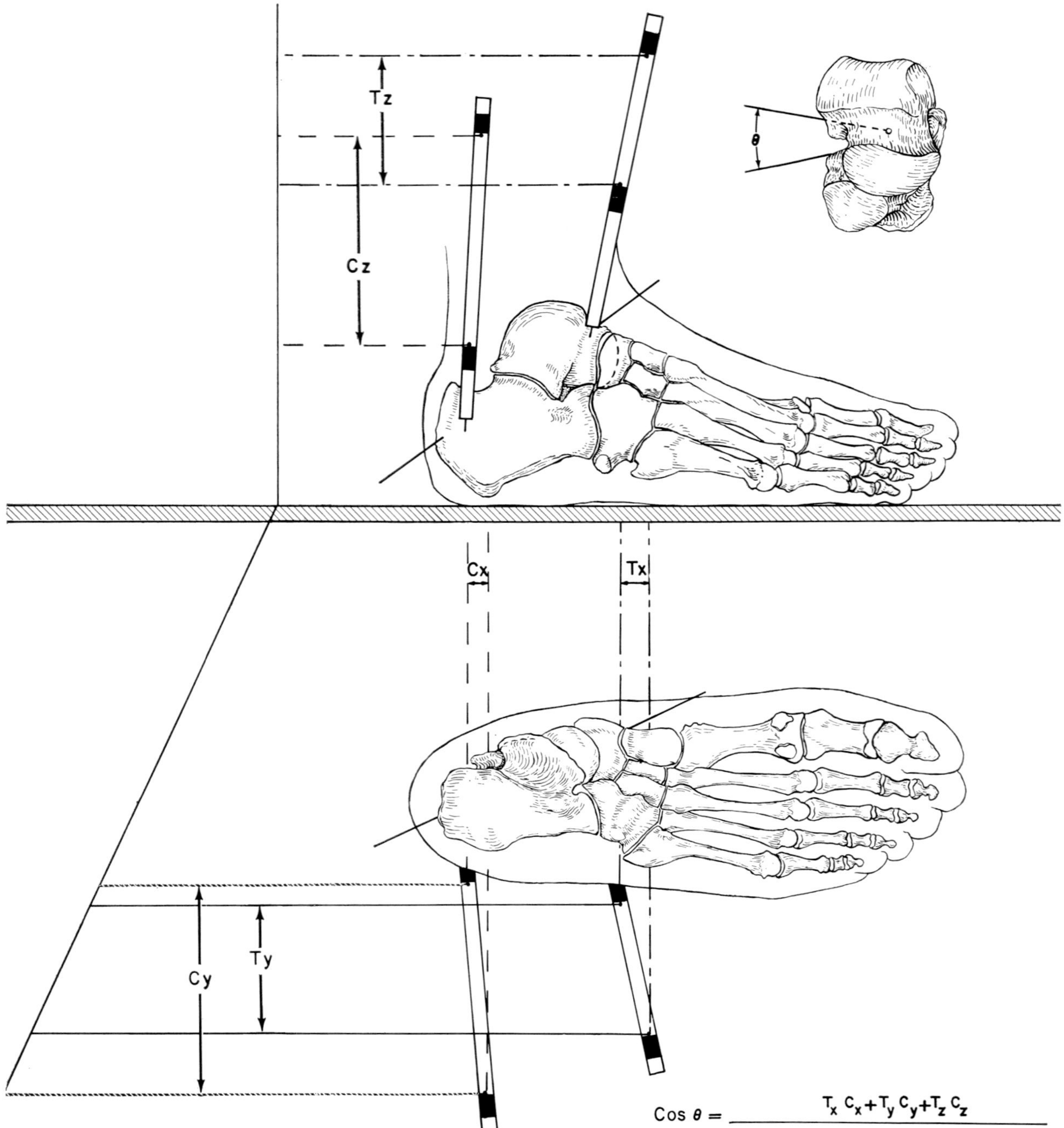

$$\cos\theta = \frac{T_x C_x + T_y C_y + T_z C_z}{\sqrt{T_x^2 + T_y^2 + T_z^2}\ \sqrt{C_x^2 + C_y^2 + C_z^2}}$$

Figure 56. X, Y and Z projections of talar and calcaneal targets as measured for each 35-mm frame of motion picture film, for which calculation of the angle θ is done. (From *Clinical Orthopaedics and Related Research*, 50, 1967).

dicular to the subtalar axis. The possibility of error due to change in the position of the subtalar axis is neglected in these calculations. However, our observations are that this change is so small that it has no significant effect on the values found in this study.

Roentgenograms of the foot of each subject studied were made after the test. In the lateral view, small pinholes in the lateral aspect of the neck of the talus and the lateral aspect of the tuber of the calcaneus were clearly demonstrated.

RESULTS

The data from the five additional subjects studied by the bone-pin technic were reduced by the same method used for the original study of three subjects.[38] The method enables one to calculate the angle formed between two skew lines.

As demonstrated by the data from the original three subjects reported in 1953, subjects may be classified as in Tables I and II. The recently studied five subjects were chosen at random. The original three subjects were selected for the purpose of studying the behavior of the subtalar joint in the flatfoot, the moderately pronated foot and the moderately cavus foot.

TABLE II
RANGES OF SUBTALAR MOTION*

Subj. No.	*Range of Motion*
1	23.58°
2	11.20°
3	28.00°
4	23.10°
5	9.93°
6	12.54°
7	14.42°
8	18.13°

* Eight subjects.

In the second study, some improvement in target visualization during data analysis was obtained, since the camera-to-subject distance was 21 feet. In the original study, this distance was 50 feet. The use of the shorter distance introduced some error, but mathematical analysis proved this to be less than 1 percent.

As in the first study, successive curves of the plotted angle differed. Although the range of values of the plotted angle were well within the range of significant variation and each curve was therefore considered to be significant, the information obtained was that successive walking cycles are not duplicated. Although stability of the pin targets was checked carefully and trial measurements of known angles were made, the subject demonstrated no tendency to duplicate individual curves. The data does reflect a definite range of angular change for each subject. There was also a definite individualistic grouping of each subject's complete curves for angular change during walking. Although the curves for an individual were never identical, individual subjects demonstrated their own distinctive types of curves (Fig. 59).

The position of the metatarsophalangeal joints, considered to be 180° during the period when the entire foot is on the walking surface, is conveniently recorded in the lateral view. This angle and the curve representing the position of the ankle are shown in Figure 59A–H. These curves are taken from the walking cycle represented by the fourth curve for calcaneotalar motion. It is evident that dorsiflexion at the metatarsophalangeal joints is reflected sharply in the calcaneotalar curve just above it. Since the metatarsal heads are oriented obliquely, external rotation of the tibia would be necessary when the heel rises were it not for the subtalar joint. The sharp change in direction of the subtalar rotation at this crucial point of the walking cycle (Curve 4, Fig. 59A–H) is evidence of nature's device for controlling horizontal rotation in the lower extremity.

Although subjects did not precisely duplicate their curves for subtalar rotation during the stance phase and each subject demonstrated an individual pattern, certain events were common to all subjects.

In most of the curves, very early in stance

there was brief internal rotation of the calcaneus on the talus. This was followed by a longer period of external rotation—or pronation. This interval represents the period of maximum weight bearing. As the metatarsophalangeal angle changes abruptly in each curve of each subject, the calcaneus simultaneously reverses its rotation from external to internal (Fig. 59A–H). Thus the stance phase may be divided into three periods regarding subtalar motion.

DISCUSSION

Studies demonstrating horizontal rotation of the segments of the lower extremity,[115,151] using bone-pin targets, indicate increasing magnitudes of horizontal rotation in the more

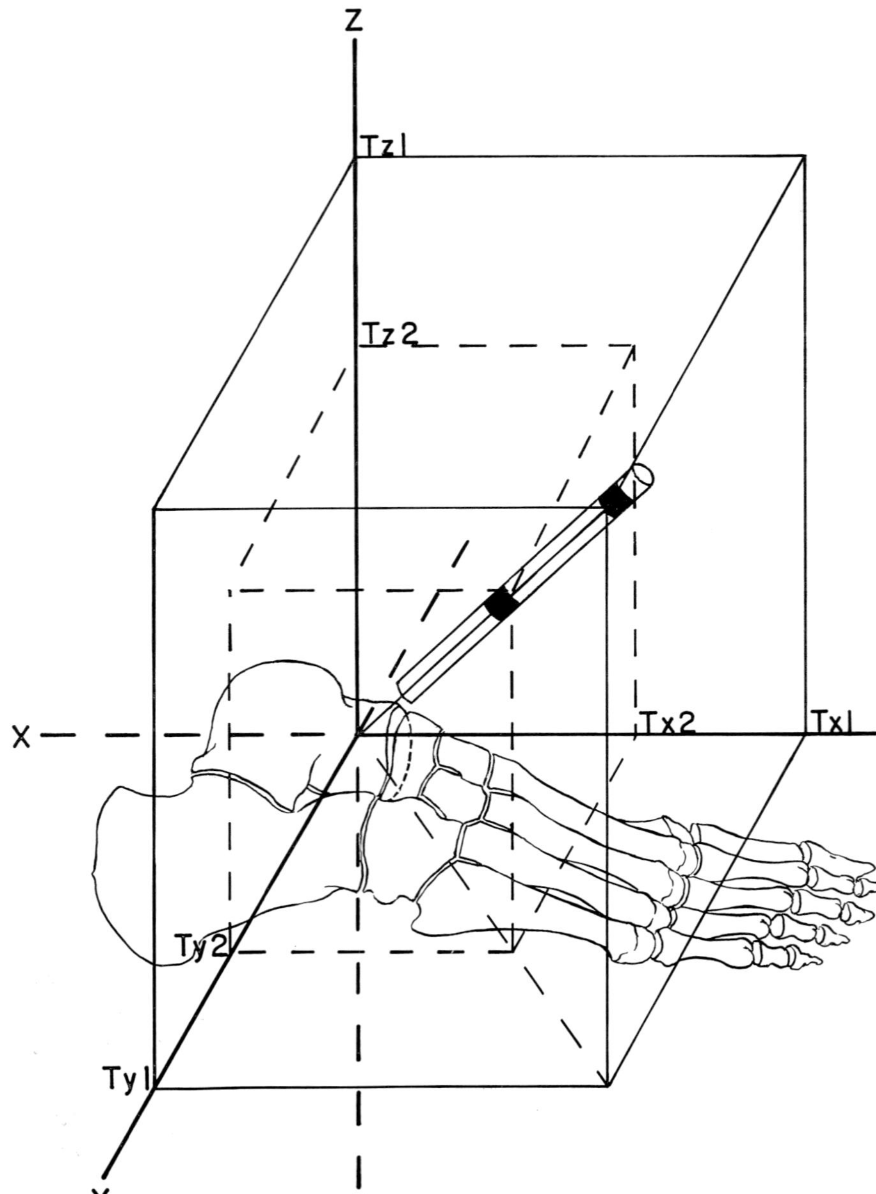

Figure 57 *left.* X, Y and Z projections of talar target.

distal segments. Thus the pelvis rotates 3 to 4°, the femur 7 to 8°, and the tibia 10 to 11°. These values are averages taken from the data of 12 normal subjects.[115] The data was taken, for the most part, during the stance phase of walking.

Regarding the study of the ankle and the subtalar joints, one must seriously consider the apparent rotation of the tibia, since the foot is fixed on the walking surface during a significant portion of the stance phase of walking.

As the tibia is rotating in the horizontal plane about its long axis, even when the foot is fixed on the walking surface, one might well expect the skeletal system to take advantage of the subtalar joint, thus making use of its horizontal component of motion in order to adapt the rota-

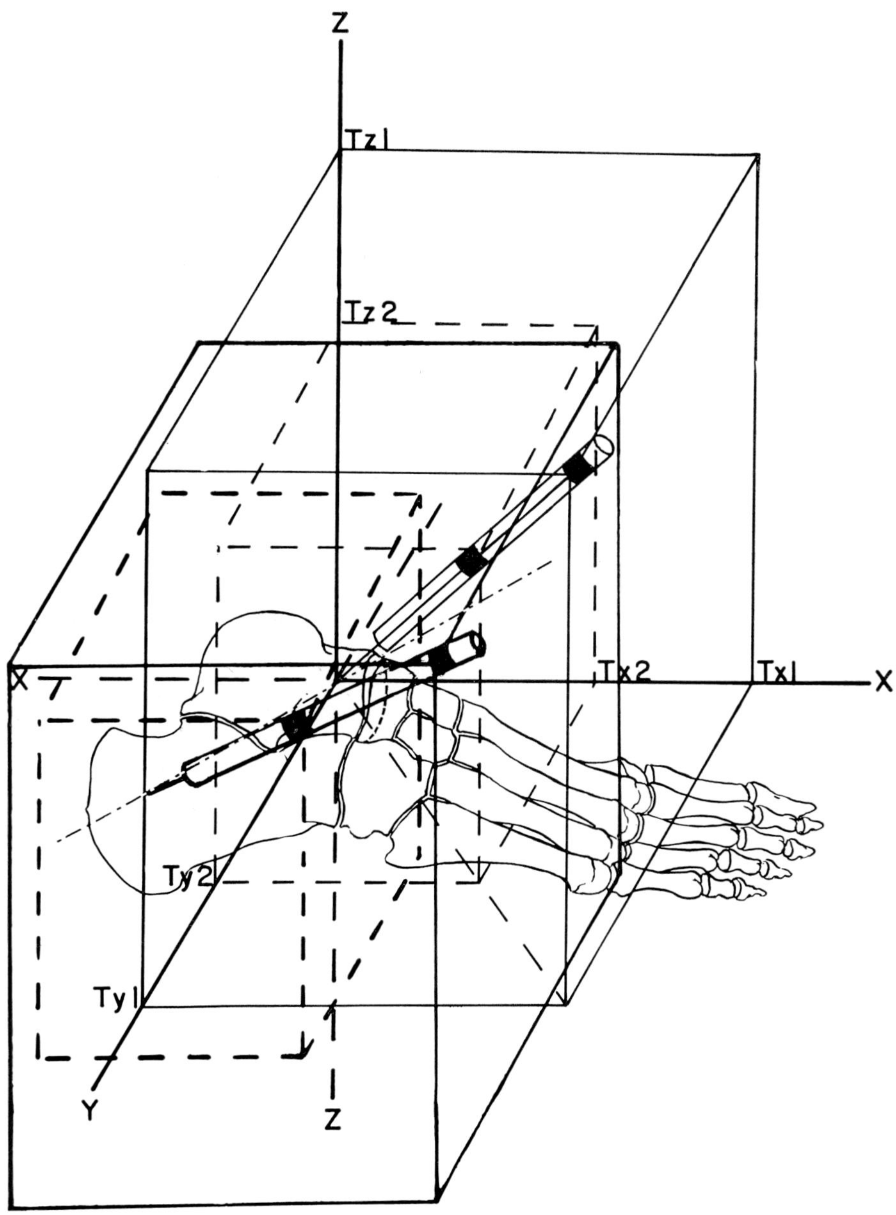

Figure 57 *right*. An overlay of the X, Y and Z projections of the calcaneal target is added. (From *Clinical Orthopaedics and Related Research,* 50, 1967.)

Cx	Tx	Cy	Ty	Cz	Tz	Cos θ	θ
-0.646	-0.775	1.780	0.855	0.199	0.544	0.877	28.693
-0.601	-0.732	1.713	0.889	0.200	0.580	0.883	28.024
-0.543	-0.675	1.745	0.908	0.226	0.602	0.884	27.830
-0.560	-0.669	1.761	0.959	0.293	0.668	0.897	26.279
-0.513	-0.618	1.723	0.959	0.304	0.684	0.898	26.137
-0.490	-0.575	1.734	0.984	0.339	0.717	0.904	25.349
-0.462	-0.550	1.745	0.968	0.369	0.725	0.903	25.425
-0.416	-0.518	1.742	0.993	0.362	0.734	0.904	25.359
-0.397	-0.499	1.763	1.012	0.380	0.743	0.908	24.809
-0.386	-0.435	1.742	0.973	0.406	0.765	0.905	25.128
-0.392	-0.465	1.787	0.978	0.370	0.734	0.902	25.632
-0.394	-0.469	1.754	0.984	0.389	0.736	0.908	24.746
-0.383	-0.457	1.762	0.965	0.366	0.716	0.904	25.361
-0.398	-0.464	1.765	0.970	0.361	0.710	0.906	25.087
-0.384	-0.505	1.761	1.014	0.340	0.713	0.905	25.167
-0.386	-0.480	1.766	1.009	0.352	0.711	0.909	24.613
-0.390	-0.459	1.753	0.970	0.364	0.735	0.901	25.735
-0.406	-0.452	1.746	0.963	0.386	0.725	0.909	24.692
-0.388	-0.461	1.753	0.995	0.386	0.737	0.910	24.521
-0.400	-0.462	1.757	0.989	0.396	0.759	0.906	25.004
-0.399	-0.471	1.750	0.986	0.414	0.762	0.908	24.728
-0.394	-0.472	1.752	1.004	0.403	0.755	0.911	24.395
-0.401	-0.478	1.762	1.003	0.404	0.752	0.911	24.378
-0.435	-0.460	1.760	1.002	0.394	0.769	0.909	24.653
-0.430	-0.486	1.743	0.984	0.429	0.808	0.903	25.487
-0.432	-0.493	1.753	0.991	0.424	0.773	0.910	24.521
-0.444	-0.500	1.743	0.981	0.403	0.809	0.897	26.294
-0.441	-0.462	1.733	0.978	0.447	0.819	0.906	25.096
-0.399	-0.434	1.735	0.986	0.478	0.827	0.910	24.438
-0.393	-0.428	1.732	0.991	0.481	0.841	0.909	24.612
-0.375	-0.423	1.727	0.996	0.500	0.858	0.910	24.513
-0.371	-0.404	1.732	0.987	0.490	0.855	0.907	24.915
-0.365	-0.365	1.723	0.997	0.500	0.887	0.906	25.056
-0.334	-0.350	1.725	1.008	0.507	0.895	0.907	24.934
-0.313	-0.344	1.761	0.999	0.498	0.900	0.898	26.049
-0.285	-0.293	1.741	0.967	0.510	0.912	0.893	26.746
-0.241	-0.247	1.747	0.969	0.505	0.945	0.883	27.946
-0.230	-0.212	1.762	0.942	0.494	0.964	0.868	29.749
-0.196	-0.127	1.743	0.919	0.457	0.960	0.854	31.399
-0.167	-0.083	1.750	0.885	0.463	0.985	0.837	33.181
-0.098	0.000	1.779	0.921	0.390	0.998	0.819	35.051
-0.082	0.138	1.756	0.937	0.411	0.958	0.834	33.510
-0.050	0.221	1.758	0.925	0.458	0.985	0.830	33.863

Identification DRIR5S

Heel Strike

The #7040 IBM Computer calculates the Angle θ at each 1/50 second interval of the STANCE PHASE of Normal Walking:

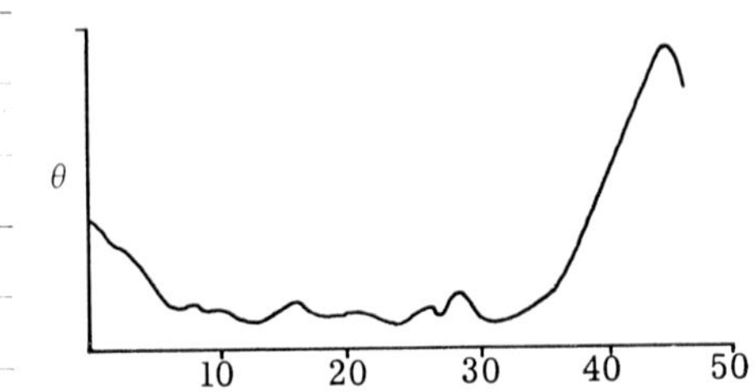

$$\theta = \cos^{-1} \frac{T_xC_x + T_yC_y + T_zC_z}{\sqrt{T_x^2 + T_y^2 + T_z^2}\ \sqrt{C_x^2 + C_y^2 + C_z^2}}$$

Toe Off

Figure 58. Sample of data processed by IBM 7040 computer, programmed according to the equation for θ. Each row of figures consists of the six projections necessary for the calculation of θ. Each row represents one frame of 35-mm film. The entire body of data represents one stance phase. (From *Clinical Orthopaedics and Related Research.* 50, 1967.)

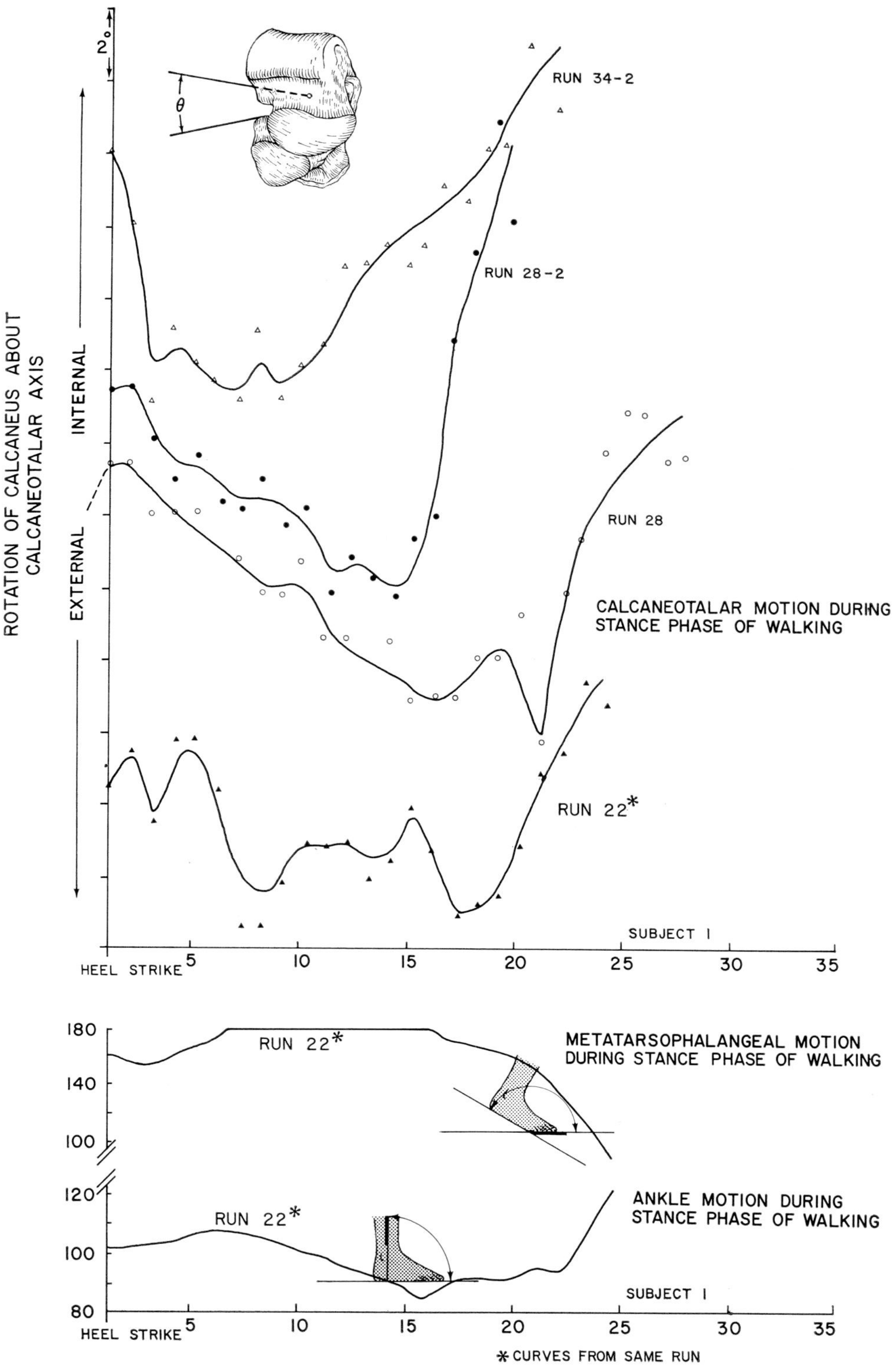

Figure 59A–H. Four curves, representing each of eight subjects, show the angle θ during the stance phase of walking; for each subject, the last of the four curves has below it curves representing (*upper*) the metatarsophalangeal angle and (*lower*) angle of the ankle. (From *Clinical Orthopaedics and Related Research, 50,* 1967.)

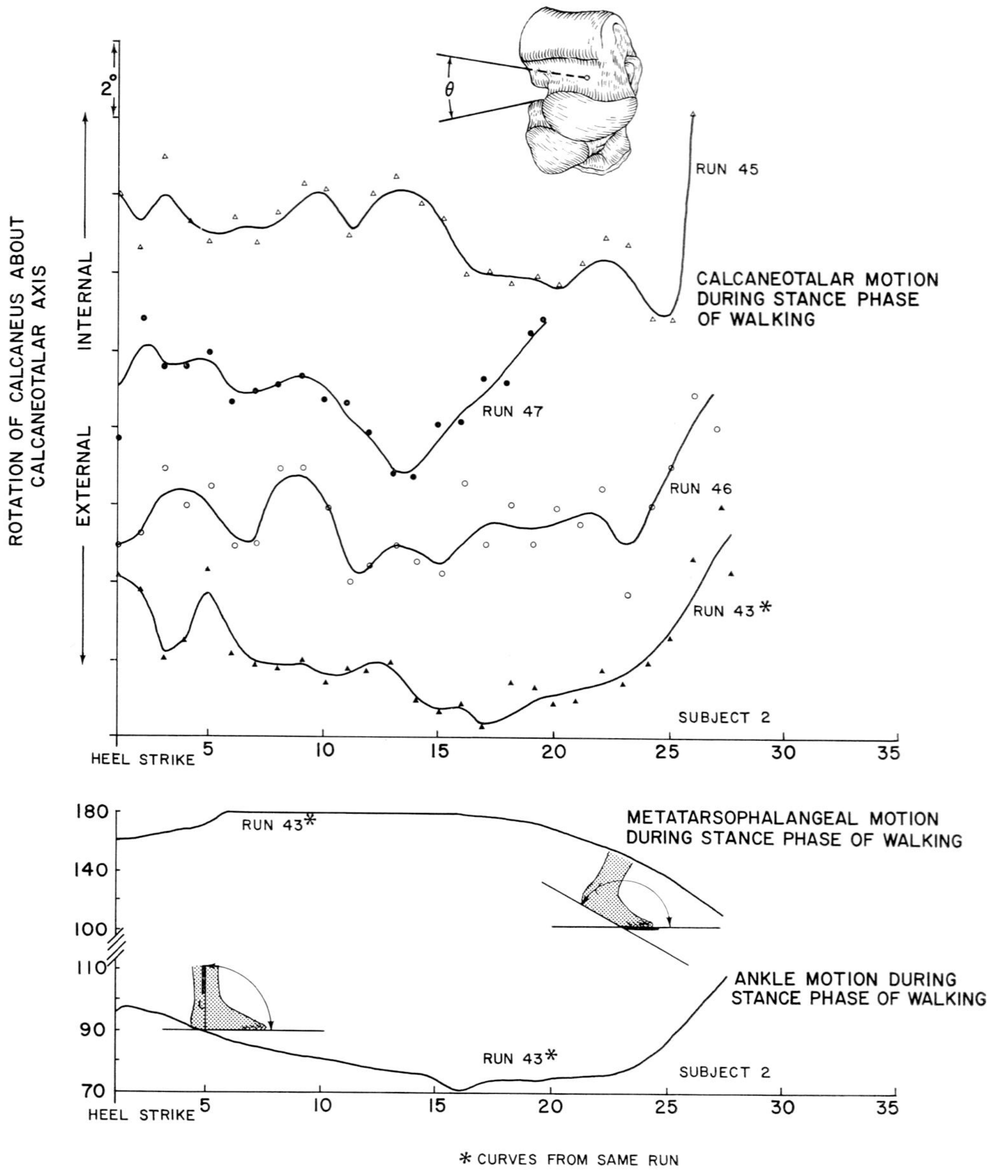

Figure 59B. Subject 2.

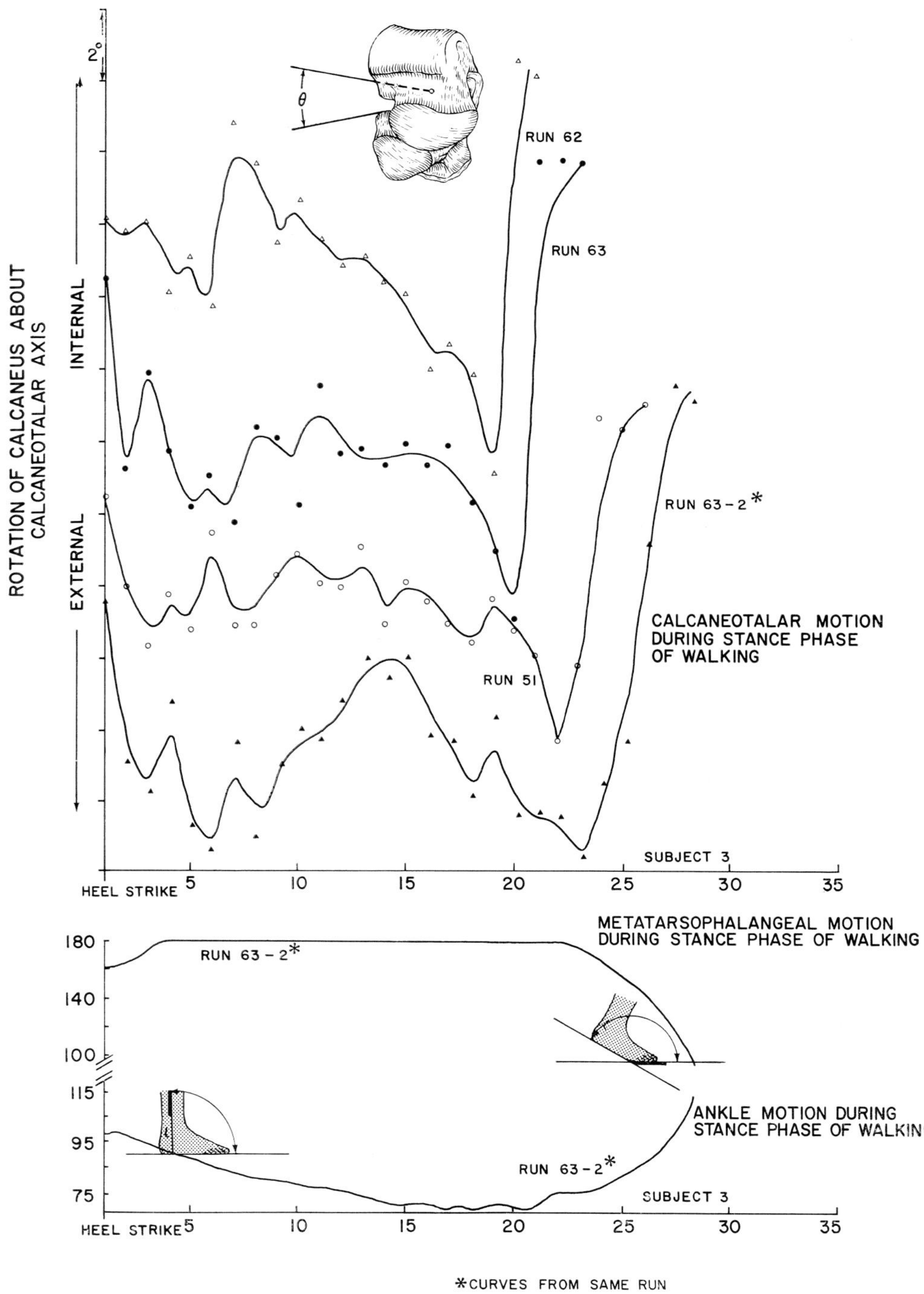

Figure 59C. Subject 3.

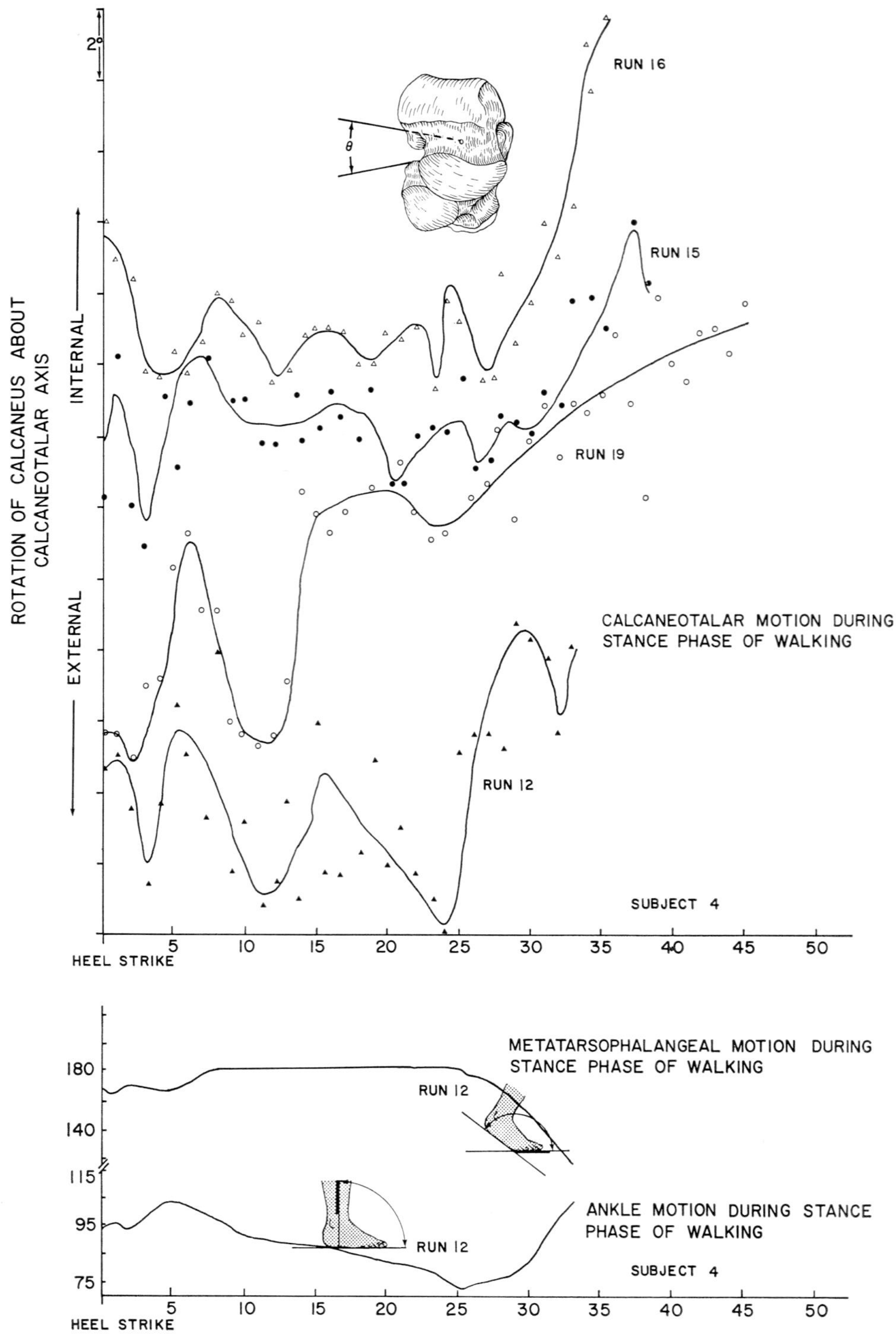

Figure 59D. Subject 4.

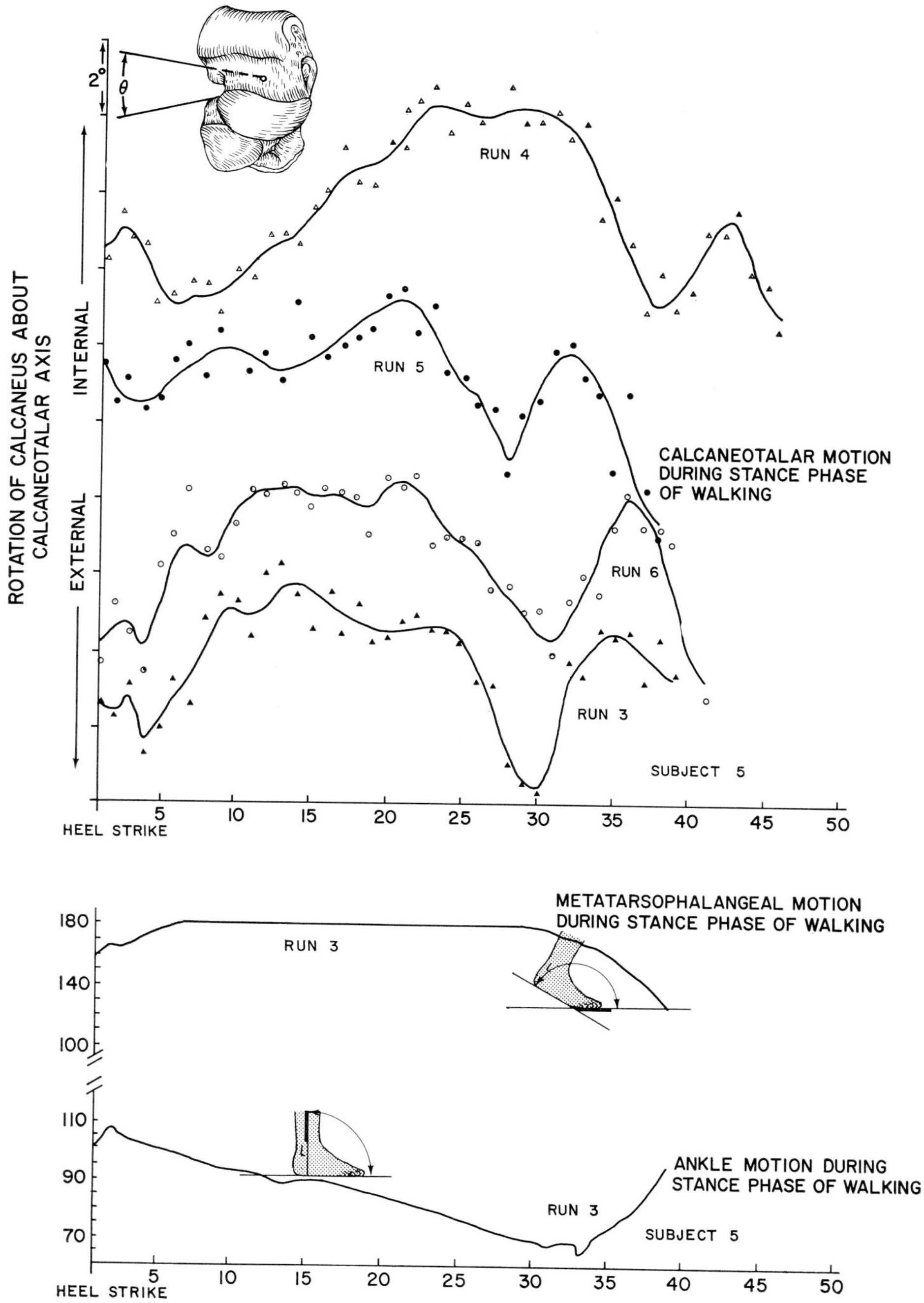

Figure 59E. Subject 5.

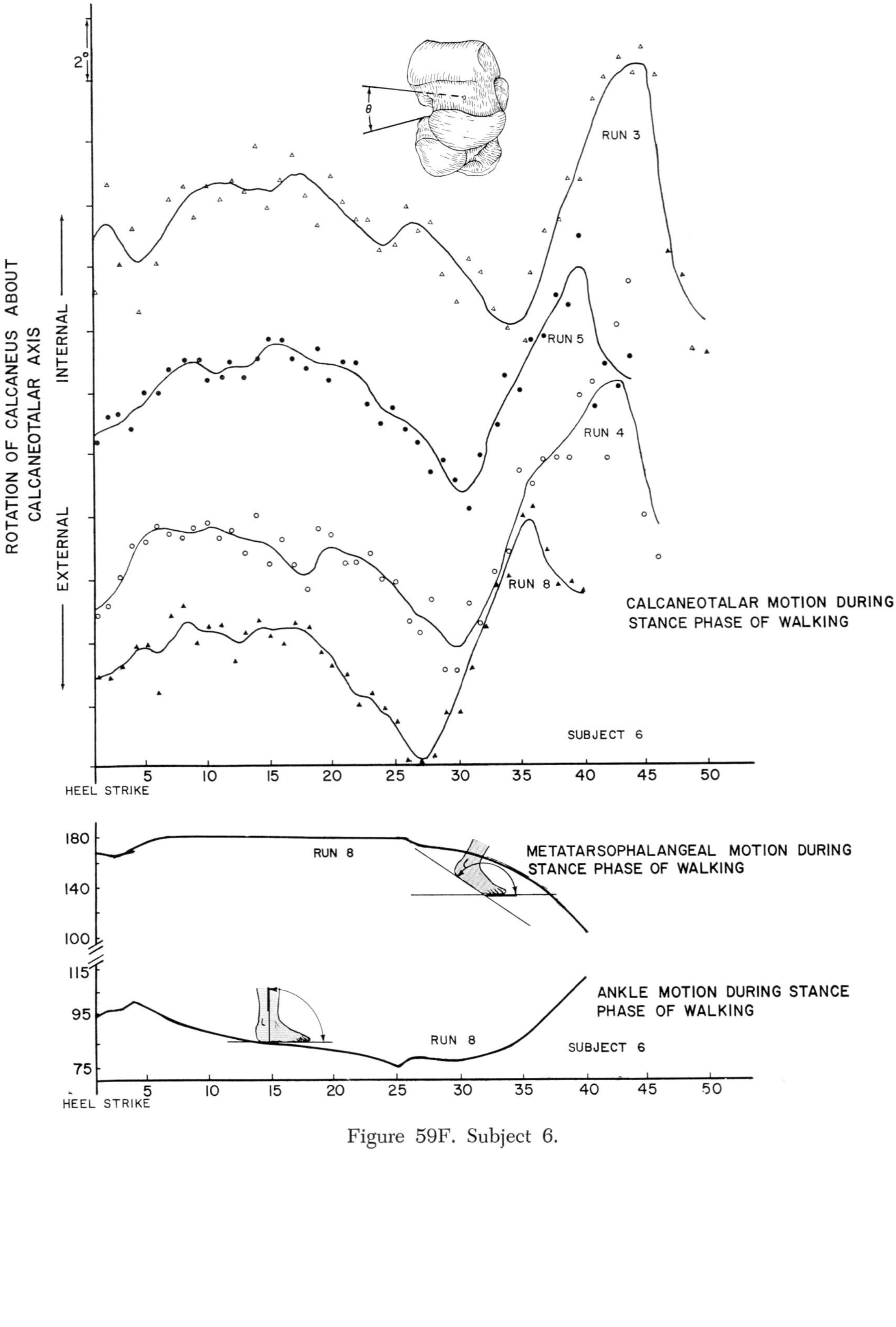

Figure 59F. Subject 6.

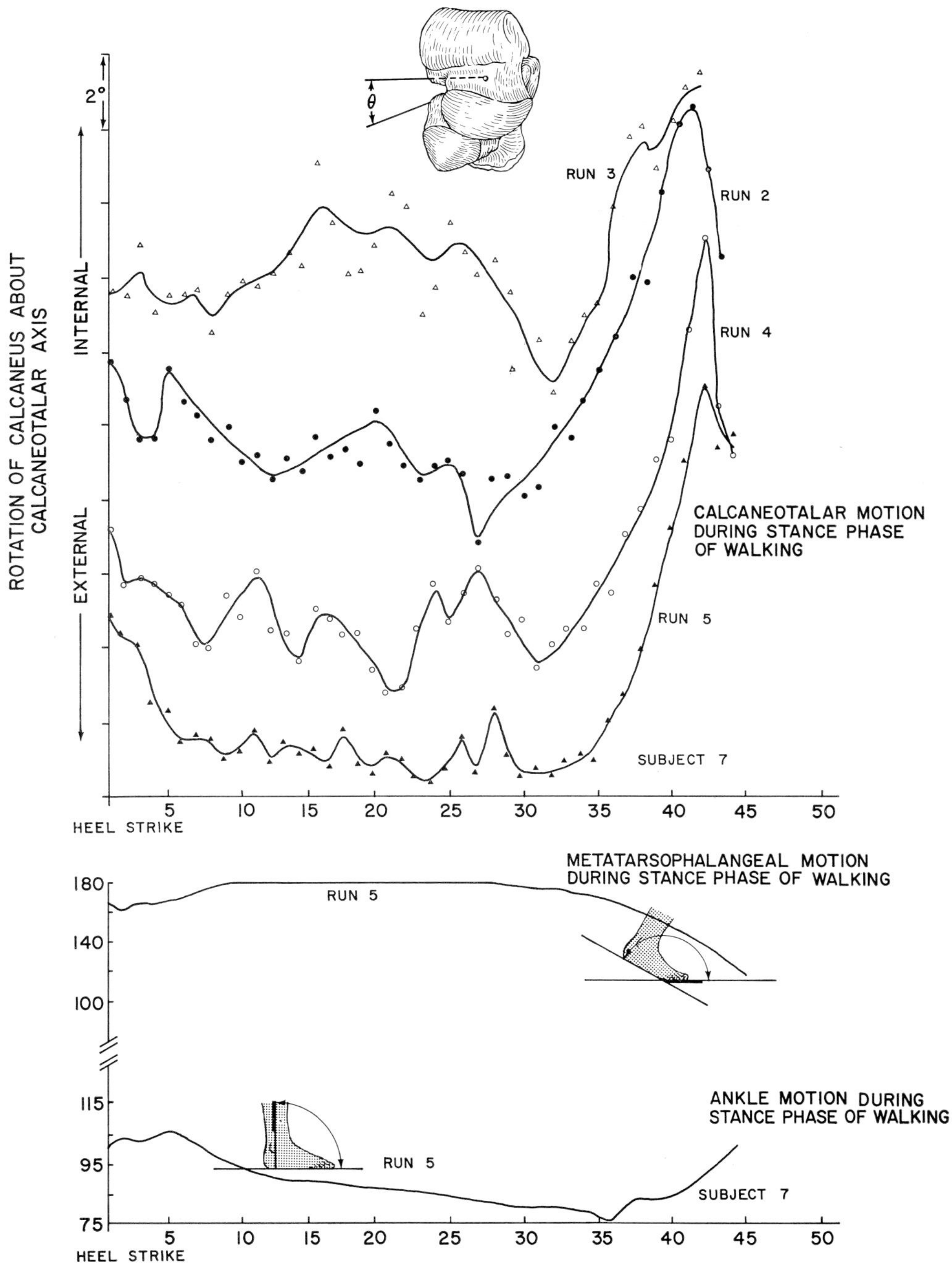

Figure 59G. Subject 7.

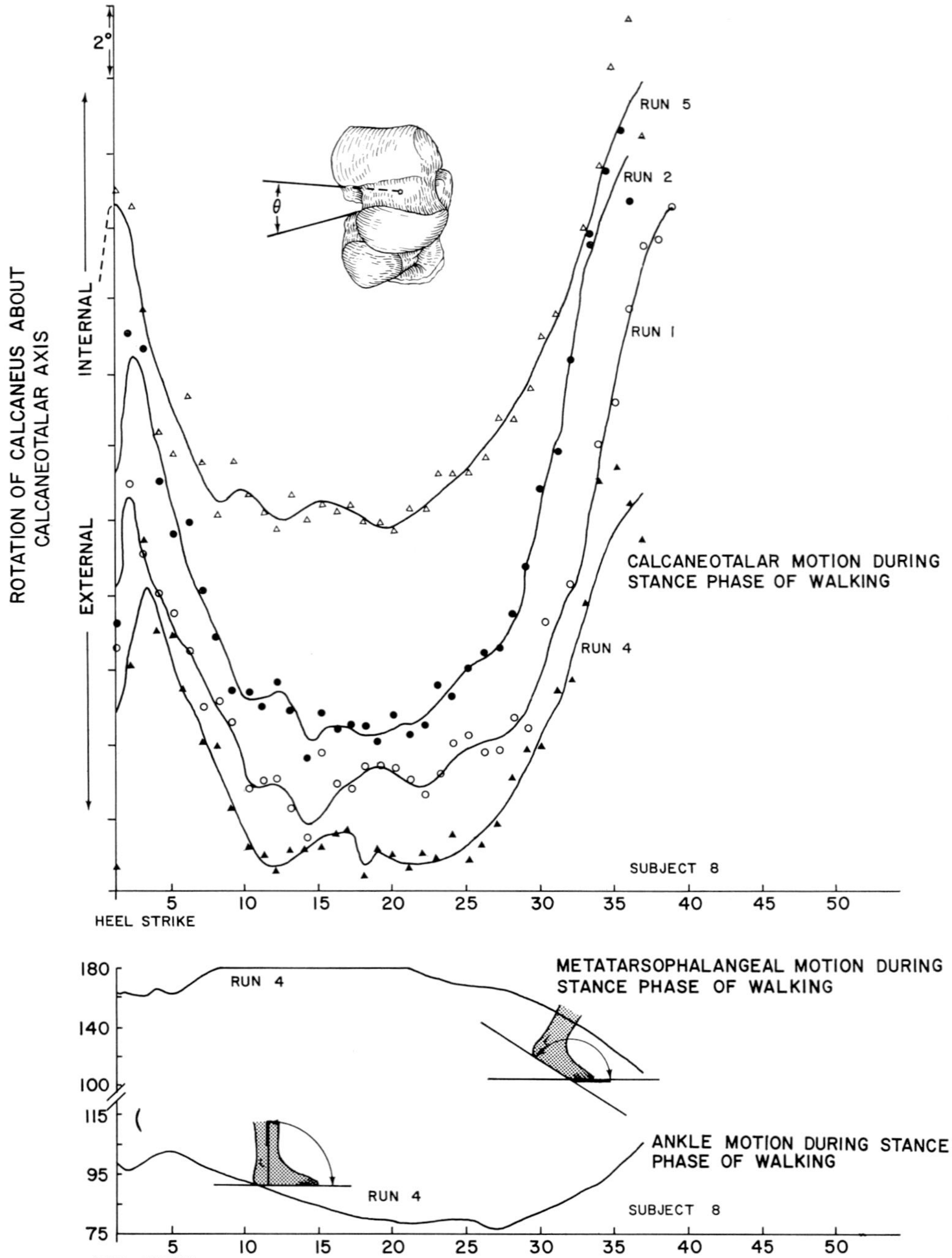

Figure 59H. Subject 8.

tion of the tibia to the foot. The average magnitude of horizontal tibial rotation during stance is approximately 10 to 11° during that part of the stance phase when the foot is fixed on the floor.[115,151] Since the foot must provide this amount of horizontal rotation, by what mechanism does it accommodate for this motion? Is this performed by the ankle joint, the subtalar joint or a combination of these two joint systems?

If the subtalar joint accommodates this horizontally rotating (tibial) terminal segment as it completes the phase of internal rotation, then often reversing to external rotation during the time that the foot is fixed on the walking surface, its component in the horizontal plane would equal the horizontal rotation of the tibia known to take place at that time. Since the amount of subtalar motion occurring at this interval is the main finding of this particular study, its horizontal component has also been estimated. The average vertical inclination of the axis from the horizontal plane is 42°. The component of rotation in the horizontal plane may be considered to be greater than one half of the rotation recorded about the subtalar axis.

If we observe that additional horizontal rotation between the leg and the foot is provided by other joints, such as the ankle joint, we are able to account for a large percentage of the motion. Since there is a known horizontal torque between the leg and the floor, we do not expect to account for all of the horizontal rotation.

CLINICAL APPLICATIONS

The findings of this study have a number of clinical implications. The traditional conception and treatment of flatfoot have consisted principally of attempts to lessen the medial deviation of the axis. Various supports, such as the use of an outwardly slanting shoe surface, have been devised to limit the range of external rotation of the calcaneus. Others, by use of a support under the medial tarsals, are intended to increase the angle of the axis with the horizontal, as well as to decrease the amount of medial deviation of the axis.

In order to appreciate the consequences of loss of the subtalar motion following arthrodesis, one must be aware of the type and the magnitude of motion at the subtalar joint. Since horizontal rotation of proximal skeletal segments requires the horizontal component of subtalar motion to make possible rotation of the leg on the fixed foot, this motion meets resistance after triple arthrodesis or subtalar bone block has been performed. It has been observed that the skeletal system increases its demands for the ankle joint to participate in this motion. A review of numerous cases demonstrates a high rate of change in the shape of the trochlea of the talus. Often there is evidence of degenerative arthritic change to an advanced degree (Fig. 61). This finding is particularly apparent in some cases a few years after subtalar bone block has been performed. Since this arthrodesing procedure may be performed at a considerably younger age than triple arthrodesis, destructive findings in the ankle joint are more likely (Fig 61 *Top left*).

The above is not intended as an argument against these arthrodesing procedures, nor are severely destructive changes uniformly found. It is not felt that this late complication necessarily suggests a contraindication to the procedures. It is felt that the above represents added knowledge of prognostic value to those who perform the procedures.

The range of rotation of the calcaneus about the subtalar axis is influenced by muscles with strategically placed insertions. These insertions and the course of their tendons have a profound effect on subtalar motion. The peroneal muscles, their tendons passing lateral to the axis, *externally* rotate the foot about the talus.

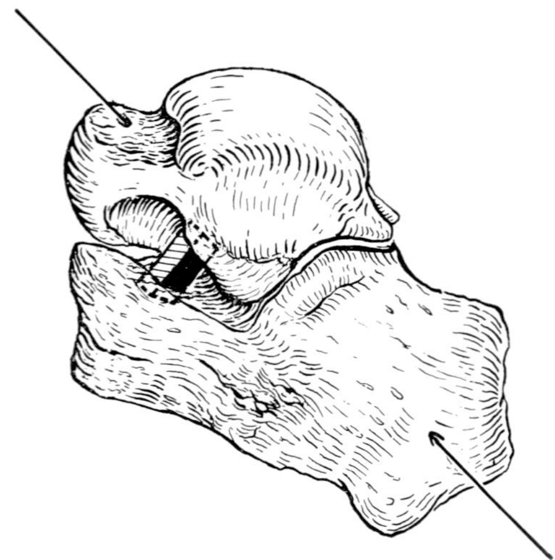

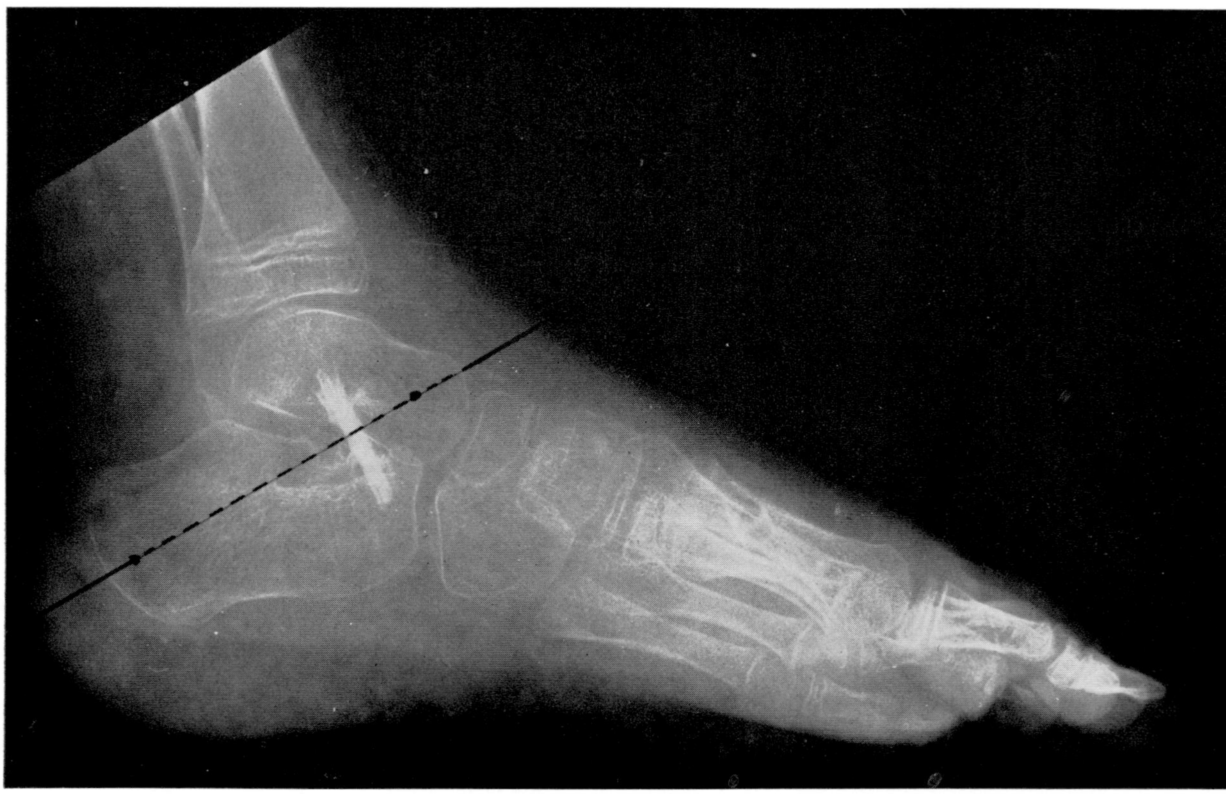

Figure 60 *top, bottom.* The subtalar bone block oriented in its recommended position perpendicular to the subtalar axis. (From *Clinical Orthopaedics and Related Research, 50,* 1967.)

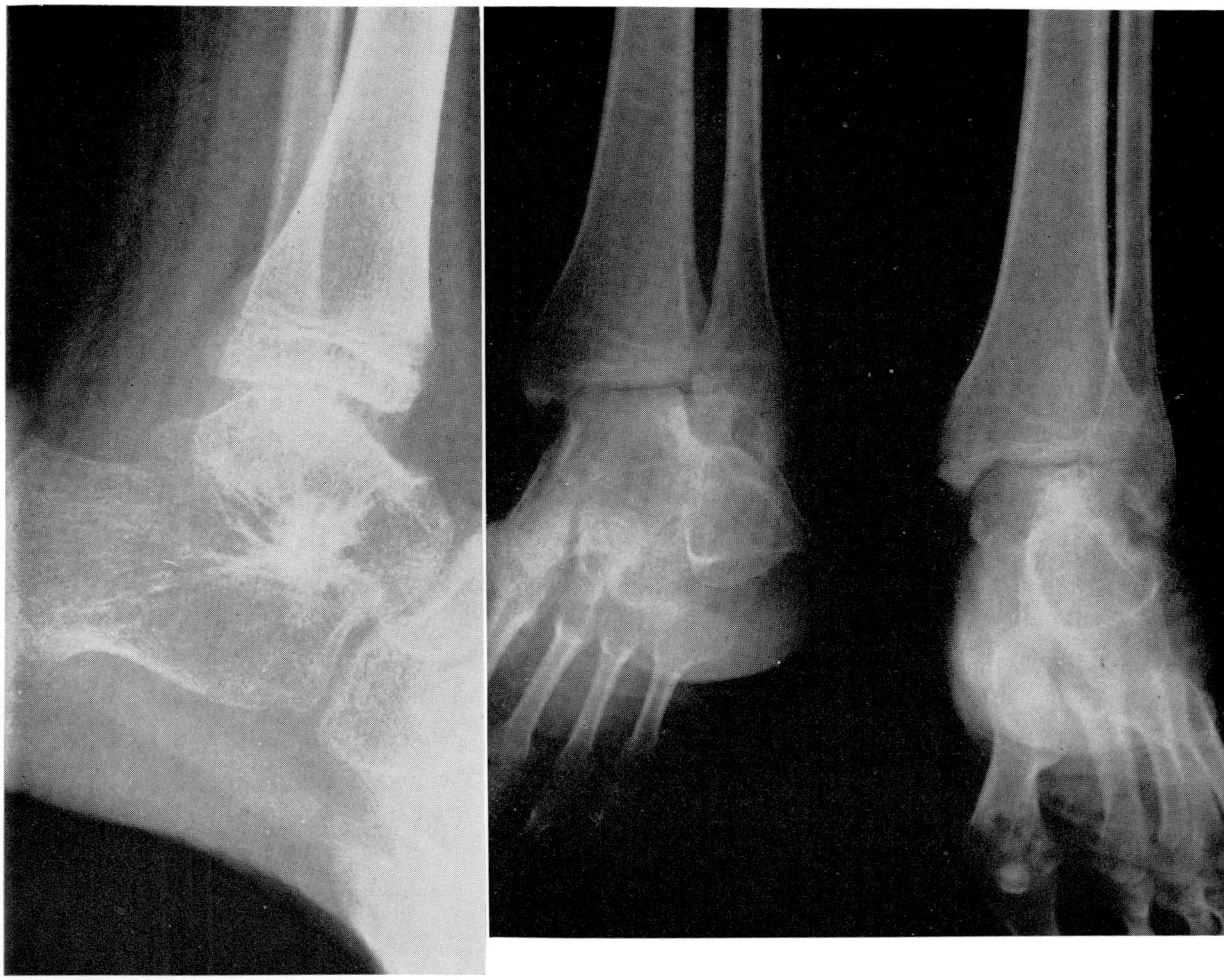

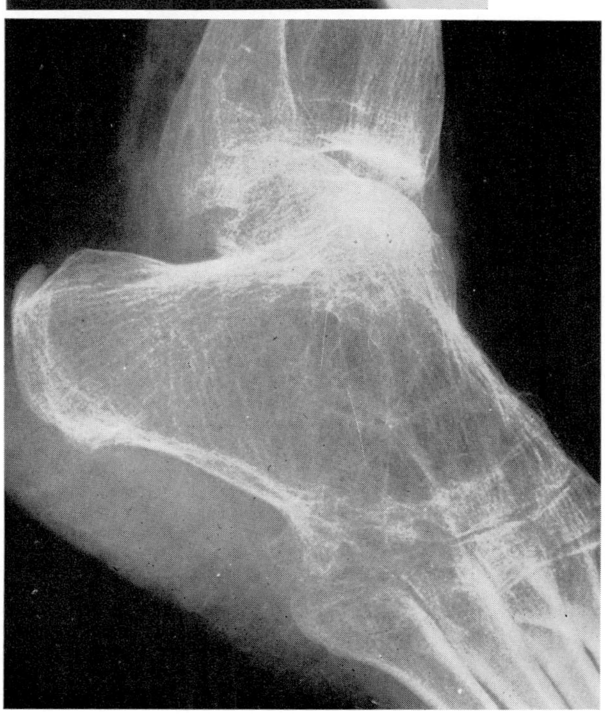

Figure 61. *Top left,* degenerative changes in the ankle joint four years after the procedure shown in 60 *Bottom. Top right,* degenerative changes in ankle joint at age 35. Triple arthrodesis was done at age 14. *Left,* marked traumatic degenerative ankle joint changes with partial dislocation at age 36. Triple arthrodesis was done at age 14.

The same is true of the long extensor tendons which pass far lateral to the axis (see Figs. 44, 46).

The anterior tibial muscle is shown by electrical stimulation to produce sharp internal rotation as well as dorsiflexion. The posterior tibial muscle is well placed to produce decisive internal rotation of the foot about the axis of the subtalar joint.

The position of this axis in the foot is thus of great importance in the study of the action of muscles whose tendons pass in relation to it. The problem of exactly where to place the insertion of a tendon to be transferred is solved by knowledge of the position of the axis and the state of other muscles passing it.

Certain skeletal structures are important in establishing the extent of *internal* and *external* rotation about the subtalar axis. The sustentaculum tali terminates *internal* rotation when it is approached by the medial aspect of the trochlea of the talus. External rotation is terminated by the abutment of the lateral aspect of the talus on the anterolateral process of the calcaneus. This is a familiar mechanism to all orthopaedic surgeons who observe it routinely during triple arthrodesis.

The anterolateral process of the calcaneus is developed in man to a far greater degree than in any other primates, as observed by Elftman.[61] Surgical alteration of this process was proposed by Chambers. This surgical procedure was intended to limit external rotation in the adolescent flatfoot. The advanced development of the anterolateral process of the calcaneus in man is compared to that of the gorilla in Figure 36A.

The familiar fracture of the lateral aspect of the talus[84] may no doubt be explained by its auxiliary function of assisting in terminating pronation. As the lower portion of the lateral articular surface of the talus strikes the anterolateral process of the calcaneus, a mechanism for the fracture of this structure exists. It is likely that the anterolateral process of the calcaneus may suffer the fracture on other occasions (Fig. 36A).

On the other side of the talus, forcible internal rotation or supination may explain the isolated fracture of the sustentaculum tali.

SUMMARY AND CONCLUSIONS

The motion of the calcaneus about the subtalar axis during walking has been calculated in eight subjects. The method was based on observation of metallic targets inserted into the lateral aspect of the calcaneus and the lateral aspect of the neck of the talus. The total range of active pronation and supination was also observed.

The talocalcaneal joint is responsible for the major part of pronation and supination. The more distal bones of the foot and the ankle joint make insignificant contributions to this motion.

The foot may pronate or supinate with respect to the leg. Active motion of the tibia on the fixed foot may also make use of the subtalar axis. Normally, this motion—horizontal rotation of the leg on the fixed foot—may utilize only the horizontal component of the subtalar motion. This component may be calculated. Its percentage of the total subtalar motion of a walking cycle depends on the position of the axis in the foot.

Loss of subtalar motion denies the leg the use of its horizontal rotary component. The horizontal torque between the leg and the foot increases unless another source of horizontal rotation becomes available. In many patients who have had triple or subtalar arthrodesis, deformity develops in the ankle joint. The talus becomes looser in the mortise. The traumatic arthritic change found in the ankle joints of such individuals is thus explained.

Although the subtalar axis is normally deviated an average of 16° medial to the long axis of the foot, this angle increases during pronation and decreases during supination. The position of the axis in the foot determines the action of muscles whose tendons pass it on their way to insert in the foot. Those passing lateral to the axis are pronators; those passing medially are supinators. It is possible for change in the position of the axis during pronation or supination to alter the function of a muscle. For example, the anterior tibial muscle, inserting just medial to the subtalar axis, may lose its power to supinate when the foot is pronated. Its insertion is moved to the lateral side of the axis. These findings assist in deciding where to place transferred tendons.

The inferior lateral prominence of the lateral articular process of the talus abuts on the anterolateral process of the calcaneus. This terminates pronation. This very likely accounts for fracture of the lateral portion of the talus. Anatomical restoration following this fracture is highly desirable. Open reduction is often indicated.

Supination terminates as the posteromedial portion of the talus meets the sustentaculum tali. This may be the source of the occasional isolated fracture of the sustentaculum tali. Unlike fractures of the tuber of the calcaneus, this fracture may often be reduced anatomically by open surgery. As an integral part of the talocalcaneal articulation, anatomical restoration of this structure is also most desirable (its upper surface constituting the middle calcaneal facet of the talocalcaneal joint).

The phylogeny of the foot is of special interest regarding the loss of the prehensile hallux and the spherical contour of the head of the talus as one considers the monkey (*Rhesus macaque*), the arboreal gorilla, the lowland gorilla and man. In the monkey and the arboreal gorilla, the articular portion of the talar head is closer to a perfect sphere; the hallux is prehensile. In the lowland gorilla, the hallux has almost reached the adducted position (Fig. 54), and the head of the talus is less regular in its roughly spherical portion. In man, the hallux is normally apparently adducted to a close to optimum position. The talar head is parabolic in cross section. Vertical and horizontal sections represent parabolae with different equations. The structure resembles an elliptical paraboloid, suggesting a method of "locking" with the articulating navicular. The more spherical talar head of the arboreal gorilla (Fig. 54) explains the mechanism for a greater range of motion necessary in the prehensile organ.

The human foot, with its more stable talonavicular joint, indicates that with pressure applied, the joint has a means of "locking." The position of the talonavicular joint rises to a position superior to the calcaneocuboid joint, thus developing the longitudinal arch. All of these factors indicate that the human foot is developing into a more efficient lever.

Man—the only primate with a longitudinal arch—also possesses the only foot that does not break at the transverse tarsal joint as tension during walking is applied through the gastrosoleus musculature. Man, with his more efficient lever for application of the gastrosoleus system, thus has a far greater ratio of triceps surae mass to total body weight than do any of the other primates.

We have seen that motion at the subtalar joint may be of considerable magnitude. Since the axis is inclined approximately 45° in the vertical plane, the motion has a horizontal component of approximately two thirds of the total motion. This part of horizontal rotation provided by the subtalar joint is denied the patient when triple arthrodesis is done. Thus we see in Figure 61 *left* a roentgenogram of a subject with long-standing triple arthrodesis where a dome-shaped talus resulted. Undoubtedly this is due to the fact that the ankle joint was required to perform more horizontal rotation than previously in an effort to provide at least part of the lost function of the arthrodesed subtalar joint. Other applications of the knowledge of this subtalar axis are demonstrated in Figures 41, 44 and 60. In Figure 60 *Bottom*, where subtalar bone block has been performed on a young individual, knowledge of the position of the subtalar axis leads one to place the blocking graft perpendicular to it. This makes the procedure more effective, and healing is thought to take place more rapidly with less chance of nonunion. In Figures 41 and 44, two muscles, the anterior tibial and the extensor digitorum longus, have different positions and insertions with regard to the subtalar axis. One is an internal rotator and the other is an external rotator. Thus, knowledge of the position of the axis is essential for prediction of performance of tendon transfer of muscles. The surgeon can place the new insertion where it will have the desired effect concerning rotation about the subtalar axis.

THE AXIS OF THE LOWER EXTREMITY

In order to appreciate fully the action of complex muscular systems such as the adductors, one must consider the mechanical axis of the lower extremity (Fig. 38). This axis influences the action of both the adductor longus and the adductor magnus, determining their participation in transverse rotation at the hip joint. The principal distal attachment of the adductors into the posterior aspect or linea aspera of the femur suggests that they are external rotators. A more important consideration, however, is their relationship to the mechanical axis of the lower extremity (Figs. 39 and 40). The axis of the lower extremity is undoubtedly a line passing from the center of the head of the femur to the point of contact of the lower extremity. If the lower extremity is not in contact with the walking surface, the mechanical axis passes through the head of the femur and the center of gravity of the dependent lower extremity. The latter is, of course, a changing point and the axis, therefore, is a constantly changing one (Fig. 38). The action of the adductors is therefore an especially complex consideration when the lower extremity is not in contact with the walking surface. After insertion of dual electrodes into either of these muscles, one can easily produce weak external rotation and adduction by electrical stimulation. Upon observing the electromyographic recording, however, one is impressed by the absence of electrical activity in each of these muscles on external rotation and their abundant activity on voluntary internal rotation or adduction. This paradoxical finding indicates the futility of attempting to determine the action of muscles solely from their response to electrical stimulation. Undoubtedly, the body uses muscles as groups, and some cannot even be used individually. According to Duchenne,[56] "a single muscle has no function as such," and according to Jackson,[116] "the brain knows nothing of muscle, only of movement." Thus motions and not muscles apparently are represented in the higher centers of the central nervous system.

The complexity of single muscle function and of muscle groups is illustrated by the above-described adductor group. The most distal portion of the adductors, the adductor magnus, inserts into the linea aspera and medial femoral epicondyle. Its origin is also broad, extending from the ischial tuberosity well along the inferior ramus of the ischium (Fig. 40).

Anteriorly, the adductor longus (Fig. 39) takes origin just behind the gracilis muscle and inserts into the linea aspera. The adductor brevis originates just behind the adductor longus muscle, along the inferior rami of the pubis and ischium.

The origin of the adductor group thus extends posteriorly from the tuberosity of the ischium forward along the inferior ischial and inferior pubic rami almost to the symphysis. The insertion of the group covers the extent of almost the entire length of the femur.

The broad extent of both the origin and insertion of such a group makes analysis of motion about axes difficult. There is variation of even the three parts of one component, the adductor magnus, regarding their ability to rotate the limb about the axis of the lower extremity (Fig. 40).

Simultaneous multichannel studies will provide further information regarding the adductor muscles and their components.

Studies of the normal hamstring muscles reveal further interesting findings. By stimulation or voluntary contraction, the short head of the biceps is a consistent flexor of the knee. The long head, however, in many individuals is inactive on knee flexion but very active on extension of the thigh or external rotation of the lower extremity.[35] These muscles do not contribute substantially during normal walking, according to findings by the electromyogram (Fig. 91).

Another demonstration of the importance of mechanical axes and centers of rotation in the analysis of muscular function is seen in the activity of the previously mentioned iliopsoas muscle. As this muscle passes downward to the lesser trochanter of the femur, it passes directly over the head of the femur. (Fig. 37). There has been disagreement concerning the role of this muscle in internal and external rotation of the lower extremity. In order to study the functions of the components of the iliopsoas muscle, electrodes were placed in the iliacus muscle and, during separate examinations, in the psoas muscle, via a posterior (sympathetic block) route. Electrical stimulation, in either case, invariably produced pure flexion without internal or external rotation. The electromyogram revealed no activity on voluntary internal rotation but slight activity on external rotation of the hip. There was abundant activity on flexion of the hip and, surprisingly, on abduction. As demonstrated in Figure 37, when the hip is abducted, a portion of the tendon of this muscle passes lateral to the center of the head of the femur. The muscle then participates vigorously in abduction.[35] During walking, electrical activity in the iliopsoas muscle is virtually absent. It is used during the first step only or while the subject ascends stairs, at which times it shows abundant activity.

The test would be of greater value as an orthopaedic study device if the electromyogram could be directly related to the tension produced by the muscle. The amplitude of the signal may vary somewhat when one test is compared to another because of variations in surrounding body fluids or orientation of the electrodes in a given muscle. During a specific test, however, it has been observed by Lippold *et al.*, and Inman *et al.*, that the integrated electromyogram has a linear relationship to the developed tension of the muscle if the contraction remains isometric.[100,119]

Isotonic contractions have been found to be related to developed tension if the muscle shortens at a constant rate.[20] (Shortening slowly at a constant rate might be considered a series of isometric contractions.) A considerable amount of uniformity has been observed, using the cathode ray oscilloscope, where the same setting of the preamplifier is required each time for normal muscles in order to have the amplitude of the strongest contraction just contained on the oscilloscope screen. This setting for normal muscles has been found on almost all of the tests done to be ideally placed at 0.5 mv/cm (Fig. 90). Polio muscle, which often produces high electrical amplitude at a reduced rate, usually requires a setting of 1 mv/cm (Fig. 5 and Chapter V).

Electromyographic records have been made from the extraocular muscles,[104] using multichannel equipment. This requires rather high amplification, and many interesting findings have been reported. It should be remembered that extraocular muscle, possessing a much finer degree of movement than the average muscle, has a much higher ratio of axons to muscle fibres. Each motor unit thus consists of a small number of muscle fibres. It has been found that the extraocular muscle is never at rest. Some electrical activity can always be elicited, even during sleep. The procedure has been carried out with flexible internal electrodes, using only sedation and local anesthesia.[104]

Using multichannel galvanometer recording equipment and the sound motion picture with synchronized sound track, one can make an admirable display of the many muscles of the upper or lower extremity involved in performing certain acts. These activities can thus be audibly and visually observed simultaneously. Each channel is alternatingly switched to the sound track of the motion picture camera. The above analyses of the action of the iliopsoas muscle and the adductors, for example, rule out prevailing concepts cited in textbooks of anatomy where individual muscle action is judged by origin and insertion alone.

In the study of normal muscle function, individual muscle action has been demonstrated during various activities. However, such action or function should be thought of as participation in a group activity in which a number of synergists take part. Thus no claim can be made for individual isolated muscle function. Single-channel equipment has been used in this study when the electromyogram is recorded on the sound track of the motion picture film. This, however, gives an excellent self-synchronized display of both movement and the muscle's electrical activity. One can in this way very readily appreciate the individual participation of each muscle in certain activities. Multichannel equipment undoubtedly has many great advantages in the study of muscle function, as the tracings of several muscles may be simultaneously recorded. Since each muscle requires a pair of internal electrodes for this type of study, it is possible that normal movement may be influenced by the use of too many internal electrodes at one test.

Only a single channel can be recorded at one time on the sound track of the motion picture film, and if muscles are studied individually, attention can readily be given to the factors of gain, sensitivity, speed of recording material and shape and characteristics of wave form. The diphasic character of the electromyogram produced by this method may be displayed on the oscilloscope screen as a vertical oscillation. If the horizontal amplifier is employed, then the oscillating wave is swept across the screen and a time factor, or baseline is introduced. As the impulse is swept across the screen, however, a very small part of the activity is not seen or recorded, since each sweep must end and then be resumed on the other side of the oscilloscope screen. When using continuous recording equipment, only the vertical oscillation is displayed. The sweep or time dimension is supplied by constantly moving recording material, such as 35 mm photographic paper or film. Since the rate of the recording material can be varied, one can easily arrange to operate the oscillograph camera at the same rate as the sound motion picture camera. This produces a conveniently synchronized system which records without interruption all of the electrical activity taking place.

The events taking place during the subject's activity, such as walking, with the proper designation of stance phase, swing phase, etc., or of voluntary contraction, e.g. extension of the knee, can then very easily be transferred to the 35-mm material. The electrical activity produced by the muscular contraction can thus be associated with the activity of the subject, as demonstrated in Figures 90, 91 and 92. In this way, phasic activity during walking may be precisely recorded (Fig. 89). Although the action potential may vary somewhat from day to day for a given muscle, due to variation in electrode position, standardization is made possible by the use of this particular equipment. Thus a voltage gain of approximately 2,000 is used for each test. The oscilloscope sensitivity may be set at the same setting of 0.5 mv/cm—a setting which one almost always finds optimal for recordings from normal musculature in the lower extremity using 0.003-in. copper wire electrodes with ⅛-in. tips. The electrodes are placed approximately 1 cm apart for functional studies where it is imperative to confine the recording to a single muscle (Fig. 6). Constant recording speed may be employed, avoiding the introduction of inconsistent forms of the action potential.

Several dynamic and structural changes are known to occur in muscle as contraction takes place. These include changes in (a) length, (b) cross section, (c) viscosity and (d) tension.

When contracting muscle shortens, as during isotonic contraction, the maximal rate of change is proportional to the length of muscle; i.e. the maximal velocity of shortening is proportional to muscle length.[144] The maximum developed tension is proportional to the muscle's cross section. During isometric contraction, the pattern of the integrated electromyogram can be shown to be proportional to the developed tension.[20,21]

Chemical changes result in loss of heat during isometric contractions. The isometric contraction involves no physical work. During isotonic contraction, work is performed; some heat of shortening is produced. This heat is independent of load (Fenn effect[149]). A transmembrane potential change passes down the membrane. This electrical potential change, which is due to rapid changes in differential permeability of the complex membrane to potassium, sodium and other ions, is the so-called action potential commonly recorded by electronic techniques. The action potential is almost coincident with the development of tension. Tension begins in the motor unit just before the termination of the primary spike of the action potential (Fig. 176). The action potential, however, terminates prior to the cessation of tension.

The amplitude and form of the recorded action potential depend upon several factors. It varies with the type of electrodes used (Fig. 4) and the adjustment of the preamplifier of the oscilloscope. It also varies according to the distance between electrodes placed in the muscle. The action potential produced represents the activity of the motor units in the immediate area surrounding the two electrodes. Individual motor unit action potentials are not usually seen with this type of recording device, since vigorous muscular activity produces abundant electrical activity that results in superimposition of many motor units, one upon the other, and loss of identity of the individual motor unit—the normal interference pattern. The total electrical changes in the area examined, however, are represented by the recording. Since the electrodes are in contact with many fibres, subunits and motor units, the recording represents an average of many values differing in sign as well as amplitude. The latter, for the most part, is a function of distance of some of the fibres (and motor units) from the electrodes (Fig. 53).

Electromyographic analysis of the phasic activity of normal hamstring muscles and quadriceps components has been very useful in the analysis of hamstring replacement transfers to the patella for quadriceps weakness. The familiar picture of persistent hip and knee flexion contracture, with weakness of the quadriceps and the patient attempting to advance the center of gravity of the body anterior to the mechanical axis of the knee, can be changed dramatically by forward transfer of hamstrings in some cases. The hip flexion contracture and lordosis may be a necessary result of the knee flexion contracture and quadriceps weakness, especially if the gluteus maximus and gastrocnemius muscles are weak, since the patient must place the center of gravity of the body in front of the knee axis to prevent collapse. Here, plaster wedging of the knee alone seldom produces permanent correction and may have to be followed by measures to balance the muscular distribution about the joint—in this case, hamstring transfer to the patella. Hamstring transfer may succeed dramatically in maintaining the correction, eliminating the necessity for the flexed position of the hip and the lordosis. Often one of the hamstring muscles may be found preoperatively to have a considerable amount of stance-phase activity (Fig. 91). Such a muscle might be a very desirable choice. However, continuing swing-phase activity of some hamstring transfers may also be useful in extending the knee joint just before the onset of the stance phase. In some cases, the transfer actually produces extension throughout swing, resulting in the "stiff-legged" gait.

Figure 92 demonstrates the individual phasic activity of the quadriceps components. It is observed that their amplitudes are low and their duration of contraction short. Contraction begins normally in the quadriceps components just prior to the onset of the stance phase. Activity is concluded just before the mid-portion of the stance phase. In all of the records in this study, the factors of recording speed, oscilloscope sensitivity, and gain are the same. Using this standard procedure one consistently observes that during normal walking the rectus femoris plays a smaller part than the other quadriceps components. Hamstring transfers are frequently made because of marked instability with bilateral quadriceps paralysis or because of external rotation of the tibia with genu valgum in the presence of a strong outer hamstring. Occasionally, the hamstring transfer into the patella succeeds in duplicating the quadriceps "stance-phase" activity during normal walking (Fig. 95). In other cases, the muscle retains its swing-phase activity during level walking but may spontaneously change to the stance phase as the subject ascends stairs.[34,40] The patient is usually able to call upon the transfer when necessary, since the muscle responds more effectively under voluntary control.

The hamstring muscle transferred to the patella may be disappointing in its failure to produce complete knee extension in the absence of a quadriceps mechanism. One wonders if it is possible for a single muscle, such as the semimembranosus or the long head of the biceps femoris, in itself, to produce complete extension and a full range of motion of the knee. Such a transfer represents our best efforts to substitute for the quadriceps a complete group or system of muscles of varied lengths. In our experience, complete extension is possible if one sacrifices some flexion in such transfers, as the biceps femoris to the patella. However, if stretching of the transfer is permitted and complete flexion produced, extension invariably suffers. Thus the best hamstring transfer, though capable of producing complete extension, is not capable of duplicating the action of the normal quadriceps mechanism to the extent of offering its contraction (extension) throughout the complete range of motion. This led to a study of the activity of the various components of the quadriceps mechanism during extension of the knee.[36]

One might postulate that different components of the quadriceps normally come into play at different stages as extension proceeds. In fact, it has long been assumed that the vastus medialis produces its maximal activity at the terminal portion of extension in order to prevent lateral dislocation of the patella.[56] Careful electromyographic recordings, however, fail to substantiate this concept. Records were made from the components of the quadriceps mechanism during voluntary extension of the knee, with the subject sitting. Although these records were made separately and not with multichannel equipment, they were repeated many times with consistent findings. It is clear that the three vasti begin their activity as extension is begun, with their activity increasing in intensity as voluntary extension is completed. A difference is often observed in the activity of the rectus femoris. Although this is the longest of the four components, its activity normally begins later. It often fails to participate in extension until the knee is almost completely extended to 180 degrees. Rectus femoris activity begins in most cases at a position of approximately 165 degrees. The contractions of the quadriceps components are isotonic until the knee reaches 180 degrees, at which time the contractions become isometric. This does not necessarily include the rectus femoris. With the subject supine, the rectus femoris assumes the same pattern of activity as the vasti.

Using two dual-beam oscilloscopes, the Visicorder amplifier and the Visicorder 906B galvanometer recorder, the electrical activity of the components of the quadriceps group may be accurately and simultaneously recorded. As knee extension takes place, with the subject sitting, the single-channel study was substantiated. Again, no rectus femoris activity was evident until knee extension was almost complete. However, subsequent records with the subject supine demonstrated participation of the rectus femoris similar to that of the remaining three components of the quadriceps muscle.

In the study of locomotion, excellent records of individual muscular activity during walking, both in normal and in postoperative orthopaedic reconstructive procedures, may be produced. This method is an excellent aid in the evaluation of the tendon transfer. With the electrodes, it is possible to penetrate deep muscles, such as the posterior tibial or the psoas major, and to record their activity when such activity might not be apparent by other means. Preoperatively, the method may be of value in assessing the state and time of contraction of a muscle under consideration for transfer. By adapting the apparatus to the sound camera, an appreciation of normal muscle function is readily obtained from the resulting self-synchronized audiovisual recording.

A difference in the time of action of the deep and superficial flexors of the fingers can be clearly shown by multichannel equipment of the type shown in Figures 16 and 131. The differential technique with insertion of two wires into each muscle is essential to confine the recording to individual muscles. Electrical stimulation for verification of electrode position is essential. (See Chap. VI.)

Chapter IV

ACTION POTENTIAL COUNT RATE RECORDINGS AND ABSOLUTE VALUE INTEGRATION

The recorded action potential of muscle has found many applications in the medical sciences. It is useful in neurological diagnosis for determining the time of activity of muscle, e.g. phasic activity, and for appreciation of the general level of muscular activity. A significant quantitative representation of the electromyogram has eluded investigators seeking a valuable relationship between some form of the electromyogram and muscle tension. The so called "integrated" electromyograph curve at times has shown promise of producing this long-sought correlation.[20,101,100] This expression of the electromyogram, while introducing a neat-appearing analog curve, has no numerical equivalent directly related to the tension developed by the muscle. In its present form of common use, it can be shown to be misleading if not erroneous. If it were possible to count all of the rapidly occurring action potentials found in the electromyographic record of normally contracting muscle, an analog curve with direct numerical equivalents would result. Such a curve should be very similar to the analog tension curve produced by isometric contraction (Fig. 62). The aggregate action potential count of the isotonic contraction should be proportional to the work performed by the muscle (Fig. 63).

Certain electronic counters readily yield action potential counts or values proportional to the counts. One may ascertain the count per unit time or the total count for a single contraction. It is the purpose of the study described in this chapter to examine the various forms of these counts and to explore their significance, particularly with regard to their relationship to isometric and isotonic contractions in normal human muscle. Although the density of the normal interference pattern of the electromyogram increases with the strength of the contraction, a numerical relationship is not demonstrated by the "integrated" electromyogram. In this study, action potentials are electronically counted. The counts thus obtained are compared to the tension developed by the muscle under various conditions.

Since the normal electromyogram consists of a series of oscillations of varying amplitude and of more or less diphasic form, special devices must be used in order to appreciate their aggregate value. Such a summation is necessary in comparing the electromyogram to anything with a continuous numerical value, such as tension. One may add the total amount of electrical activity, rectifying the inverted forms, thus producing an average "integrated" curve or employ a method for counting the total number of action potentials. The so-called "integrated" electromyogram produces a summation curve, each point on which indicates the total electrical activity at that instant. Many of the conventional resistor-condenser "integrators" commonly used in electromyographic work in reality produce an average curve representing the envelope of the electrical activity of the electromyogram. This curve itself may be integrated by measuring the area included under it (Fig. 64). Several excellent studies have been performed in which the "integrated" curve[20,21,100,101,119] has been compared to that of the tension continuously developed by the mus-

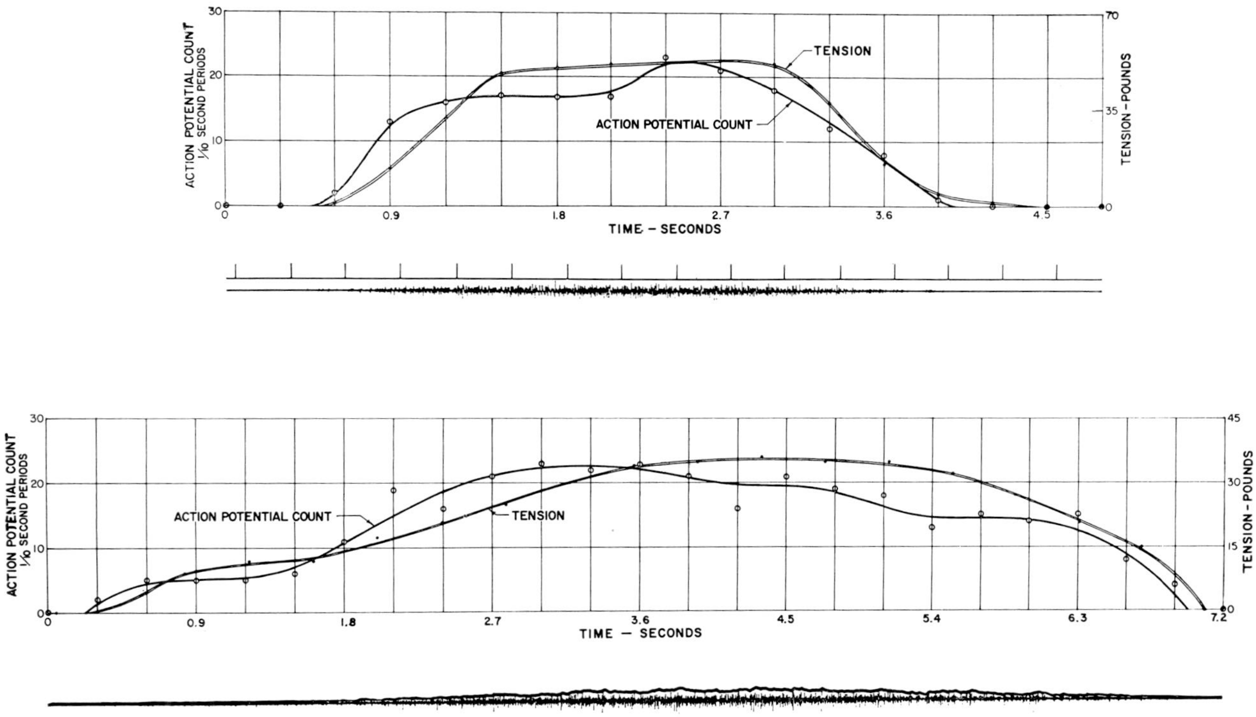

Figure 62. Comparison of the analog curves of tension and action potential count; two isometric contractions. In the lower chart, the averaged electromyogram is shown. (From *Journal of Bone and Joint Surgery,* October, 1960.)

cle. Relationships have been demonstrated between the "integrated" electromyogram and tension under special conditions, such as during isometric contraction or during isotonic contraction where velocity was controlled. The alternative method of producing a quantitative expression for the electromyogram in the form of a count of action potentials might be expected to yield values proportional to the percentage of the total possible motor unit activity from a given volume of muscle. If the motor unit conforms to the all-or-none law, the total number of action potentials may be of greater significance than total electrical activity which is a function of amplitude and duration, as well as action potential count.[41]

As was explained in Chapter II, the appreciation of action potential counts depends on what is being counted. What does each action potential represent? Since recorded action potentials vary greatly in size, each could not represent a single motor unit or a like number of synchronous motor units. By setting the counting

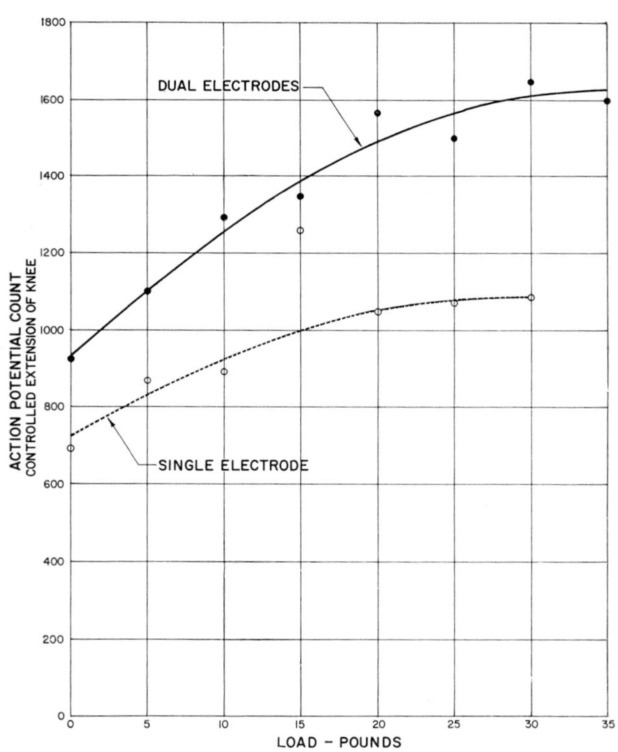

Figure 63. A series of isotonic contractions with control of time of contraction and variation of load.

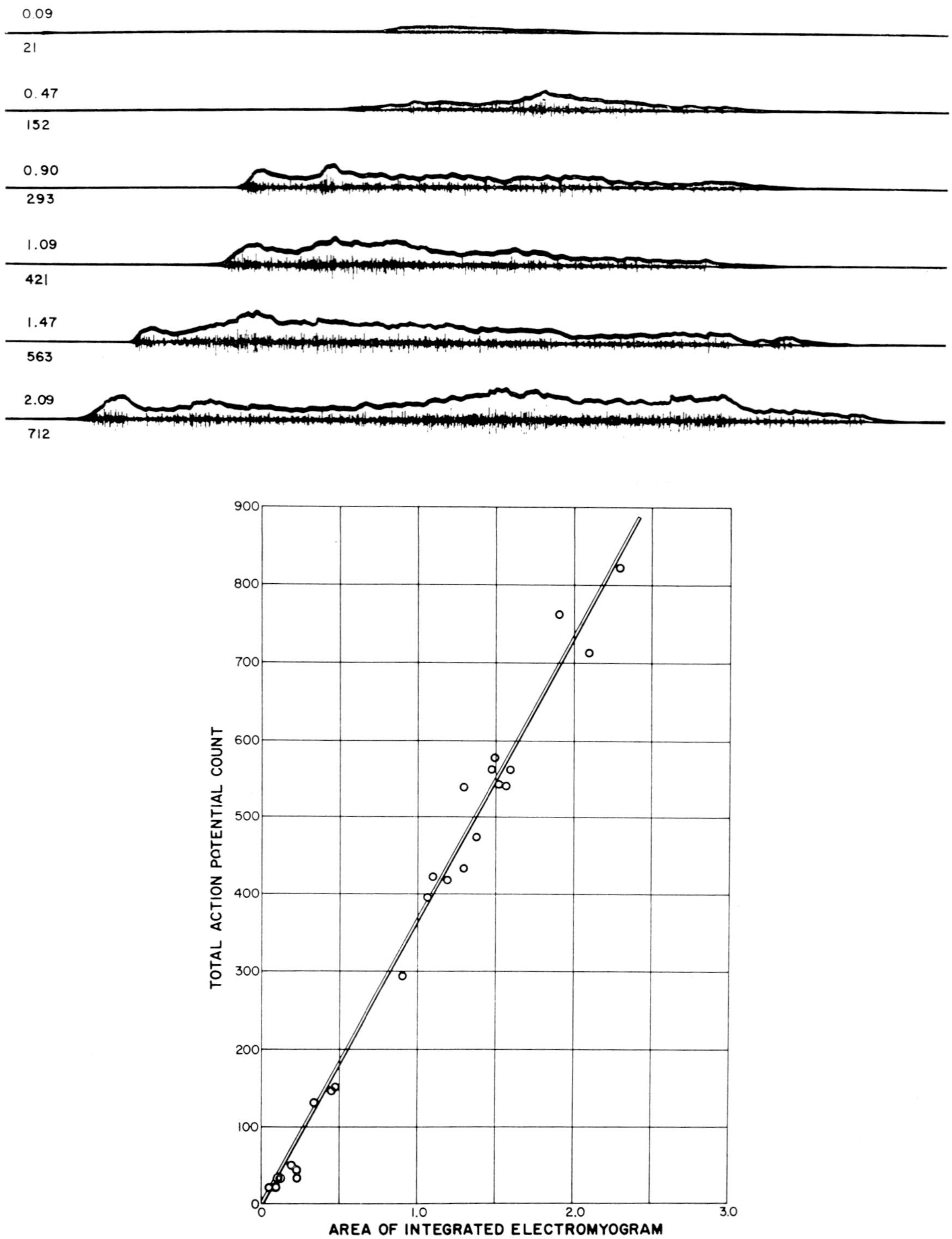

Figure 64. Comparison of the area of the "integrated" averaged electromyogram to total action potential count. The areas under the averaged curves were determined by planimetry. (From *Journal of Bone and Joint Surgery,* October, 1960.)

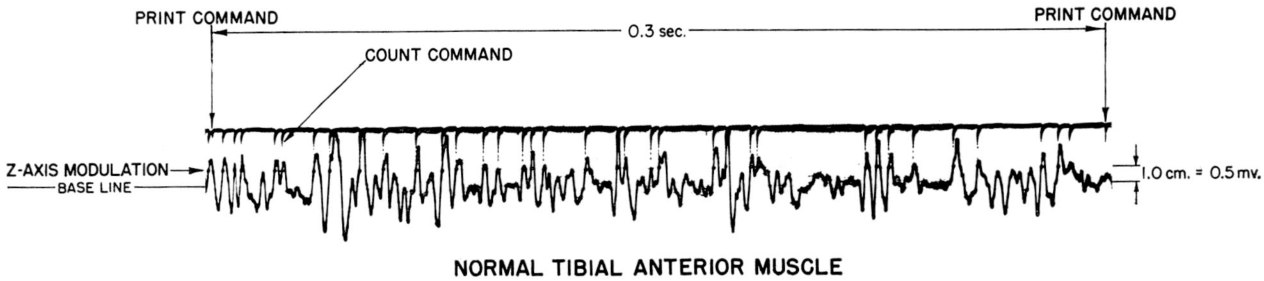

Figure 65. Indication of count level selected as demonstrated by the Z axis and "count command."

threshold, it is quite easy to count action potentials of only a minimum or greater value (Fig. 65).

Using the single coaxial electrode, it is often possible during weak contractions to confine the count to the recording of the main spike of the single motor unit. In most of our counting procedures, however, the dual internal wire electrodes resulted in a signal far superior statistically for counting, integration, etc. For example, the Q/C relationship, described later in this chapter, remained linear when the dual internal wire electrodes were used. This curve was rather irregular, as tension was increased in step-like fashion when the coaxial electrode was employed. Also, counts of action potentials from the single rigid probe bore no close relationship to analog tension curves. Counts from the dual internal wires, plotted in analog form, were quite similar to the tension curve (Fig. 62). This reliability is undoubtedly explained by the larger amount of muscle tissue examined when using dual internal electrodes, by synchrony, and by the resulting characteristically reliable average of the predominant frequency of aggregate action potentials (Figs. 1 and 2). This is found in moderate and strong contractions.

A factor which must be considered in the counting process is synchronization. This phenomenon, often found throughout the muscle, lends more significance to the count (Figs. 1 and 2). Since large numbers of motor units can often be demonstrated to fire in synchrony, the isolated action potential bears quantitative significance, as will be elaborated upon later in this chapter. In Chapter II, a method for restricting counts more closely to a few isolated motor unit action potentials is described (Figs. 29 and 30).

A continuous oscilloscope trace of action potentials from contracting human muscle under various conditions was recorded. The action potentials were counted by the electronic counter which was fed by the vertical signal output of the oscilloscope. In making such counts, it was possible to record full 10-sec intervals or to record continuously for such intervals as 0.1 sec, or 1.0 sec. A second oscilloscope trace simultaneously recorded the conventional electrical average curve of the electromyogram. The "average" or "integration" curve was recorded in this study only for purposes of comparison to the analog count rate curve which we consider a superior source of information (Fig. 62). The application of the real or absolute-value integrator is described in another section of this chapter.

Internal electrodes were used with differential amplification. The subject was grounded to the instrument which was permanently grounded to as stable a ground as was available. The two electrodes were made from 0.003-in. nylon-coated copper wire. In this application, one-half inch of the electrode tip was bared. They were oriented longitudinally in the muscle a considerable distance from the point of entry of the nerve. Longitudinal orientation of electrodes quite close together is desirable in order to restrict the recording to one muscle only (Fig. 66).

There are certain difficulties in obtaining accurate readings of the absolute tension developed by muscle. Since muscles in the body are

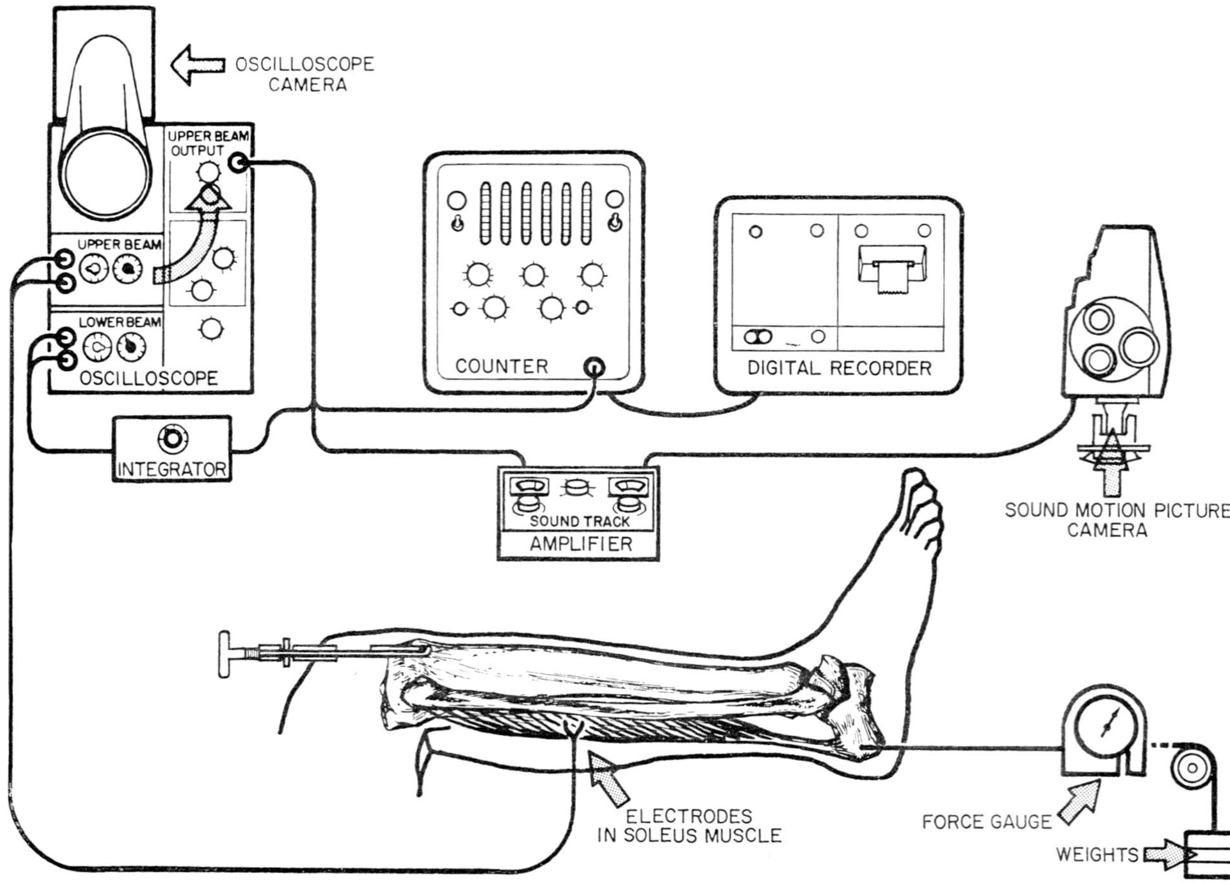

Figure 66. Apparatus in place for recording from muscle. Simultaneous records are made of tension, action potential count and the "integrated" (averaged) form of the electromyogram. (From *Journal of Bone and Joint Surgery,* October, 1960.)

attached, usually to bone, their ends are inaccessible for tension studies. Animal experimentations on muscle tension have usually consisted of laboratory measurements made on stimulated, usually isolated muscle, a situation hardly comparable to tension developed by voluntarily innervated muscle in the unanaesthetized subject.

Our concepts of the magnitude of tension of which muscle is capable of developing has resulted from indirect methods of study. In experimental animals, it is possible to measure the tension of muscle indirectly stimulated, after the tendon has been isolated. Indirect methods, such as calculations of the probable tension developed by the gastrosoleus group attached at the tuber of the calcaneus using the foot as a lever, lead us to believe that human skeletal muscle is capable of developing great amounts of tension. A degree of tension of approximately 500 kg has been suggested as the capacity of this muscle when body motion is accelerated. In the present study, a technique is employed by which the tension of normal, voluntarily contracting muscle can be measured directly, as though the muscle end or its tendinous insertion were free (Fig. 66). This is accomplished by immobilizing the tibia with a Steinmann pin through the tuberosity of the tibia and another through the tuber of the calcaneus near the attachment of the triceps surae. Tension developed between these two pins represents the tension in the soleus muscle when the knee is flexed. In this experiment, the subject was trained to suppress the peroneal, posterior tibial and flexor muscles. By attaching a force gauge to the pin inserted through the tuber of the calcaneus at the muscle's insertion, the tension developed by the soleus muscle could be recorded. The greatest magnitudes of these voluntary contractions were in the order of 35 kg.

This absolute tension does not agree with theoretically calculated tension.[105] Also theoretically, great tension is required of the triceps surae in order to raise the heels from the floor as the subject stands on his toes. However, the center of gravity of the body must shift forward, thus lessening the requirements on the triceps surae.

However, during the test, inhibition may have been a direct factor, since, as claimed by Rodriquez,[146] strength of the quadriceps muscle can be greatly increased by isometric exercises. It is possible to attain such an increase in strength without an increase in diameter of the thigh. Furthermore, the contralateral unexercised quadriceps increases in strength. Such results have been attributed to overcoming of central inhibition.[146]

The tension developed by the muscle was continuously recorded from a precision force gauge (Fig. 67). The force gauge was calibrated as shown in Figure 68. Its readings had a linear relationship to load. Rigid attachment to the force gauge was assured by the use of a calcaneal Steinmann pin representing the insertion of the soleus muscle (Fig. 66). In order to insure absolute immobilization of the tibia, a second Steinmann pin was used. This pin was placed through the tibial tuberosity and secured by a Zimmer clamp to the apparatus. The ability of the soleus muscle to develop tension could be readily appreciated by this direct method.

In laboratory preparations, muscle can shorten a greater amount than when limited by the joint ranges of the skeletal system. The bone-pin technique used in this study permitted shortening and lengthening within the muscle's physiological dimensions only. This experimental arrangement (Fig. 66) yielded the following information:

1. Continuous electromyogram (Fig. 62).
2. Action potential count, with print command and count signal (Fig. 65).
3. Electrically "averaged" electromyogram (Fig. 62), lower record.
4. Continuous readings of muscle tension developed as indicated by the force gauge (Fig. 62).

Figure 67. The Dillon force gauge, useful for direct isometric tension readings.

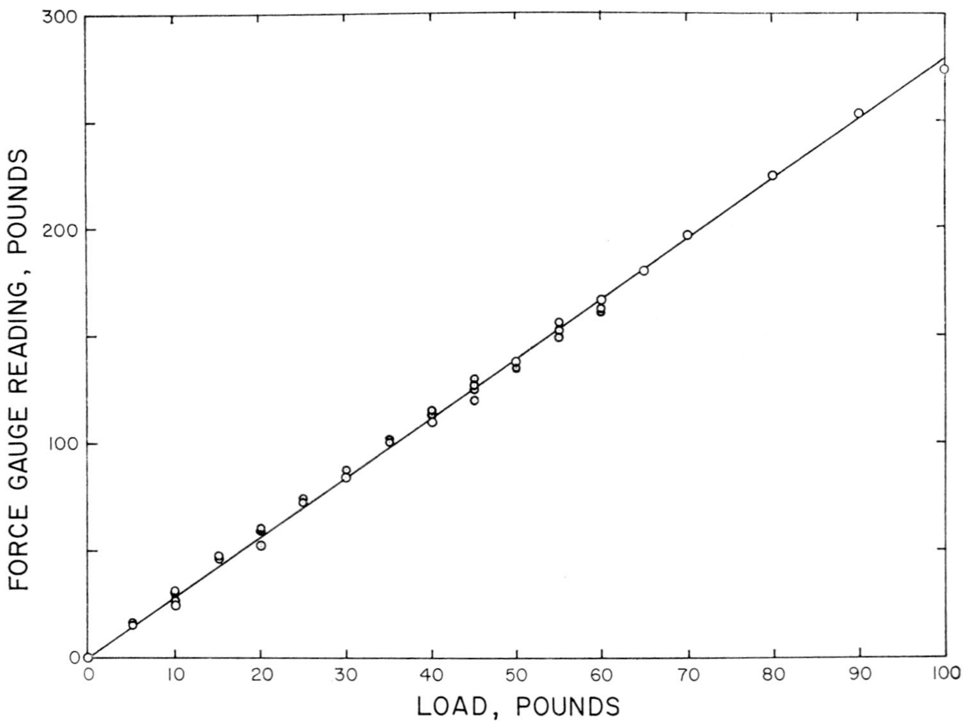

Figure 68. Calibration of Dillon force gauge.

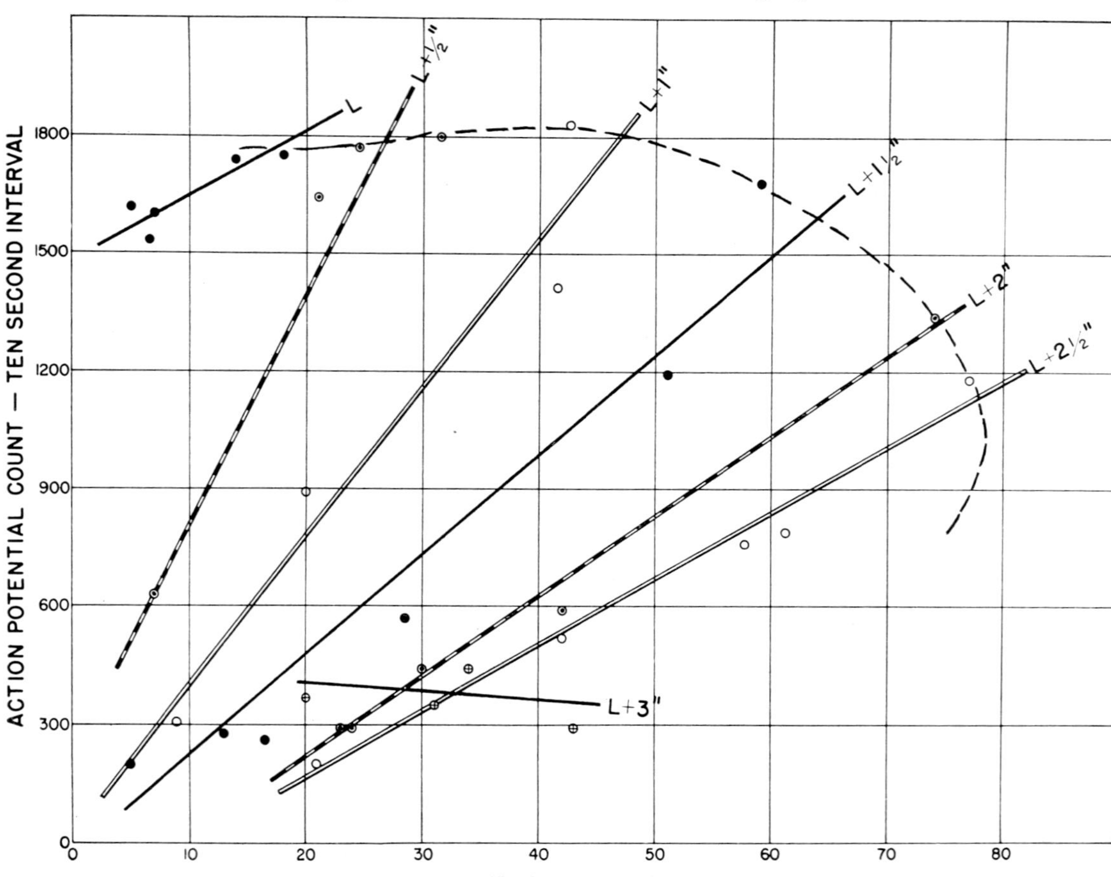

Figure 69. Isometric contractions of 10 sec duration. Tension is recorded by the force gauge.

Data were taken for single isometric and isotonic contractions for 10-sec intervals and for continuous contractions at 1-sec and 0.1-sec intervals (Figs. 62, 69, 70). The knee was flexed to 90 degrees to eliminate any possible participation of the gastrocnemius muscle. It was believed that the soleus muscle (tension) was virtually isolated by the use of the calcaneal-pin technique. The calcaneal pin was placed near the muscle's insertion. Relatively young, active individuals were used as subjects. The procedure was carried out six times. Fifty to sixty contractions were recorded at each test. Isometric tests were made at various muscle lengths and isotonic tests were made. With practice, the peroneal and posterior tibial muscles were easily suppressed.

In one experiment, contribution of the peroneal musculature was definitely excluded by common peroneal nerve block. The force gauge used measured a considerable amount of force with a negligible amount of excursion, thus permitting measurements of virtually isometric contractions. Continuous readings from the force gauge were made by the use of the 16 mm sound camera, the sound track of which recorded the electromyogram in synchrony with the oscilloscope and oscilloscope camera. For exact synchronization of the period of activity counted and recorded by the digital recorder, it was necessary to introduce a signal from the "print command" of the counter. By this signal, the precise instant of the aggregate count for the period of time recorded was indicated. When counting for 1.0-sec intervals, the counter yielded a count for every other full second. When recording 0.1-sec continuous counts, the figure printed by the digital recorder represented every third 0.1-sec interval. The count signal indicated the precise instant at which each individual count was made. The "count command" thus definitely designates which oscillations are included in the total count.

When recording from normal muscle, the oscilloscope sensitivity chosen for this study was 1.0 mv/cm. In addition to this setting, it was necessary to adjust the sensitivity of the electronic counter. Using the standard input of the counter at maximum sensitivity, the pulse threshold counted was 0.2 V. This minimum threshold value may be expressed in terms of the original signal:

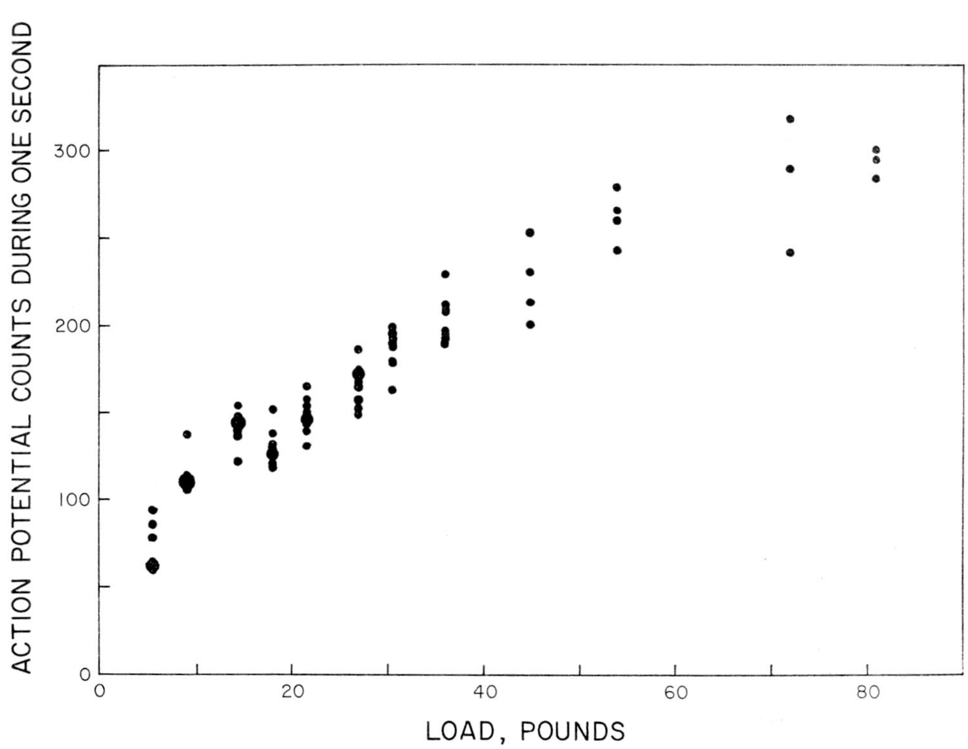

Figure 70. Isometric contractions of 1 sec duration with variation of load.

$$\frac{0.2}{2.0} \times 1.0 \text{ mv} = 0.1 \text{ mv}$$

with 0.2 volts representing the counting threshold of the electronic counter, 2.0 representing the vertical signal output of the oscilloscope in volts per centimeter of oscilloscope deflection and 1.0 mv is the oscilloscope sensitivity used. The magnitude of the minimum counting threshold was thus 0.1 mv.

Thus, the counter included very small oscillations in the count. It was recognized that this might not be the ideal sensitivity at which to count action potentials. It was further recognized that such a sensitivity might give a very misleading count in certain pathological conditions such as muscular dystrophy where there are many action potentials of low magnitude, but where the muscle is obviously weak. Settings must be determined in order for the counter to be used as standards in polio, muscular dystrophy and normal musculature. Since motor units of various normal muscles vary in size,[66] action potentials of different magnitude result. Also, the aggregate count found in normal muscle varies, since the number of motor units per unit volume varies from muscle to muscle throughout the body.

The procedures where tension was recorded were done only on normal subjects. However, action potential counts were recorded from a variety of subjects with pathological conditions such as poliomyelitis, muscular dystrophy and amyotrophic lateral sclerosis.

An alternative counting threshold was available by use of the discriminator. This adjustment of the counter made it possible to set the count threshold at selected values. By this means, counts could be produced representing either positive or negative potentials exceeding a certain value (Fig. 65). The smaller components of the motor unit complex could thus be eliminated from the count. However, many potentials low in amplitude because of their distance of origin from the electrodes were also excluded.

The rather strange phenomenon of synchronization is often found, using multichannel equipment, where even large groups of motor units are exposed to long electrodes (½ in.) using differential amplification. Synchronization has been found in 40 percent of our dual recordings made from the normal anterior tibial muscle. Electrodes in this experiment are oriented transversely in the muscle. This is a strange finding which seriously challenges the concept of recruitment of motor units at random,[158] as the strength of contraction is increased through the influence of the central nervous system. Therefore, it appears that for each intensity of contraction, the muscle as a whole has a certain rate (frequency). This frequency is prevalent throughout the muscle to the point of producing virtually identical records from different parts of the muscle on many occasions. These findings indicate that muscular tension is increased more by an increase in the rate of firing of its motor units than by recruitment of "previously idle" motor units.

There is often a predominating action potential frequency (rate) in skeletal muscle. This frequency is closely related to the degree of contraction produced. It is also closely related to the tension of the muscle, and counts of this frequency vary with the intensity of the contraction. It is representative of the rate of activity throughout the muscle.

On the other hand, "integration," or averaging, of the electrical activity represents the electrical activity of only the amount of muscle exposed to the electrodes. Therefore, it is not so closely identified with whole muscle behavior as is the action potential rate found in the recording of a single muscle site. This represents the predominating frequency to be found throughout the muscle. The study indicates that different sites in the muscle will yield comparable counts.

Synchronization is readily found in the anterior tibial muscle. However, when comparing action potentials from two pairs of electrodes in such muscles as the vastus lateralis or rectus femoris, one is not likely to find synchrony but will find

1. Identity of timing (beginning and end of activity).

2. Similarity of frequency (rate).
3. Like amplitude.

The action potential, having frequency, amplitude, duration, and fine variations in form is influenced by the adjustment and calibration of the equipment used. Standardization of results depends upon control of the many technical factors governing the size and form of the action potential. In this study, uniformity of many of these variables in technique was established. It was thought that the most important factors were:

1. Caliber of electrode and length of tip; the number of motor units contributing to the recording is subject to the diameter and length of the exposed tip.
2. Distance between electrodes.
3. Sensitivity settings of recording apparatus.
   a. Oscilloscope.
   b. Counter.
4. Uniformity and standardization of speed of recording equipment.
5. Proper setting of counting voltage threshold.
6. The use of equipment of suitable frequency recording range (Figs. 8, 20, 21).
7. Since alternating current coupled input is almost universally employed, it is imperative to standardize the input condensers and resistors; their values influence the lower frequency limit (Fig. 28). It is desirable that the amplifier impedance be much greater than the resistance of the tissue between the electrodes.

In using large internal dual electrodes, as in this study, a larger volume of muscle was sampled than when using a single coaxial electrode which may confine the recording to a very small volume. On the other hand, surface electrodes are thought to record from a large mass of muscle but with less specificity. Surface electrodes were found to yield a lower action potential count than internal dual electrodes. Accordingly, dual internal electrodes placed 1 in. apart with 0.5 in. of the electrodes bared were used. Their orientation was longitudinal, although transverse orientation gave comparable results. The behavior of this volume of muscle was considered characteristic of the muscle, as represented by the consistency found in count and appearance of action potentials of repetitive tests made in different regions of the muscle and by the striking similarity of the action potential curves often recorded from different portions of certain muscles during the same contraction.

Using dual internal electrodes and differential amplification, with the subject grounded to the apparatus, one records the difference in potential at the two electrodes. Thus, one may record motor unit action potentials from either electrode; or action potentials at each electrode may be combined. Conceivably, potentials of opposite sign may simultaneously occur at the two electrodes and thus be added, producing a larger deflection. However, there is almost always a sufficient time interval between action potentials for at least part of each to be recorded on the oscilloscope and to be counted by the electronic counter.

Since one side of the differential amplifier subtracts its value from its other identical section, one might rarely find equal simultaneous pulses arriving at the inputs, with a resulting zero deflection. At times, the input at one terminal may be zero. At this instant, a complete action potential may arrive at the other input terminal of the differential amplifier. Under these conditions, the recorded action potential would be the same as if the single-ended mode were employed. Thus at very low levels of activity, one observes the pure action potential. It is interesting, as shown in a later section of this chapter, that the same relationships are found at low as well as at high levels. This relationship, Q/C, is the ratio of total electrical activity to total action potential count during a contraction. It represents the average electrical value in microcoulombs of the single action potential.

Figure 82 demonstrates the fairly linear relationship of counts of action potentials and of integrated action potentials to tension. The linearity of the relationship is retained in the lower ranges of the curve.

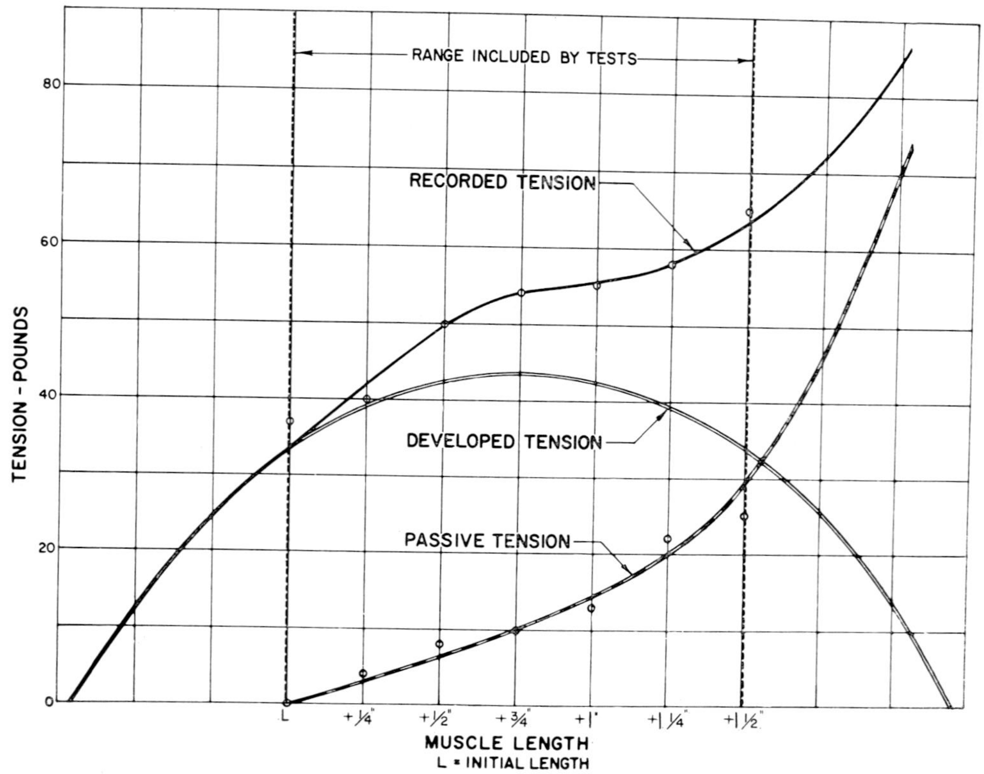

Figure 71. Values for recorded isometric tension and passive tension placed according to their presumed positions on the theoretical length-tension curve.[143]

Total counts of 10-sec isometric contractions were plotted against tension demonstrated in Figure 69. Since these contractions were studied at various muscle lengths, several graphs were prepared. As the contractions increased in tension, a number of readings were recorded at each length. The data for each length demonstrated a fairly linear relationship between total count and tension. In this diagram, it must be noted that in order to place the muscle at progressively greater lengths, it was necessary to stretch it or apply tension to it. This tension has been subtracted from the recorded tension in the values used in this figure. The values given for tension are therefore the developed tension for each contraction. The portion of the theoretical length-tension diagram,[143] in which the muscle lengths are comparable in range to those used in Figure 69, is demonstrated in Figures 71 and 72.

The continuous 0.1-sec interval counts are demonstrated in Figure 62. These are records of isometric contractions. The relationship of count rate per unit time to tension is demonstrated. The similarity of the two analog curves is noted. It is also noted that the count rate curve precedes that of tension. Since the count rate given for each point plotted on the curve is instantaneous and is indicated by the print command (Fig. 62), there is no lag due to apparatus. The actual delay between tension and count rate then, as indicated, is physiological. The counter, unlike the "integrator" (averager), does not have significant time delay. The resulting analog curve of count rate per unit time is somewhat similar to the average curve. However, any point on the count rate analog curve has numerical value. This value represents the action potential count rate at the point under consideration for the volume of muscle immediately surrounding the electrodes. In Figure 62 (lower graph), the analog count rate curve may be compared to the force curve and to the "integrated" electromyogram. Since no print command is recorded here, synchrony is not considered absolute. However, the 16-

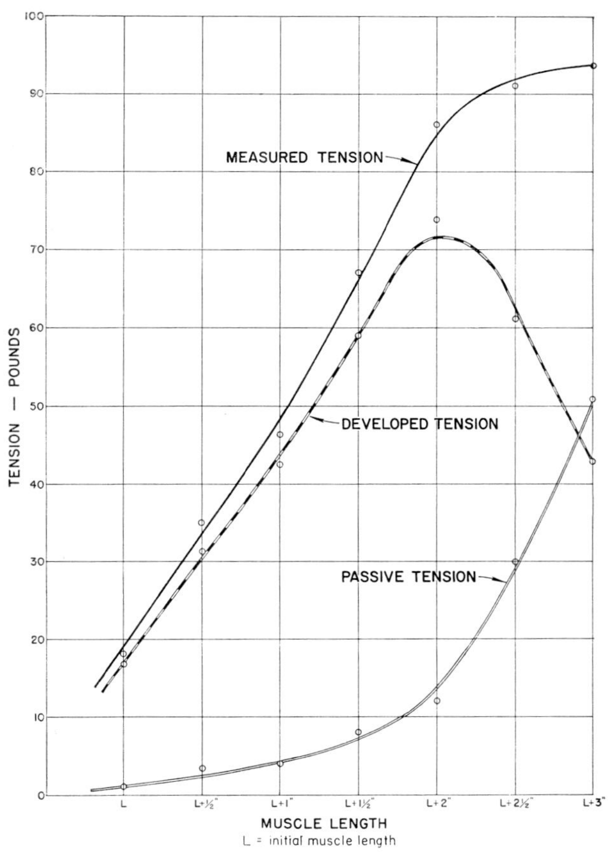

Figure 72. Values for isometric tension and passive tension plotted as in Figure 71. Here, a greater range of muscle lengths was studied.

mm sound camera used can be demonstrated to operate at precisely 24 frames/sec (by use of a known frequency from the counter). Since its sound track consists of an audio recording of the signal counted, the analog curves of action potential count rate and tension may be very accurately matched. One can thus compare count rate, tension, electromyogram, and integrated electromyogram for a complete contraction, as in the lower graph of Figure 62.

Isotonic curves of contractions with progressively increased loads were recorded (Fig. 63). Here, tension was not recorded and the pin technique was not used. An attempt was made to have the subject perform the contraction as uniformly as possible. Approximately the same velocity of contraction was used for each contraction. As the load was progressively increased, action potential count rate increased in regular fashion. The time of each contraction was the same. The leveling off of the curve as the load is increased indicates that as the contraction approaches its maximum, there is an increase in synchrony of motor units. This synchrony results in a lower count than would be anticipated. The density of the normal interference pattern also increases as the strength of the contraction increases.

When the muscle, with bone pins in place, was allowed to contract under free isotonic conditions (Fig. 73), no relationship of tension to action potential count or to the averaged electromyogram was observed. Tension was measured by the force gauge.

Using planimetry, the areas of some of the averaged curves of complete isometric contractions were obtained. The area thus obtained represents actual integration of the electromyogram. Several consecutive isometric contractions of gradually increasing intensity were made, the total action potential count recorded, and the area of the average curve plotted against total action potential counts (Fig. 64). These contractions were of moderate intensity and performed under similar conditions. The linearity of this relationship for normal muscle is evident. Throughout the body, with its muscles having motor units of various size,[66] graphs of different slopes would be expected. A different slope of the resulting line is also anticipated for pathological muscle (Fig. 79). Here, a real integrator is employed. Such an integrator actually indicates total electrical activity for a contraction or for a given portion of the contraction; the total action potential count is taken from the same portion of the contraction. This was made possible by the use of a switch which turned the integrator and counter on and off at precisely the same time.

Investigations were done in which the average electromyogram and total count were recorded in such pathological states as poliomyelitis, muscular dystrophy and amyotrophic lateral sclerosis (Fig. 79). It was not possible to record tension in these studies. Instead, maximal contractions were used. These are to be considered preliminary studies. They were performed both with dual internal wire elec-

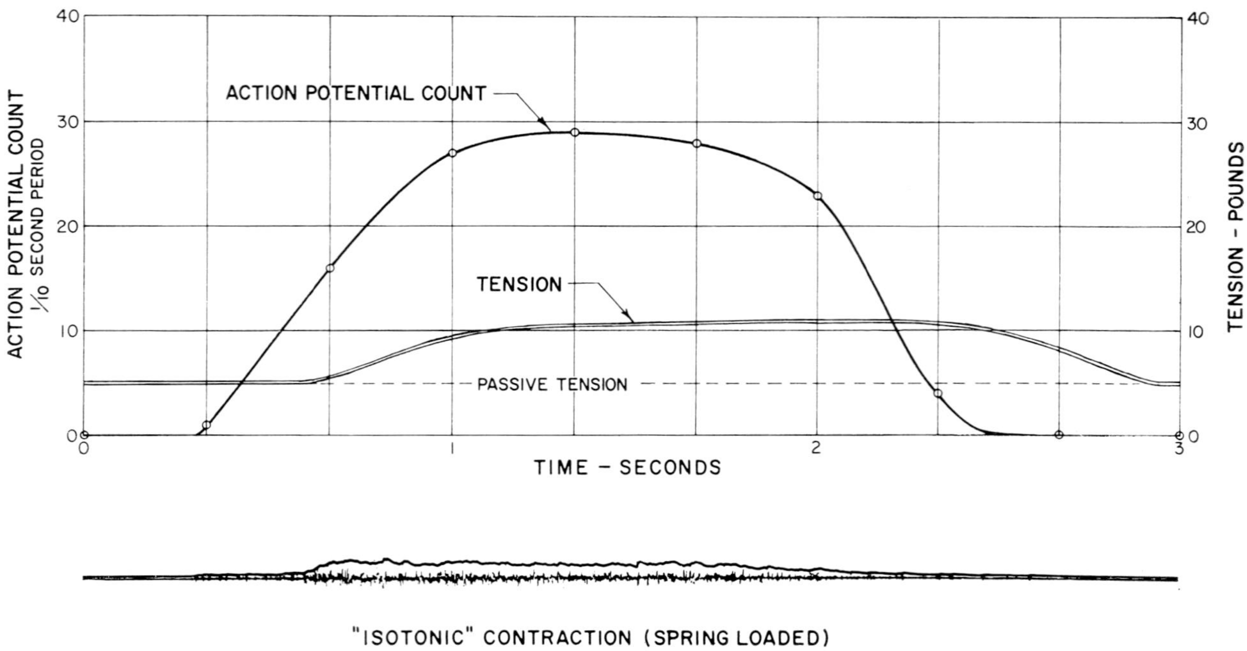

Figure 73. An isotonic, uncontrolled contraction. No relationship is found between action potential count and tension.

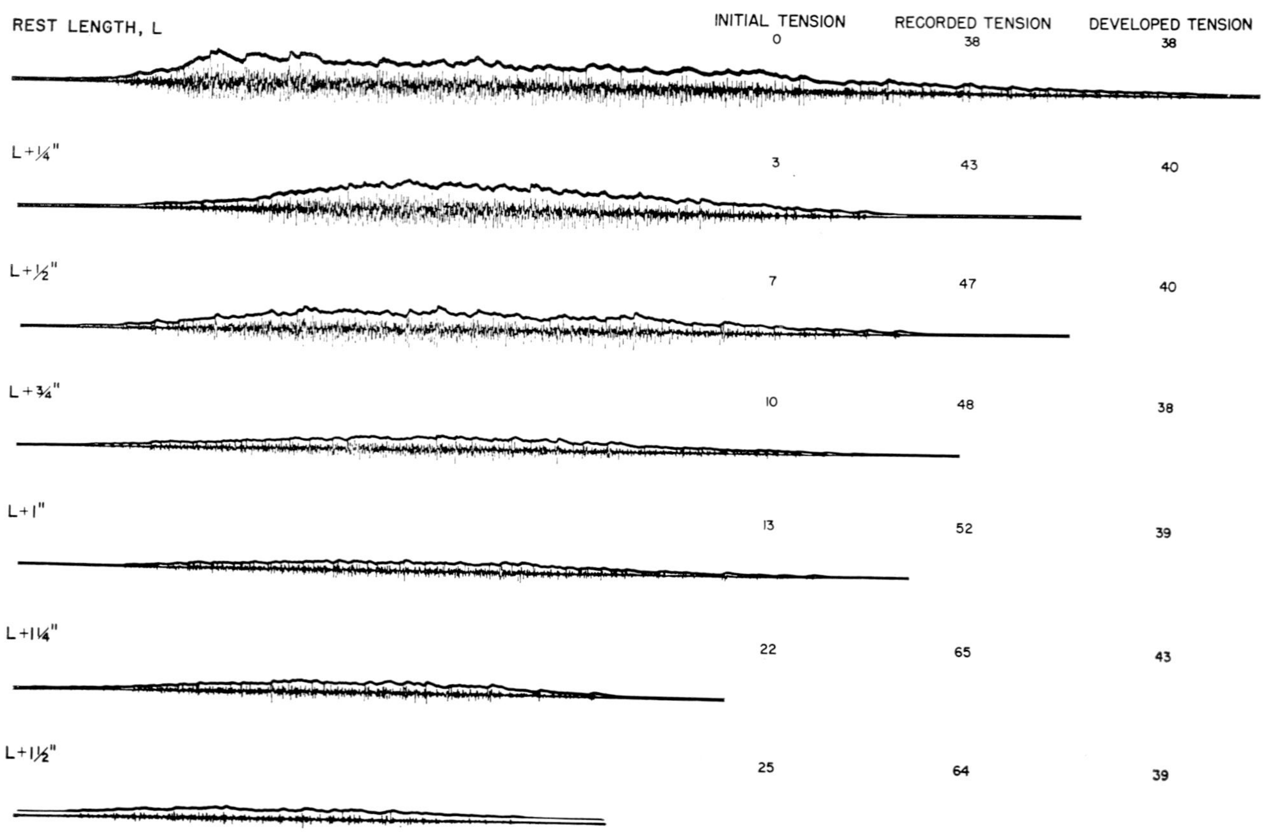

Figure 74. The action potential and averaged electromyogram at different muscle lengths. Passive and developed tension are given for each contraction. (From *Journal of Bone and Joint Surgery,* October, 1960.)

trodes and with the single coaxial electrode. Such studies served as a basis for comparisons to normal muscle, which were used later in this report.

The electronic counter used in this study accepts and counts pulses of extremely fast rise time. It is capable of counting 1,200,000 pulses per second. Muscle action potentials occur at a much lower rate. Since the action potential is produced by one or more motor units and each of the fibres in the volume tested contributed, the question arises whether the counter might consider each fibre's action potential as an individual pulse. This possibility can be eliminated by proper threshold setting of the electronic counter. Thus, if the possibility exists that firing of the fibres of the motor unit is not absolutely synchronous, individual fibre amplitude would not be of sufficient magnitude to be counted. The threshold of action potentials counted can be conveniently selected by use of the discriminator.

The 10-sec isometric contractions and the analog displays of the 0.1-sec interval counts demonstrate a linear relationship between count rate per unit time and tension. At given

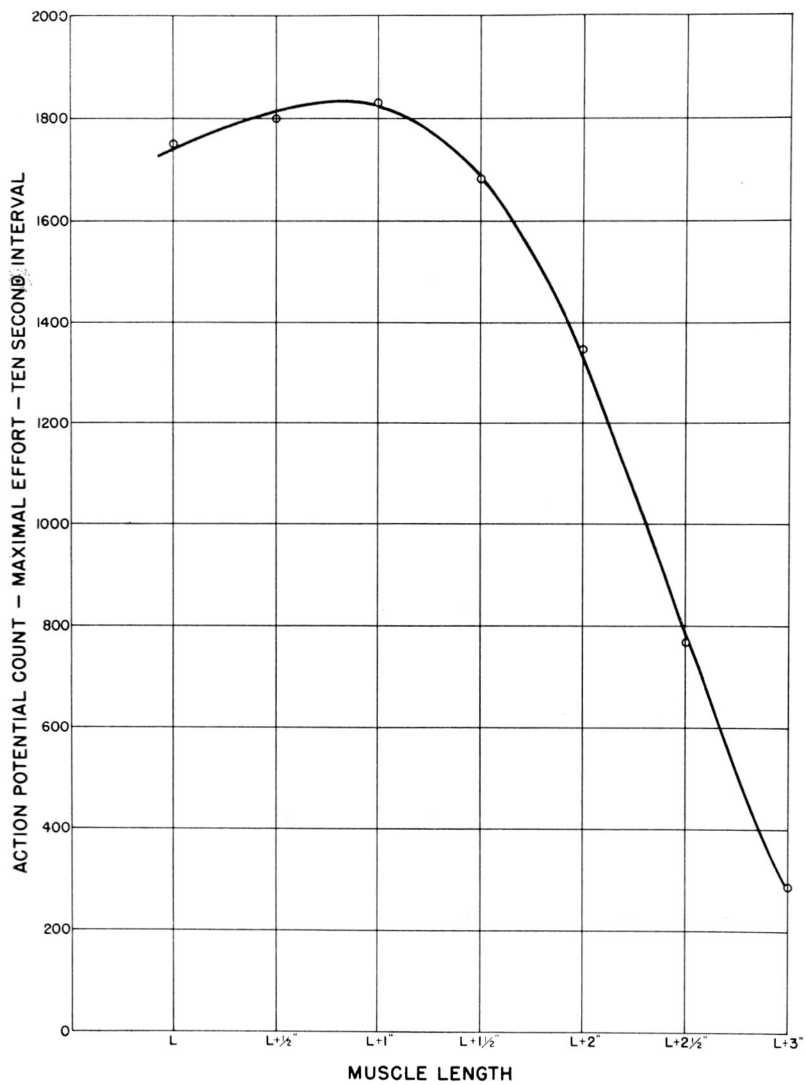

Figure 75. Total action potential counts for 10-sec maximal contractions as muscle length is increased by increments of ½ in. (From *Journal of Bone and Joint Surgery*, October,1960.)

muscle length, the maximum action potential count rate can be predicted.

The impulses from the tendon afferents are thought to be responsible for reducing the electromyogram under conditions of stretch.[117] Muscle spindles[77,78,79] and other proprioceptive organs, such as the Golgi apparatus, influence the contraction. Thus, any measurements of tension in muscle must be made under certain conditions and those conditions precisely recorded. The isotonic contractions described by Bigland and Lippold[20] and Lippold[119] were made under specified conditions. The muscle was required to perform a task in which shortening was at a measured velocity. In isometric tests, initial length and tension are of similar importance. Figure 74 shows the effect of initial length and passive tension on action potential.

Obviously, the behavior of muscle is influenced by the load to which it is subjected. Contractions of the soleus muscle during walking involve very little change in length, and great tension may be possible though the action potentials are of moderate amplitude. High action potential counts may develop in the nonloaded muscle with very little resulting tension, since the contraction is unopposed.

The bone-pin technique used to record tension eliminates the very difficult task of correcting for the leverage system of joints. Unless the muscle end is free or bone-pin techniques are used, muscle tension cannot be measured without elaborate leverage system calculations.

Action potential count rate is a function of recruitment and rate of motor unit contraction. The counter is unable to identify the source responsible for increase in total count. Which is the more important source, recruitment of new units or increase of the rates of units already in a state of contraction? Does recruitment of a new motor unit produce more change in total tension than increase of the rate of one already in a state of tetanus or near tetanus?

Isometric count-rate–tension relationships are linear (Fig. 69). Each muscle length is a separate consideration. Paradoxically, the action potential counts are lower, as the muscle is placed at greater length where the *recorded* tension is greater. At still greater lengths, the *developed* tension finally becomes smaller as the right side of the length-tension curve is reached (Fig. 69). If tension is increased by recruitment and by increasing the rate of motor unit contraction, how is it possible to develop greater tension with a lower count rate (Figs. 69 and 75)? Apparently, the motor unit action potential does not have the same significance at different lengths. It might be said that within certain limits, muscle is more effective at its greater lengths—where great tension is accompanied by a low motor unit action potential count. However, the muscle must first be placed at such a length by its antagonist or by an outside force such as gravity. Work must first be done on the muscle before its greater tension output at greater length is possible (Fig. 72). The tension necessary to stretch the muscle to the length at which the test is made must be recorded (Figs. 71 and 72). If this value, the "passive tension," is subtracted from the "recorded tension," the "developed tension" is determined.[143]

The single electrode with a very small tip or the coaxial electrode may be used to obtain recordings from a few motor units and sometimes from a single motor unit. The motor unit near its maximum contraction is expected to be in complete tetanus, at which time its action potential count rate curve reaches a plateau (Fig. 63).

It may be said that we obtain different information by the use of different internal probes:

1. *Coaxial electrode, an increase of count by increase of individual* motor unit (or at least small groups of motor units) contraction rate.
2. *Dual electrodes, an increase of count by increase of individual motor unit rate (frequency)* and *by recruitment* (to the extent that recruitment actually occurs).

Thus, action potential counts are proportional to tension until the maximum tetanus frequency of the motor units is reached. The isotonic contractions of Figure 63 reach this stage and the curve levels off. The isometric contractions of Figure 62 are of lesser magnitude. Here the action-potential-count–tension curve is linear throughout. Relationships to tension for isotonic, as well as isometric, contractions are expected if the contraction is regulated and does not exceed a certain limit. No relation is expected for small loads or rapid spring-loaded contractions (Fig. 73).

The isometric contraction (Figs. 62 and 69) performs no work, since there is no change in muscle length. Here, tension is directly related to action potential count or rate:

$$C \sim T$$

The isotonic contractions of Figure 63 produce work, since the muscle shortens and lifts a weight:

$$W = T \times D$$

The results of Figure 63 show a significant relationship between total action potential count and work:

$$C \text{ (Count)} \sim W$$

Since distance as well as duration of contraction was constant, the action potential count is proportional to tension:

$$C \sim T$$

The terminal portions of the curves of Figure 63 level off, probably because of increasing motor unit synchronization as the strength of the contraction increases. The tension of *isotonic* contractions, as well as the tension of *isometric* contractions, is proportional to action potental count under the above-specified conditions.

SUMMARY AND CONCLUSIONS

Muscle action potential counts may be conveniently recorded by electronic counters. Total counts for a chosen time interval may be recorded, or a continuous count (analog display) may be recorded. Total counts and analog displays for muscular contractions make possible a numerical representation of action potentials produced by a given volume of muscle. The "integrated" (averaged) curve of total electrical activity may thus be replaced by the action potential count of the electromyogram having a total or continuous numerical expression. For muscle function analysis and locomotion studies, the continuous action potential count with established threshold values adds significantly to the information gained from the conventional electromyogram.

The time constant of conventional electrical "integrators," due to dependence upon capacitors, is overcome. A real but very short delay in the rise of tension following the appearance of the rising action potential count is observed (Figs. 62, 73). The electromyogram, its count rate and averaged form decline well in advance of tension.

Total counts for 10-sec periods of activity increase in linear fashion as the strength of the isometric contractions increases. A skeletal pin technique has been used for more accurate measurements of individual muscle tension. The range of muscle length studied is limited by physiological joint ranges of motion. Thus, only the central portion of the theoretical length-tension curve, the physiological portion (the only range actually used in the human body) has been studied (Figs. 71 and 72).

The continuous action potential count is related to tension during isometric or isotonic contractions as long as the muscle is loaded and the rate of shortening controlled.

The electronic counter appears to have great possibilities of application in the study of neuromuscular disorders. Since different adjustments of both the oscilloscope preamplifier and the counter are necessary, results are not directly comparable to records from normal muscle. However, normal dystrophic and poliomyelitis muscle have definite characteristics of action potential, rate and amplitude (Fig. 80). Thus a motor unit action potential classification for types of muscle is possible.

Normal	moderate amplitude	high frequency (rate)*
Dystrophy	low amplitude	high frequency (rate)*
Poliomyelitis	high amplitude	low frequency (rate)*

* Since muscle action potentials are not identical and do not ordinarily occur at regularly spaced intervals, they have "rate" rather than "frequency."

THE AVERAGE ELECTRICAL VALUE OF THE ACTION POTENTIAL, Q/C—TOTAL ELECTRICAL ACTIVITY DIVIDED BY TOTAL ACTION POTENTIAL COUNT

The fundamental functional composition of skeletal muscle is the motor unit (Fig. 76). Each motor unit is a bundle or a group of smaller bundles of fibres innervated by a single axon representing a single anterior horn cell. Thus each axon terminates in many end plates and usually each muscle fibre has a single end plate. In the normal human, there is great variation as one proceeds caudad to cephalad in the number of muscle fibres in each motor unit (Fig. 77). For example, the medial head of the gastrocnemius muscle, has been demonstrated by histological studies to contain an average of 1,634 muscle fibres per motor unit.* The first dorsal interosseous muscle in the hand contains approximately 340 fibres per motor unit. This knowledge is based on the counts of Feinstein *et al.*[66] The counts were made by histological methods. The total number of motor axons terminating in the muscle was first counted. Representative portions of the cross-section of the muscle were then chosen. Counts were made of the number of muscle fibres. After arriving at the total number of muscle fibres in the muscle, this value was divided by the number of muscle fibres per motor unit for many of the muscles of the body. The results reported by Feinstein *et al*[65,66] do not allow for larger gamma fibres, which should not be included in the counts of alpha motor neurons. Had it been possible to eliminate these from the axon counts, the ratio of muscle fibres to motor axons would have been greater. The desirability of establishing a method for determining motor unit size in the living subject is obvious. Such a procedure would contribute to our basic knowledge of neuroanatomy and neurophysiology. It would also give us a basis for comparison for study of pathological states.

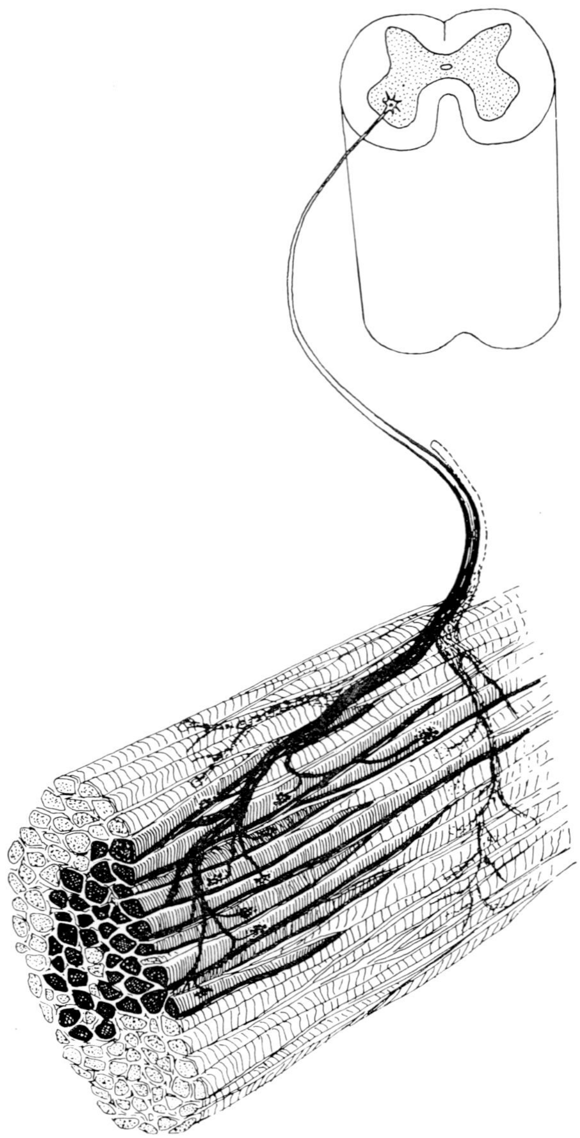

Figure 76. Diagram of the motor unit.

It is possible that by electronic means, one could calculate average motor unit size. Total integration, Q, produced by a muscular contraction has been shown to have an almost linear relationship to the total action potential count, C, of the same contraction (Fig. 78) if the values for a series of contractions are plotted. Since the striated muscles of the body are composed of motor units of different size, one would expect that the relationship Q/C would have a characteristic value, for the motor units of each muscle are proportional to the histological counts of Feinstein *et al.*[66] It would

* In one instance, this figure was 1,934 muscle fibres for the average motor unit (medial head of gastrocnemius).

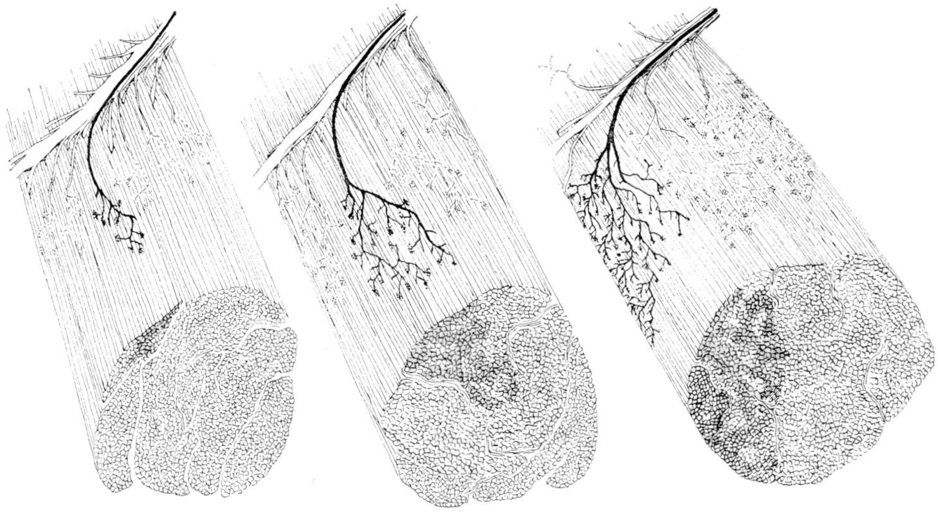

Figure 77. Diagrammatic representation of muscle. Motor units are of various sizes. Probable division of the motor units into subunits.

be most convenient in the living subject if it were possible to demonstrate this ratio by some of the simple electronic tests in use. For example, by using the absolute-value integrator to obtain Q and the electronic counter with adjustable counting threshold to arrive at the value C, the valuable relationship, Q/C, proportional to motor unit size, may be readily found.

$$\frac{Q \text{ (microcoulombs)}}{C \text{ (Total Count)}} = \text{Average Action Potential (in microcoulombs)}$$

Such a relationship (Q/C) is most obvious when one compares the muscles of the lower extremity to the finer muscles of the upper extremity. For example, the medial head of the gastrocnemius[66] is comprised of motor units containing an average of 1,634 to 1,934 fibres

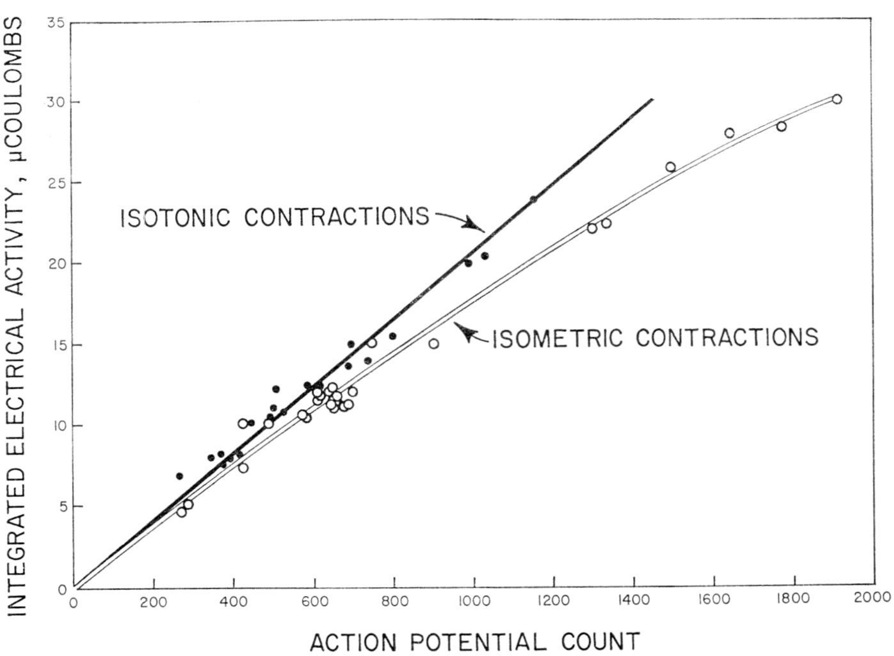

Figure 78. Relationship of total integrated electrical activity (Q) to total action potential count (C). The isotonic as well as the isometric contractions are made under controlled conditions.

each. In doing integration and counting studies on such a muscle, a higher Q/C ratio for a given contraction is anticipated than when one records similarly from such smaller muscles as the first dorsal interosseous. In the first dorsal interosseous muscle, each motor unit is thought to contain approximately 340 fibres.[66] Thus, for a given contraction, the volume of muscle surrounding the electrodes produces a certain amount of recorded electrical activity. The medial head of the gastrocnemius fires fewer but larger motor units. Such information should enhance our concepts of the histology of muscle. It may permit us to gain further insight into the knowledge of the dystrophic muscle. The rato indicates the degree of innervation and the degree of control by the central nervous system over each muscle. Characteristic ratios (Q/C) may be expected from the data obtained from the normal, postpolio or dystrophic muscle. In the dystrophic muscle where there is primary myopathy, one expects to find smaller motor units. Here the count is high, and total electricity, Q, is low. A curve of low slope is anticipated in contrast to that of certain types of polio musculature where there is a great deal of axon branching. Here the motor units of corresponding muscles are larger and the total count (C) is lower for a given amount of electrical activity (Q) (Figs. 79 and 80).

Using the absolute-value integrator and the electronic counter (Fig. 81) with variable threshold, it is possible to determine Q/C ratios quite readily for various muscles for a definite contraction or for a definite portion of the contraction. It has been our practice to obtain Q/C ratios for each muscle during different types of contractions. These contractions vary in duration. Studies have also been made where time has been constant and the load or tension varied. Each muscle demonstrated a virtually linear Q/C relationship (Fig. 82) after the several points, each representing a complete contraction, were plotted.

If desired, continuous cumulative values of Q (total electrical activity) and C (total action potential count) may be displayed on an X-Y oscilloscope. This results in a curve, the slope of any segment of which directly indicates the Q/C relationship. This procedure appears very promising for rapid exploration of this relationship in various neuromuscular conditions. The X-Y procedure facilitates the preparation of many Q/C determinations in a short time. The curves of three postpolio muscles and a normal muscle were displayed on the X-Y oscilloscope (Fig. 79A).

The setting for the counter was chosen arbitrarily at 0.6 cm. Only the positive components of the wave forms were counted. The absolute-value integrator demonstrated simultaneously by direct read-out and also by analog display, if desired, the total electrical activity, both positive and negative, at the examining electrodes during the same contraction. Several counting thresholds were tried before the set threshold was adopted at 0.6 cm. The counting threshold was selected at this critical level in order to include the principal deflection of the single motor unit and to reject the minor portions of the single motor unit action potential. A suitable oscilloscope sensitivity was chosen at 0.5 mv/cm (Fig. 65).

Since the basic pulse is diphasic, only the positive or negative deflections should be counted. Otherwise, the true action potential count is doubled. Electrical integration, however, should include all of the electrical activity, both positive and negative. Hence, the absolute-value integrator must rectify the negative portions of the wave forms and add these to the positive portions of the oscillations.

The ratio Q/C is thus a property of skeletal muscle increasing as one proceeds cephalad to caudad. The portion of the muscular contraction for both Q and C are identical. Contractions can be timed and duration indicated as demonstrated in Figure 82. If differential input using dual wire internal electrodes is employed, significant variation in Q/C in the various muscles in the normal human is not apparent. This is probably due to the fact that during strong contractions, action potentials tend to combine. A great amount of synchrony is produced. Thus, using differential input and a fairly large area of muscle, one might record the same

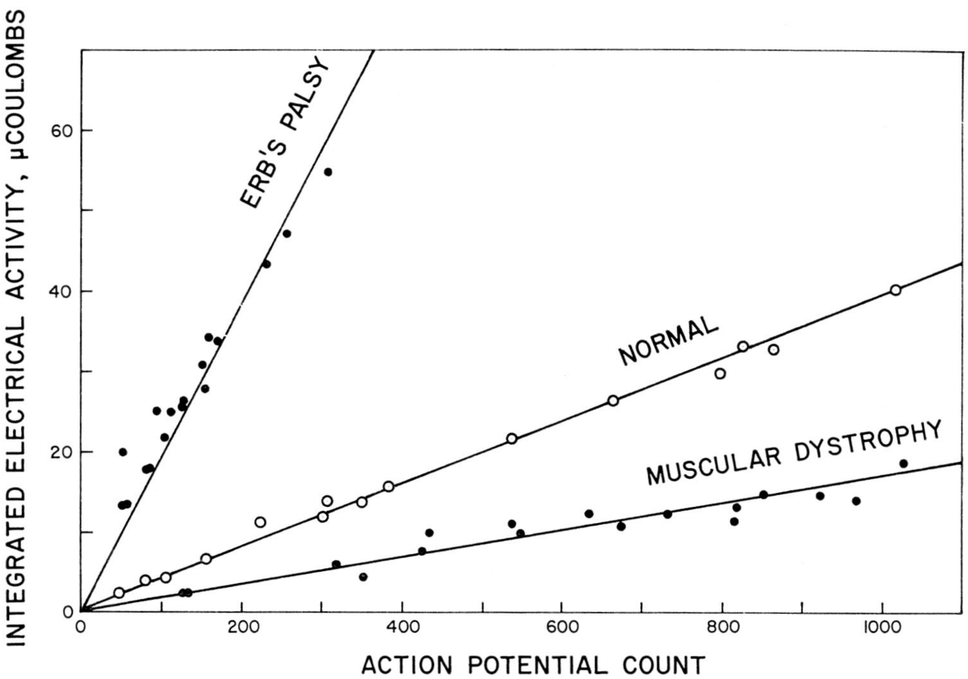

Figure 79. Q/C curves for muscle in Erb's palsy, normal, muscular dystrophy. (The convalescent poliomyelitis muscle Q/C curve has a slope similar to that of Erb's palsy.)

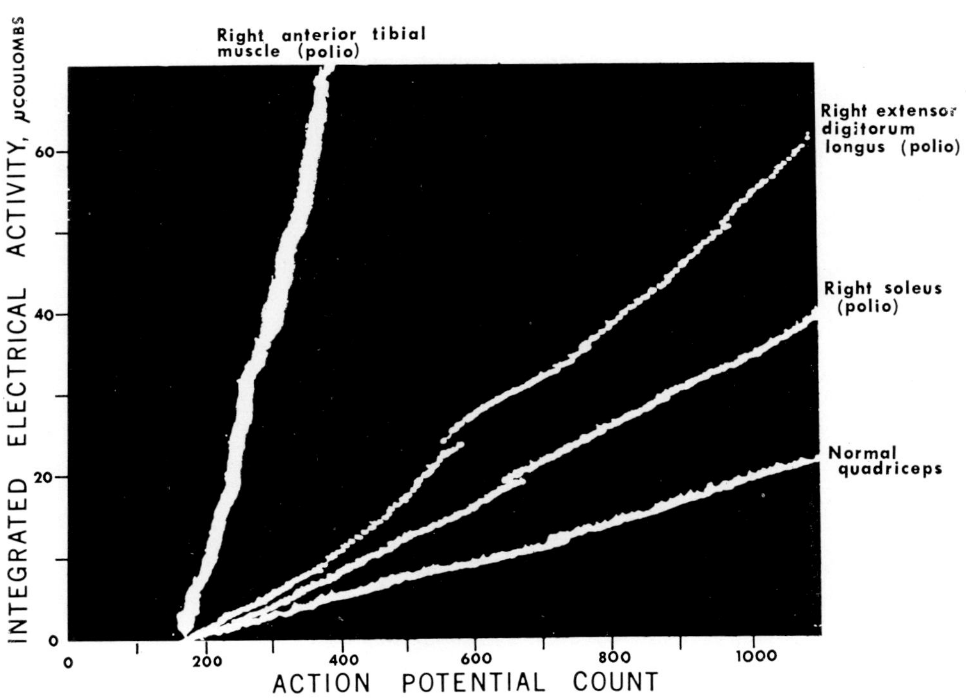

Figure 79A. Q/C curves of three postpolio muscles with varying degrees of involvement, and a normal muscle. These curves were recorded from X-Y oscilloscope. Each curve represents a single isometric contraction.

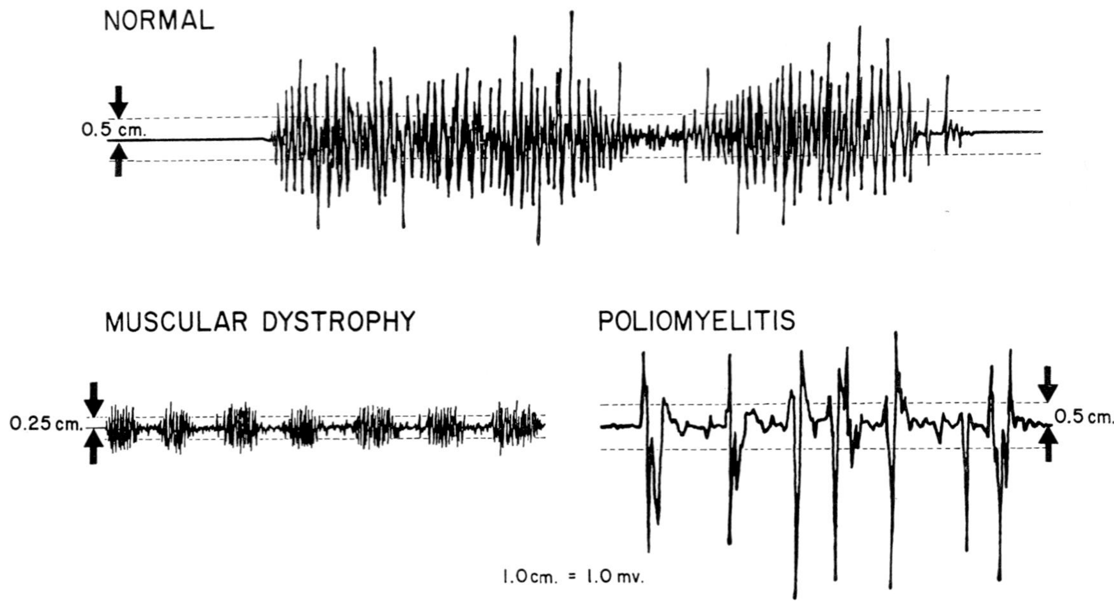

Figure 80. Sensitivity settings used for electronic counting of action potentials in normal and pathological muscle.

number of action potentials from a small muscle such as the first dorsal interosseous as from the vastus lateralis where the motor units are much larger. This, of course, is in accordance with our observation that during strong contractions, most of the motor units of the first dorsal interosseous muscle may be in a state of synchrony, losing their identity as individual pulses on the electronic recording device.

If, however, one selects a very small probe such as the unsanded 0.004-in. stainless-steel nylon-coated electrode and employs the single-ended mode, differences in the Q/C relationship in various muscles begin to appear. However, one must then accept certain disadvantages of the single-ended method. The main disadvantage is that only certain muscles can be studied in this manner, since a nearby ground placed preferably over the muscle concerned must be used. In certain muscles such

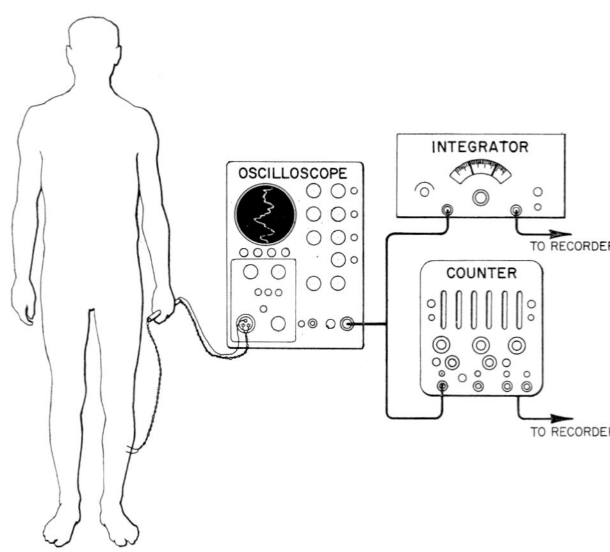

Figure 81. Apparatus in place for integrating and counting (determination of Q/C).

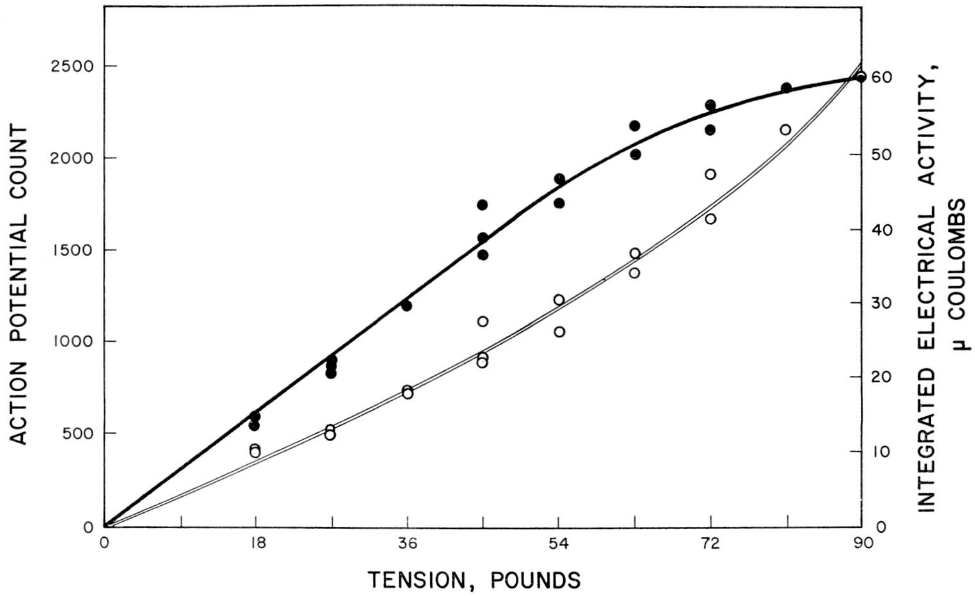

Figure 82. Relationship of total integrated electrical activity to tension, and total action potential count to tension. Each point represents a contraction of 5 sec.

as the posterior tibial muscle, this is, of course, impossible. Intervening muscles such as the triceps surae, long flexor or anterior tibial would be in a position between the surface ground plate and the recording electrode in the posterior tibial muscle. Where muscles are tested by the single-ended method, there is some activity seen on the oscilloscope as the grounded antagonist contracts. This is not the case where the differential method is employed (Fig. 6). Perhaps the best system for Q/C determination is the single coaxial electrode with a single-ended preamplifier at the source. This has been our method of choice in this procedure.

Using even the differential method (Figs. 5 and 79), great differences in the ratio Q/C were found for such lower motor neuron disturbances as diabetes, polio and Erb's palsy, compared to normal or muscular dystrophy (primary myopathy). In the lower motor neuron disease where these differences were found, one encounters abnormally large motor units due to axon branching. In primary myopathy, the motor units being fragmented by the pathological process, action potential count is high and the total amount of electrical activity is low. The expected low Q/C value is shown in Figure 79.

Q/C determinations were also made by differential and single-ended methods comparing isometric to isotonic contractions. Here the curves were close to linear, but in either case it can be demonstrated that as the dense interference pattern appears, both counting and integrating are somewhat unreliable. Figure 78 shows the linearity of the isotonic curve and the isometric curve which gently slopes off as greater intensity and length of contraction occur. These records were made by the differential method. It can, however, be said that in general, when Q is plotted against C, the relationship is close to a linear one. This is true for isometric contractions at different muscle lengths and for isotonic contractions where the load is increased by known increments. It is particularly the case where the contractions to be plotted are weak—here we are able to count more discrete action potentials and the interference pattern has not been reached. Under these conditions, there is opportunity to count and integrate series of single motor unit action potentials.

In an effort to better identify action potentials when the dense normal interference pattern occurs, making counting and integrating difficult, an attenuator which makes a restricted frequency band possible was developed. This instrument is described in Chapter II. Its frequency band of, for example, 1.2 to 10 kHz produces a far different pattern from that resulting from the conventional input of a few cycles per second to several thousand or several hundred thousand cycles per second. It was found that the upper frequency setting was not critical until it was lowered below 6,000 or 7,000 Hz (Figs. 8 and 21). The lower frequency limit, however, if set at around 40 Hz, allows the larger averaged wave forms from a relatively large volume surrounding the electrodes to be recorded. It was the purpose of the restricted frequency band to restrict the recording to a much smaller volume of muscle very close to that occupied by the electrodes. It was hoped that such a recording would reproduce the action potentials of only a few motor units and show definite difference in the Q/C ratio in pathological states and the various muscles of the normal human body. The differences between the records from these two inputs are shown in Figure 31. As expected, greater amplification is necessary when the restricted frequency band is employed, since a much smaller field contributes to the recording. The restricted high-frequency band, as intended, deemphasizes the low-frequency wave forms of the large volumes of muscle which surround the tip of the recording electrode.

Even though it was felt that the Q/C curve, particularly when recorded with differential input with conventional frequency range, is not reliable during strong contractions, a great deal of information could be obtained from the portions of the tracing made during low levels of activity. It was possible to grade the contraction while observing the resulting action potentials on the oscilloscope. The subject, for some of the tests, was instructed to contract the muscle weakly to such a degree that consecutive discrete action potentials appeared. Under these conditions, the confusion of the dense interference pattern was avoided. The action potentials were then more characteristic of the muscle examined in amplitude and duration.

An exhaustive study of electrode types was undertaken before attempting to calculate the Q/C ratio for various types of normal muscle and for pathological states. Small single coaxial electrodes, double coaxial electrodes, unsanded copper, sanded copper and unsanded stainless steel wire electrodes were studied (Figs. 4, 9, 13, 11, 12). Both the single-ended and differential methods were explored. One of the electrodes chosen to best demonstrate the relationship was the small, tough-coated, unsanded stainless-steel wire. Often the differential mode as well as the single-ended mode from the opposite side was recorded. This, together with the use of the restricted frequency band, appeared to be the best approach toward the final electrical identity of the individual motor unit wave form. Another applicator very reliable in making the Q/C determination was the single electrode with single-ended preamplifier near the muscle (Fig. 7).

When the differential input was used with normal or restricted frequency band, practically no noise was recorded. When the single-ended method was used, good recordings could be made if a source preamplifier was used. This method was practical if it was possible to locate a surface ground near the recording electrode. Even here, however, some electrical activity was evident when the antagonist contracted and this was considered an undesirable feature (Fig. 83).

Summarizing the differences between single-ended and differential inputs, one can say that the single-ended tiny probe may record potential changes in a very small area. The differential input using two electrodes is a comparison of the electrical activity at the region of one electrode to that at the other electrode. Since by the differential mode one may be assured that the recording is confined to a small volume of muscle, this is the most precise method for individual muscle function analyses.

Though one may use the tiniest wires as recording electrodes, both differential and single-

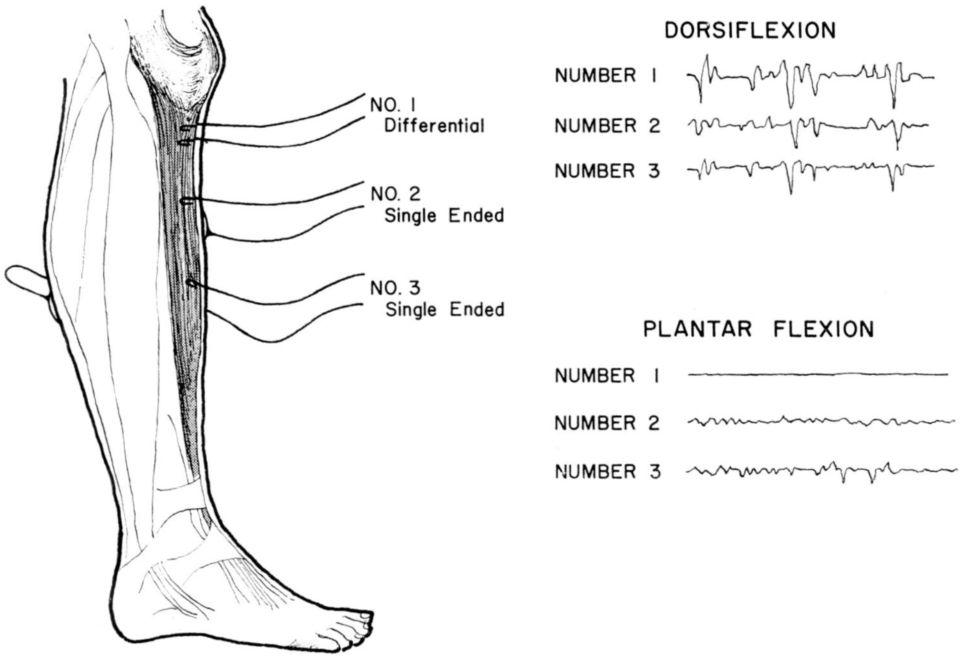

Figure 83. Postpolio muscle. Comparison of action potentials from m. extensor digitorum longus. *1*, Dual internal electrodes; *2*, single internal electrode with proximal surface ground; *3*, single internal electrode with remote surface ground. The use of the remote ground causes recording of excessive activity when the antagonist is active.

ended applications record from more than one motor unit. Just how many units are being recorded from is often impossible to say. Apparently this will always be impossible as long as we are unable to see where the electrode is placed. Since the fibres of the motor unit may not be grouped together as shown in Figure 76 but may be interdigitated with the fibres of one or more other motor units (Fig. 77), one can readily see how difficult it is to record from a single motor unit in muscle with even the tiniest probe. Even the smallest probe would tend to record from an electrical field. Because of the fact that the identity of the source could not be demonstrated clinically, the study described in Chapter VII was undertaken. Here it was hoped that after stimulating the single alpha motor axon in isolated preparations, one could record exclusively from the entire group of muscle fibres thus activated, finally demonstrating the actual single unit which produces the single motor unit action potential. Such a microstimulation study would enable us to demonstrate the amplitude, time duration and other characteristics of the individual motor unit wave form. In this way, requirements such as the optimum frequency recording band would become known. With this information, it is possible to arrive at frequency requirements as determined by Fourier transform or Fourier analysis (Chap. VII).

The Q/C linearity (Figs. 78 and 82) persisted during most types of contractions, as stated, even into large magnitudes. Perhaps this relationship was maintained as linear because of two failures:

1. Failure of the counter to add potentials already grouped and averaged together.
2. Failure of the integrator to integrate anything more than the dense interference pattern where many action potentials were combined.

The linearity at the upper portions of Figures 78 and 82 may be so because of failure of both counting and integration at upper levels.

The relationship was subjected to a further test, however. By increasing the gain approximately four times, weak contractions could be recorded, counted and integrated. At this low level, a preponderance of single motor unit action potentials was seen on the oscilloscope screen. There was linearity of the Q/C relationship. The slope of the curve was unchanged.

In making Q/C determinations, several factors must be standardized:

1. Oscilloscope sensitivity.
2. Setting of integrator.
3. Orientation of electrodes.
4. Size and length of exposed tips of electrodes.
5. Choice between single-ended and differential mode.
6. Distance between electrodes if differential mode employed.
7. Speed of recorder.
8. Proper input impedances of various units.

In recording from muscle, frequency of motor unit action potentials must not be confused with rate. The muscle fibre or motor unit has a rate of firing—the intervals between contractions are not equal. Frequency applies to very regularly occurring phenomena. The time duration of each pulse indicates the order of its frequency, since it is inversely proportional to the number of pulses that may theoretically take place in unit time. The rate for a single motor unit never becomes high because of the refractory period. Its frequency, however, may be relatively high. It is the frequency of the motor unit that determines the characteristics necessary to record its action potentials (see "Fourier Transform of Action Potential Spike," Chap. VII).

It appears that the rate of motor units can vary and that this is probably the more important way in which muscle increases its strength of contraction. This is considered to be so because of extensive synchronization often found throughout muscle (Figs. 1 and 2), even on much less than maximal contractions. It is more likely that rate can vary but that frequency of the individual motor unit remains the same. If the motor unit fires, the "all-or-none" law can be interpreted to mean that all of the electrical oscillations produced by the motor unit are alike but rate may vary.

If we could arrange the recording equipment to be "tuned in" to the frequency of single motor units, the recording could be much better defined. Excessive restriction of the lower limit of the frequency recording band, however, would tend to cause the appearance perhaps in some degree of the action potentials from single fibres or from "subunits" of the motor unit. Action potentials of the complete motor unit would be correspondingly suppressed. A permissible frequency bandpass for emphasis of the single motor unit action potential has been calculated (see Chap. VII).

As a muscle increases its strength of contraction, rates of motor units do increase, but it has not been shown that the actual frequency characteristics of the individual motor unit change. Under experimental conditions (see Fig. 160), microstimulation may sometimes produce a less-than-maximal motor unit action potential as a weak stimulus is repeatedly applied to the single axon.

Records made using a restricted frequency band are compared with records made from the same electrodes via the conventional input in Figure 31. The conventional input produces larger action potentials of greater time duration than those found in the tracing made using the restricted frequency band. It can be concluded that since the same electrodes are used to simultaneously record the action potentials of the two lower records, the large peaks shown (middle trace) arise from the conventional wide-band input. More regular spikes of faster rise time are reproduced by the attenuated circuit. The attenuated circuit, arbitrarily selected at 1.2 to 10 kHz, records from a smaller, more specific volume of muscle.

It may be that records made using the restricted frequency band have more meaning. The action potentials thus produced tend to be of more uniform amplitude and duration. They should be more suitable for counting and provide more information regarding the characteristics of motor unit action potentials of various normal and pathological muscles.

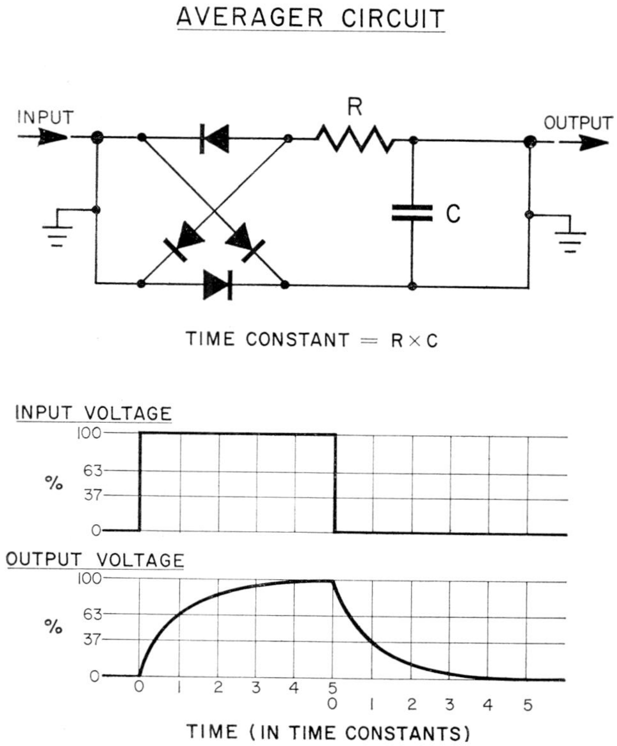

Figure 84. An average circuit with single-ended output showing distortion and time delay of input signal.

Circuit diagrams (Figs. 84 and 85) are given for the conventionally used "integrator." This instrument is, in reality, an averager and produces a neat-appearing average curve rectifying and summarizing the electromyogram (Fig. 86). On observing Figure 84, one realizes that the input square wave or other input wave form is considerably distorted by such an averaging circuit. Time delay of onset and termination of activity are evident. True integration and the resulting analog curve are demonstrated in Figure 86. This is compared to the integrated "average" curve of a conventional electromyogram. Since total integration is an absolute value, it cannot return to zero at the end of the contraction. At that time, this value is at its height. In order actually to average the

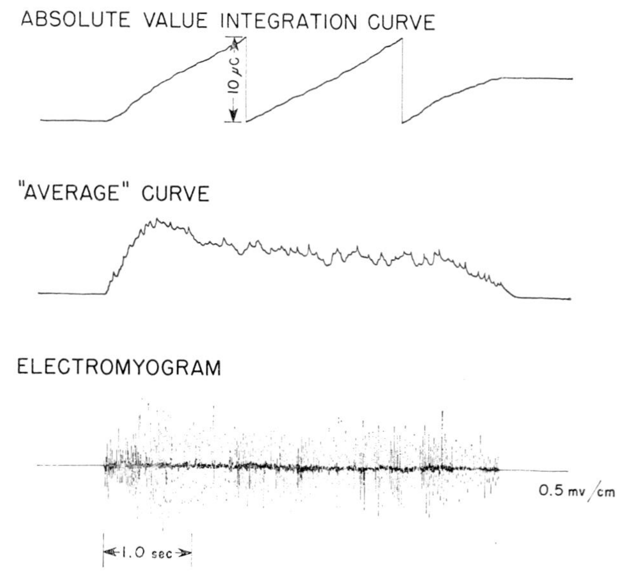

Figure 86. Comparison of analog curves produced by the Absolute-value integrator and the averager. The curves are the result of the electromyogram shown.

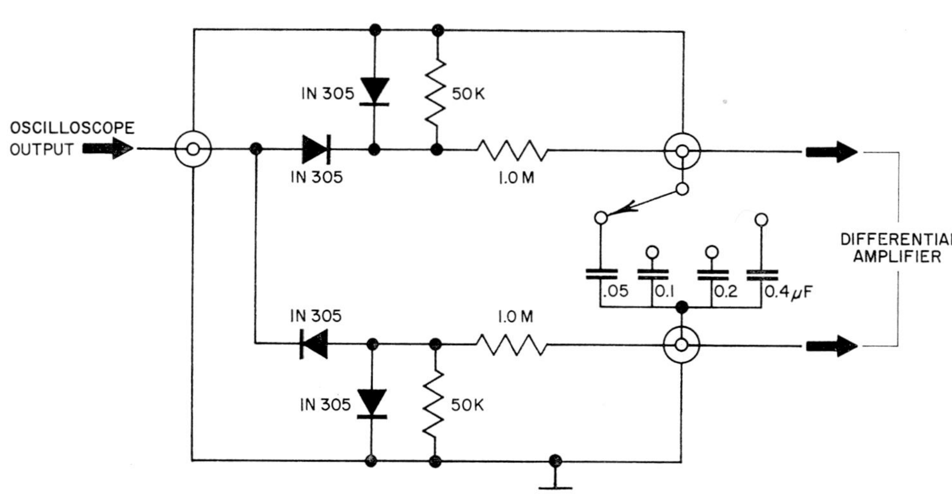

Figure 85. Circuit diagram of averager with differential output.

curve from the absolute-value integrator, one would have to differentiate the analog curve of the real integrator. However, one could integrate by taking the area under the curve produced by the averager representing the electromyogram. This was done and demonstrated in Figure 64 where a planimeter was used to measure the areas for several contractions. The values for integration thus obtained are plotted against action potential count for each contraction.

Q/C was significantly different in muscles with motor units of different size, particularly if the interference pattern was avoided. Here, strong contractions produced difficulty in counting, due to synchronization, where one records large spikes made up of the action potentials of several motor units. The average size of the motor unit, Q/C, was best expressed when weak contractions at high gain were measured.

When the value Q/C itself was plotted against load, for isotonic and isometric contractions, the relationship shown in Figure 87 was found. In Figure 88, the relationship of Q/C to isometric contractions of increasing intensity is shown.

In employing the various electronic units, as was done in several of these studies, it was necessary to observe the characteristics of the inputs and outputs of the instruments. Since the networks employed produce satisfactory data, their characteristics are given.

The Tektronix No. 502 dual-channel oscilloscope had the following characteristics:

1. Inputs (2), 47$\mu\mu$f, paralleled by 1 megohm.
2. Outputs (2), 2μf in series with 1 megohm.
3. The impedance of the electronic counter input was approximately 1 megohm shunted by 30$\mu\mu$f.
4. The absolute value integrator* had a resistive input of 450,000 ohms paralleled by 1 megohm, resulting in approximately 315,000 ohms.
5. Before each test of any kind, the resistance between the two electrodes was also measured. A new insertion was made if this value was not near 35,000 or 40,000 ohms. The resistance was again checked and recorded following the test. The data were not considered acceptable if significant change in the resistance was noted following the test.
6. If the single-ended mode was employed,

* Absolute-value integrator Model C1–112, Eldorado Electronics, Berkeley, Calif.

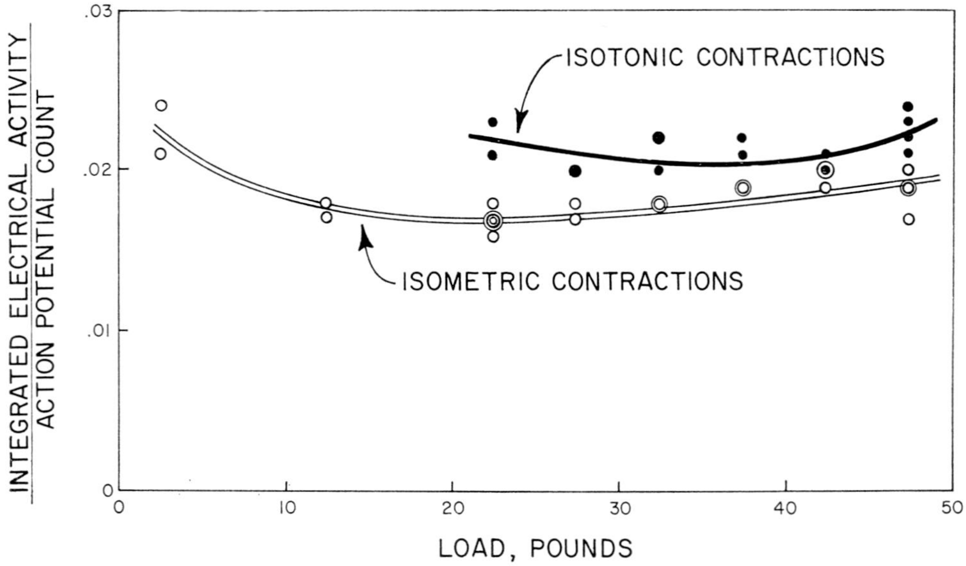

Figure 87. Variation of Q/C as load is increased; controlled isotonic and isometric contractions.

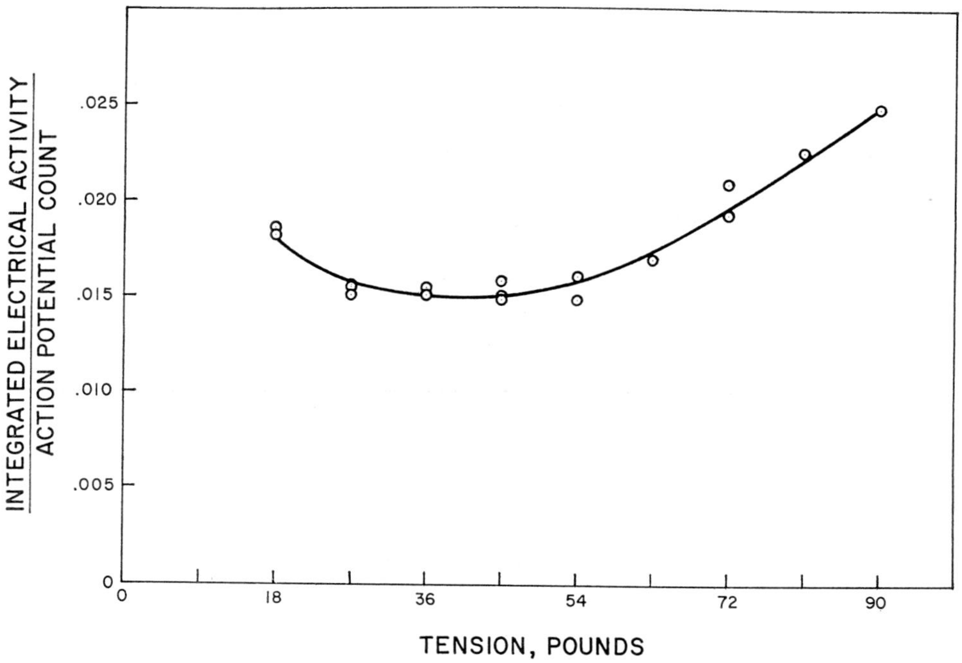

Figure 88. Variation of Q/C as isometric tension is increased.

resistance between the electrode and ground was checked before making the recording.

Q/C curves for various muscles may be readily displayed on the X-Y oscilloscope. In Figure 79A, three postpolio muscles with varying degree of involvement are compared to the Q/C curve of the normal. The greater the slope, the greater the degree of involvement, since motor unit size increases (with axon branching) and count decreases. The thickness of the line of the curve of the postpolio anterior tibial muscle indicates that a much greater time was necessary to complete the contraction. The Q/C slope is thus very sensitive to motor unit size.

Chapter V

NORMAL PHASIC ACTIVITY WITH ANALYSES OF THE BEHAVIOR OF MUSCLES FOLLOWING TENDON TRANSFER

The sound camera with synchronized sound track provides an excellent method for evaluating surgical procedures by comparison of data obtained before and after tendon transfer. A simultaneous synchronized electromyogram can be made on other media such as the 35-mm paper or film of the Fairchild camera or the now-available high-frequency galvanometer multichannel direct-reading Visicorder. One has the great advantage of visual as well as auditory instantaneous and simultaneous analysis of the patient, particularly with regard to the muscle in question. The electromyogram as represented on the high-fidelity recording does not lend itself to "bar diagram" simplification or anything less than the timed, synchronized oscillograms which demonstrate (a) time of action, (b) amplitude, (c) rate of action potentials and (d) fine variations in form of the action potential.

Should studies of more than one muscle become necessary in the paralytic or normal subject, one may employ the Visicorder* Model 906B with its 12 or 14 channels. Thus the action potentials of many muscles can be simultaneously and synchronously recorded on the rapid developing paper of the Visicorder. Each channel can be independently switched in turn to the sound track of the motion picture camera. This is particularly desirable for normal muscle function analyses and will be of great value in analysis of the function of muscles in the upper extremity where muscles such as the extensors of the wrist may contract in conjunction with the extensors of the fingers and where the temporal participation of such muscles as the interossei and lumbricals have yet to be determined. The Visicorder has the great advantage that the data can be seen at once. This, of course, is a feature found only in this type of high-frequency equipment. High frequencies have been found to be definitely necessary in preparation of these records, and frequencies up to 8,000 (flat to 5,000) cps are available when using the miniature galvanometer of the Visicorder (Fig. 16). It is thus possible in the multichannel recordings to show a moving picture of the activity performed. The sound track records one muscle at a time, while on the Visicorder paper the oscillograms of a number of muscles are synchronously displayed. Each oscillogram can be switched in sequence individually to the sound track of the sound motion picture film, a necessarily single-channel medium. The visual portion of the film records the body motions in which the muscles involved have participated.

The rhythmic, intermittent contractions of the muscles of the lower extremity during walking may be conveniently associated with the events of the walking cycle. If continuous oscillograms are recorded from individual muscles and plotted according to their occurrence during the stance and swing phase of the walking cycle (Fig. 89), a display of the phasic activity of the various muscles results.[40,70,155] Patterns of electrical activity exhibited by the muscles of the lower extremity are found to be

* Minneapolis-Honeywell.

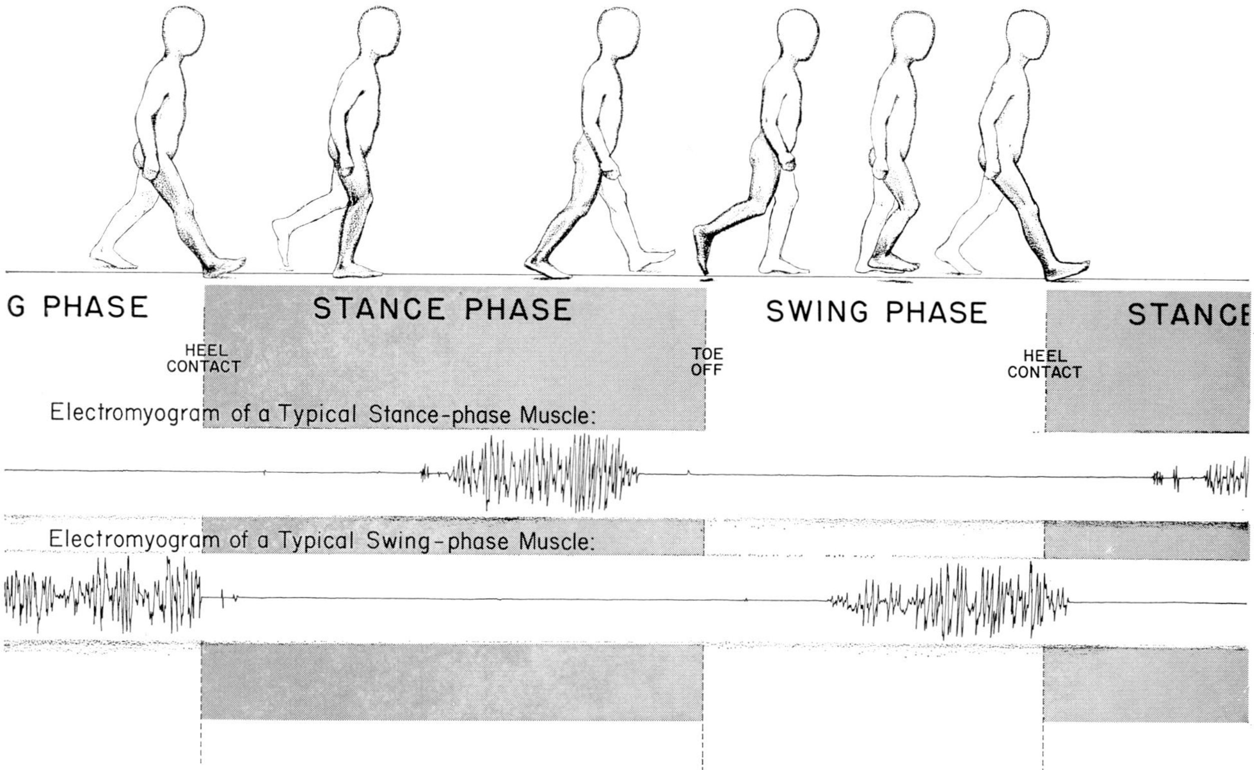

Figure 89. Conventional method for demonstrating phasic activity of the muscles of the lower extremity. (From *Journal of Bone and Joint Surgery,* March, 1959.)

quite consistent. As the subject walks, the intermittent contraction of each muscle is readily identified with its phase.[140] Thus in the normal subject, the anterior muscles of the leg are predominantly swing-phase muscles. The posterior or flexor muscles are stance-phase muscles (Fig. 90). In the thigh, the quadriceps is characteristically an early stance-phase muscle, while the hamstring muscles are basically of the swing-phase type. Examples of the oscillograms taken from the muscles of the normal thigh are shown in Figure 91. These records are based upon recordings made on many normal subjects. Variation from the normal patterns presented is very small. They agree in general with the data based on studies made at The Prosthetic Devices Research Project, University of California, Berkeley and reported as *Series 11, Issue 25, September, 1953,* "The Pattern of Muscular Activity in the Lower Extremity During Walking." The electrical activity of the muscles of the lower extremity of six normal subjects was recorded, using dual internal electrodes.[140]

Since the advent of effective vaccines for poliomyelitis, reconstructive surgery including tendon transfers is far less frequently necessary. There has been a great decline in the number of these procedures. The information gained from the study of various transfers is, accordingly, less frequently referred to. However, the basic knowledge gained is considered worthy of preservation—it appears unlikely that a demand for a study such as the following will again be sufficient to result in extensive investigative work. Knowledge of the neurophysiology and functional anatomy involving tendon transfer is still applicable in some cases following trauma and in cases involving poliomyelitis residuals in other countries. At this institution, due to the proximity of Mexican border towns, there is a considerable population of postpoliomyelitis patients with classical residuals.

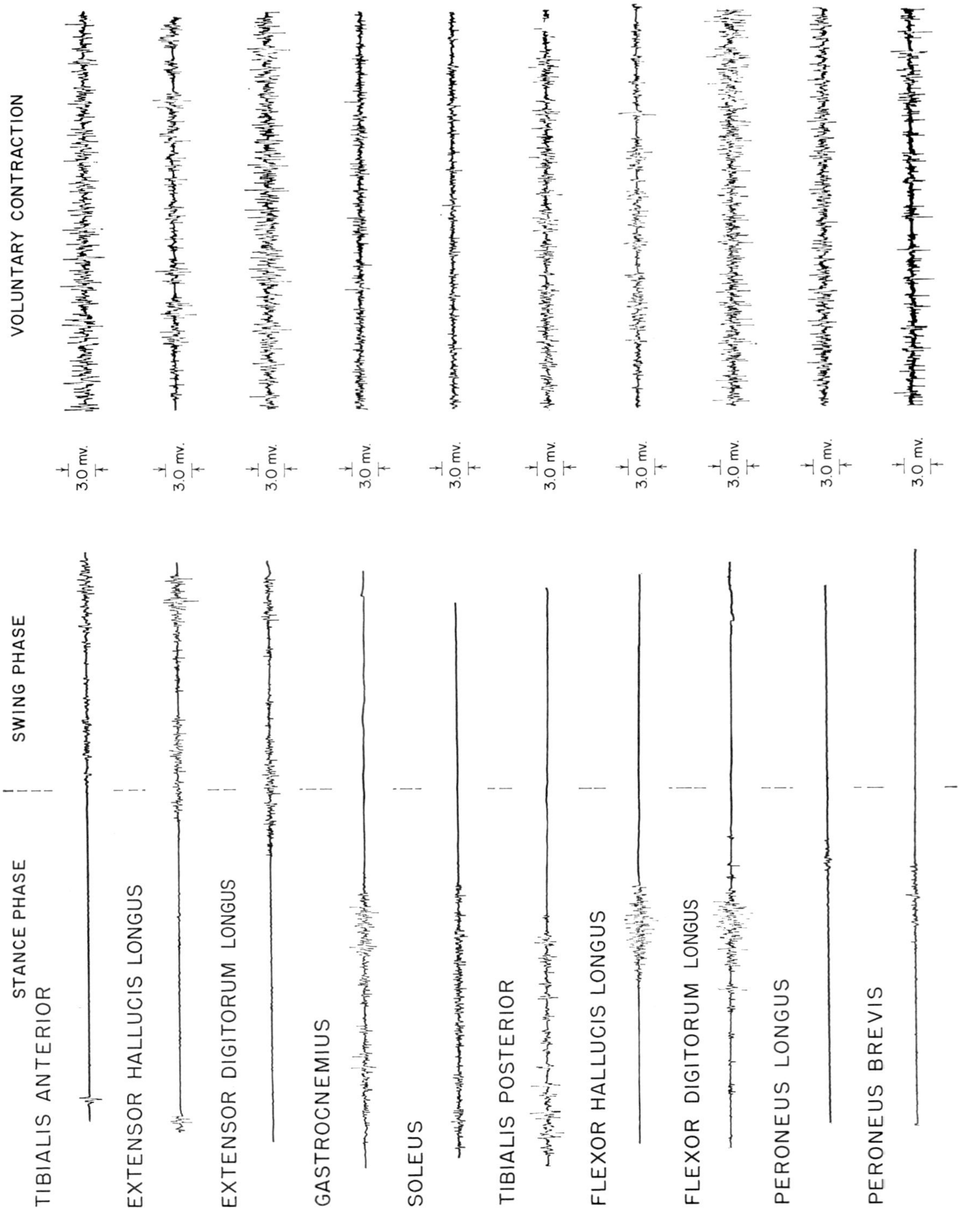

Figure 90. The normal muscles of the leg. Single typical walking cycles are shown for each normal muscle studied in the leg followed by the electromyogram of voluntary activity. Note that the oscilloscope sensitivity is the same for each study. A single subject was used for the entire series. (From *Journal of Bone and Joint Surgery*, March, 1959.)

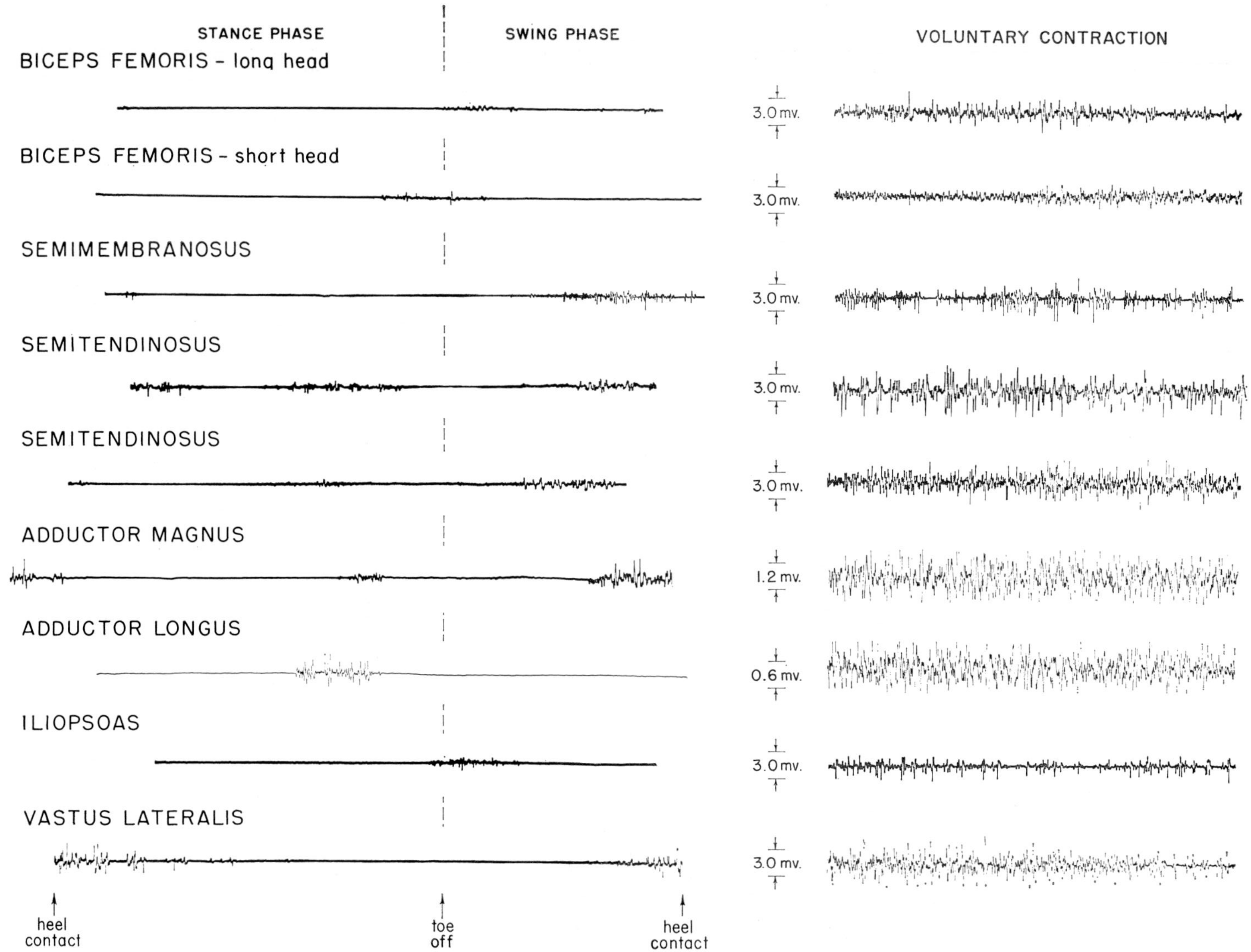

Figure 91. The normal muscles of the thigh. Typical action potentials for individual muscles of the thigh during normal walking and voluntary activity. Electrodes are unchanged during each muscle evaluation. Some variation was observed in the case of m. semitendinosus. The phasic activity of m. semitendinosus of two individuals is shown. The relatively minor role of the thigh mucles during walking can be appreciated by comparing the action potentials of walking to that produced by voluntary activity. The records shown were checked with studies of several other normal individuals. Oscilloscope sensitivity was increased for the tests on the adductor muscles. (From *Journal of Bone and Joint Surgery,* March, 1959).

Phasic activity was studied for many years by Scherb,[155] who used the method of manual palpation with the subject walking on a treadmill. Averages from hundreds of patients were compiled. He later checked the manual method with electronic methods. Apparently only a few normal superficial muscles—the gluteus medius, gluteus maximus, sartorius, biceps femoris, semitendinosus and semimembranosus—were checked by this method. There is no evidence that electrical stimulation was used to establish the position of the electrodes in the muscle belly tested. Studies were also made of some muscles after tendon transfer. These studies, however, were done by the palpation method. This technique is considered inadequate for evaluation of the action of the transferred muscle.

These basic associations of muscular contraction with the stance and swing phase of walking are also found, to a large extent, in individuals with lower extremity involvement following poliomyelitis. The phasic activity of the paralytic, however, deviates in pattern from the normal to a certain degree, depending upon the amount of paralysis. When tendon transfer in the lower extremity is contemplated, the question often arises whether one can violate the patterns of phasic behavior and expect a functional transfer. For example, if the tibialis anterior, a swing-phase muscle, is transferred posteriorly to the calcaneus, will it be able to change its phase of action and serve as a stance-phase muscle, as does the soleus muscle which it is intended to replace or reinforce? Furthermore, if transfers are done within the boundaries of phasic groupings, will these muscles retain their desirable preoperative phasic activity, and if so, to what extent? For example, if the extensor digitorum longus is transferred to the dorsum of the foot to replace a lost tibialis anterior muscle, will its preoperative swing-phase behavior persist?

If flexor muscles, such as the peroneus longus, peroneus brevis or tibialis posterior, are transferred to the calcaneus, their effective stance-phase activity continues. When these muscles are transferred forward, however, with the

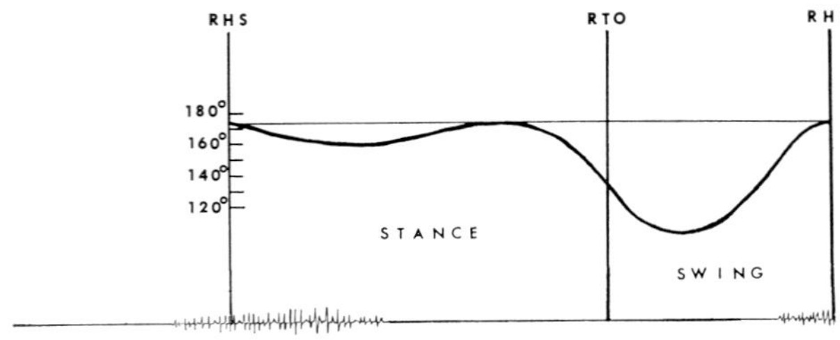

Figure 91A. Diagram of the knee angle in the sagittal plane during walking and its relation to quadriceps activity. *Upper line,* knee angle; *lower line,* electrical activity of quadriceps.

hope that they will act as dorsiflexors during the swing phase of walking, their phasic activity must undergo a conversion from stance to swing in order to be effective, functioning muscles during locomotion. Such a conversion of phasic activity would apparently require suppression of normal reflexes of reciprocal innervation and the agonist-antagonist reflexes,[158] since walking, to a great extent, is an automatic activity.

Certain transfers may be of considerable benefit to the patient with paralytic disease of the lower extremity, although no phasic conversion takes place. This is true because of the extraordinary requirement that the paralytic may place upon the muscle, whether transferred or not, if he can make any use of it at all. If the swing-phase biceps femoris is transferred to the patella, though its swing-phase action may continue, it may nevertheless be of value in producing stability at the knee by extending it just prior to the onset of the stance phase of walking. It thus prevents collapse at the knee in a different manner from that used by the normal quadriceps during walking, which provides stability at the knee by virtue of its strength just prior to and during the early stance phase (Fig. 91A). Its benefit during locomotion is then not through duplication of the function of the normal quadriceps muscle which produces knee stability in a different manner—by serving to prevent further flexion or collapse of a slightly flexed knee during the initial one-half of the stance phase of walking. An example is present, however (Figs. 93, 94, 95), where the transferred biceps femoris be-

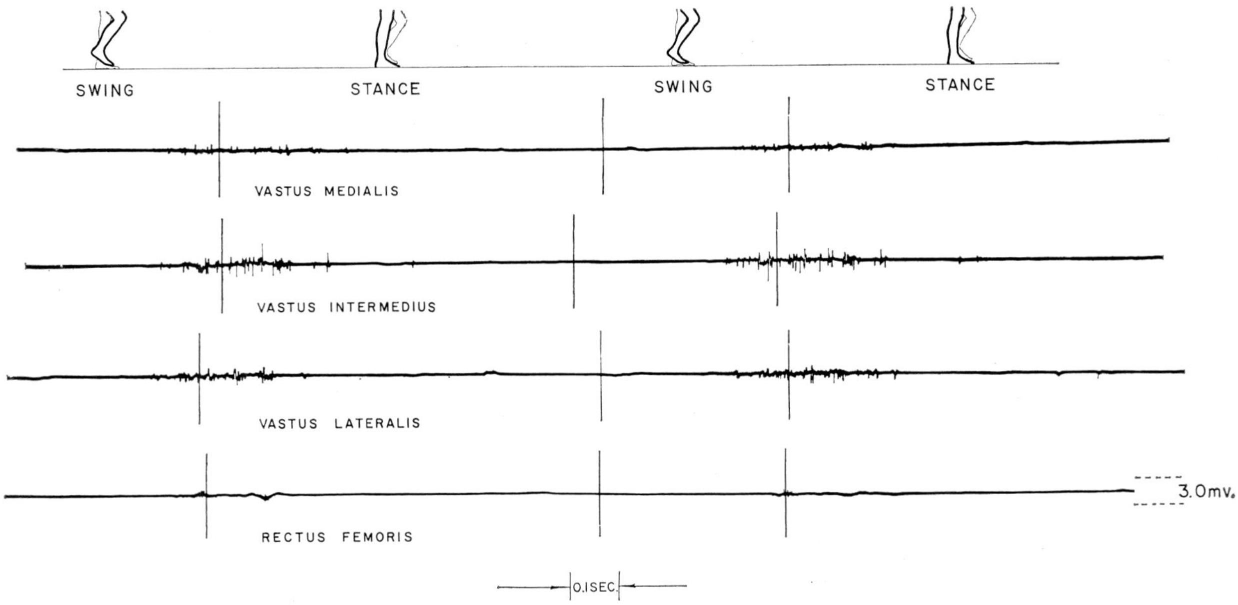

Figure 92. The phasic activity of the components of the normal quadriceps group. (From Instructional Course Lectures, American Academy of Orthopaedic Surgeons, 1959.)

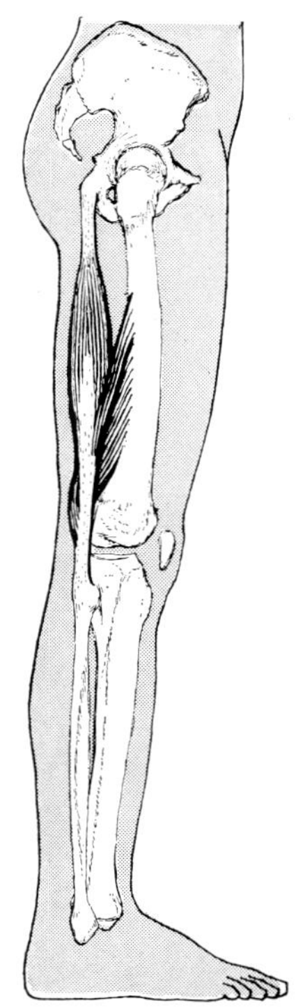

Figure 93. The two heads of the normal biceps femoris muscle.

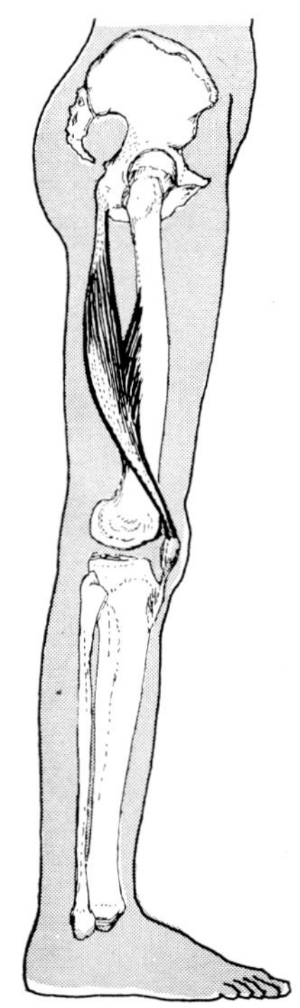

Figure 94. The biceps femoris transferred to the patella, lateral route.

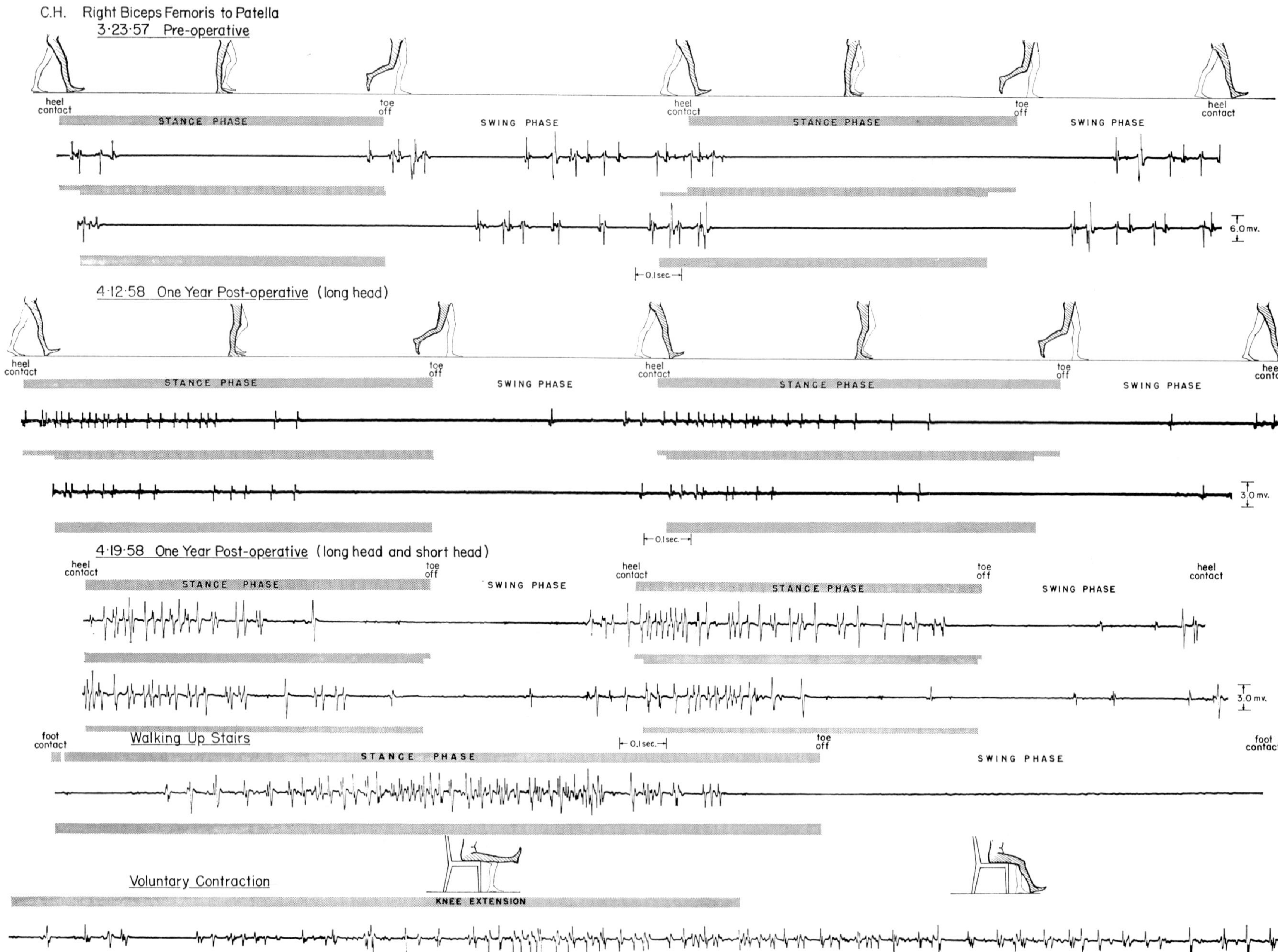

Figure 95. The patient had zero quadriceps activity preoperatively and was unable to extend the knee. Postoperatively, at four months, she could extend the knee completely. Preoperatively, the biceps femoris muscle demonstrated swing-phase behavior. Postoperatively, there is spontaneous, involuntary stance-phase behavior in the muscle. The muscle is used extensively on ascending stairs, as indicated by the amount of electrical activity. The patient preoperatively walked with a slight amount of recurvatum. Postoperatively, she walks with a slightly flexed knee. The phasic behavior and function of the transfer thus resemble that of the normal quadriceps muscle. (From *Journal of Bone and Joint Surgery,* March, 1959.)

came a stance-phase muscle under involuntary conditions. The usual value in the hamstring transfer to the patella is found in its ability to prevent flexion and collapse at the knee in the absence of the quadriceps. Such action may be voluntary but nevertheless quite useful to the patient.

It is possible that the transferred nonphasic muscle will be detrimental to the patient. For instance, if flexor muscles are transferred to the dorsum of the foot, their stance-phase activity, if retained, will oppose the action of the triceps surae. The effective tension of the calf muscles is then lessened by the tension developed by transfers.

The ability of many transfers to voluntarily perform the function for which they are intended is well known. For instance, the peroneus longus muscle transferred to the dorsum of the foot (Fig. 96) may act as an excellent voluntary dorsiflexor. The same muscle may fail completely to act during the swing phase of walking when its action is required to dorsiflex the foot and prevent foot-drop (Figs. 97, 98, 99). It can be shown in many instances, even during walking, that the patient can make a voluntary effort to use such a transfer, with the result that the muscle makes a temporary phasic conversion. This change in the muscle's phasic activity is maintained only as long as the patient concentrates on its use. Such voluntary phasic conversion is demonstrated in Figures 97 and 98. If the peroneus longus or tibialis posterior transferred to the dorsum of the foot succeeds in preventing or improving foot-drop without a demonstrable change in phasic activity, its value during the swing phase of walking is similar to that of a checkrein. It is possible to attach the transferred tendon so tightly that the elastic and fibrous portions of the structure passively prohibit plantar flexion beyond a certain limit merely by virtue of their shortness. This passive function has not been observed to be permanent. Heel cord lengthening often assists such transfers temporarily.

Occasionally, the peroneus longus, after transfer into the dorsum of the foot, may participate to some extent in the desired swing phase. Examples of such partial phasic conversion are shown in Figures 100 and 101.

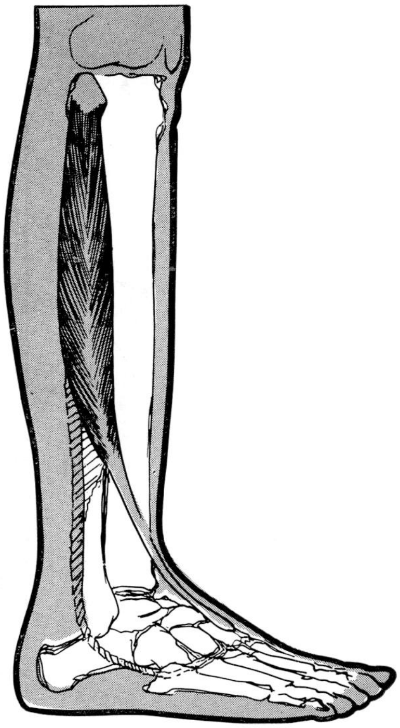

Figure 96. The peroneus longus transfer to the dorsum of the foot.

Since conservation of all remaining muscle power is a prime consideration in the paralytic lower extremity, the knowledge of what may be expected of the transferred muscle in its new site, particularly regarding its new time of contraction, is essential. The muscle does not meet all of the requirements of a good transfer, although it lessens deformity and produces a good voluntary contraction, if it does not make its expected contribution to locomotion.

Thus demonstrations of the voluntary action of a transferred muscle with the patient sitting may have nothing to do with the effectiveness of the transfer during walking. For instance, if the posterior tibial tendon has been transferred into the dorsum of the foot, photographs may be made of the patient in a sitting position using the transfer to produce some dorsiflexion of the foot. This voluntary action is of little value if the patient does not use the transferred posterior tibial muscle as a dorsiflexor during the swing phase of walking (Figs. 102, 103, 104). Since this is a deep muscle inaccessible

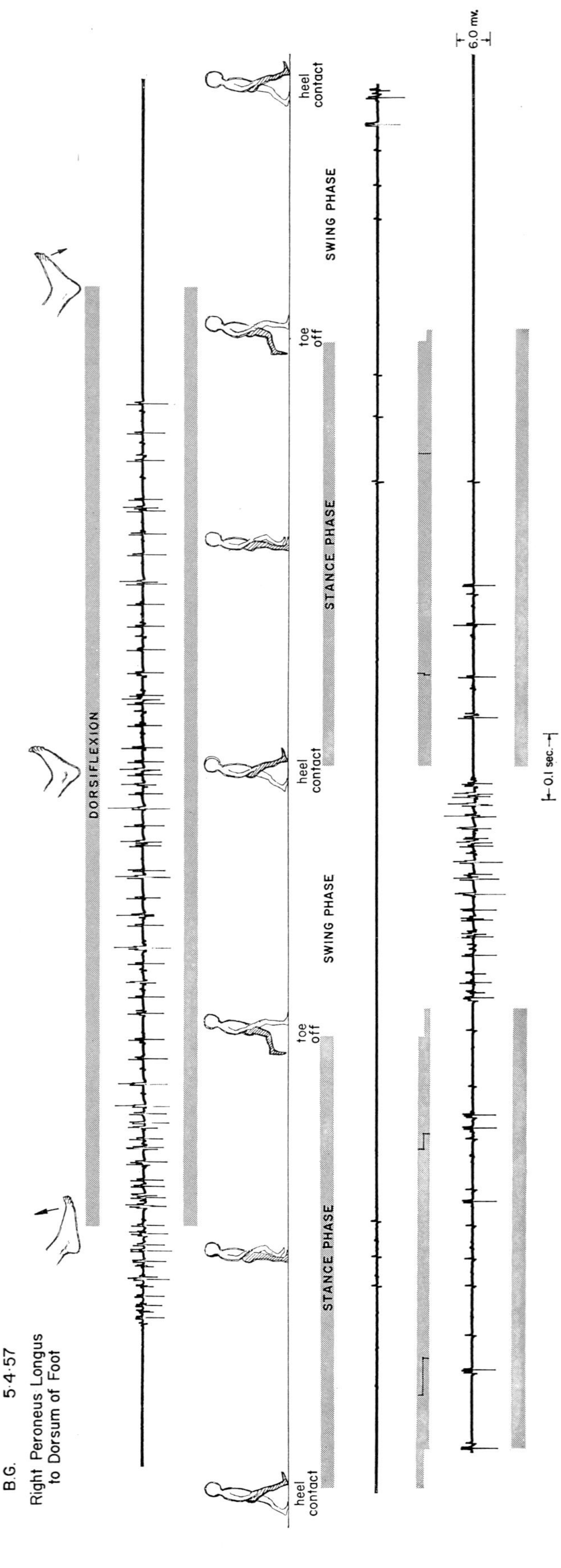
B.G. 5-4-57
Right Peroneus Longus to Dorsum of Foot
DORSIFLEXION
heel contact
STANCE PHASE
toe off
SWING PHASE
heel contact
STANCE PHASE
toe off
SWING PHASE
heel contact
6.0 mv.
0.1 sec.

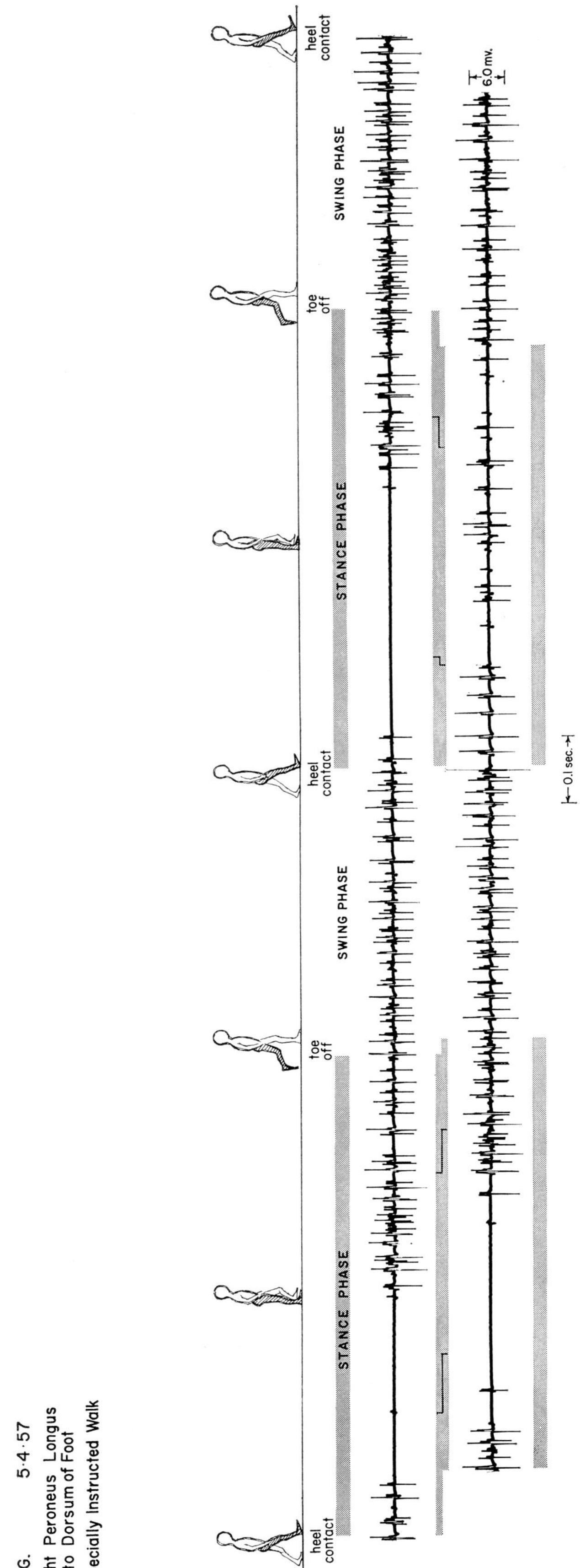

Figures 97 (*top*), 98 (*bottom*). Peroneus longus transfer to dorsum of foot. There is no definite pattern of activity and for considerable periods, the patient has no involuntary use of the muscle. Following specific instructions, however, there is a useful swing-phase action. (From *Journal of Bone and Joint Surgery*, March, 1959.)

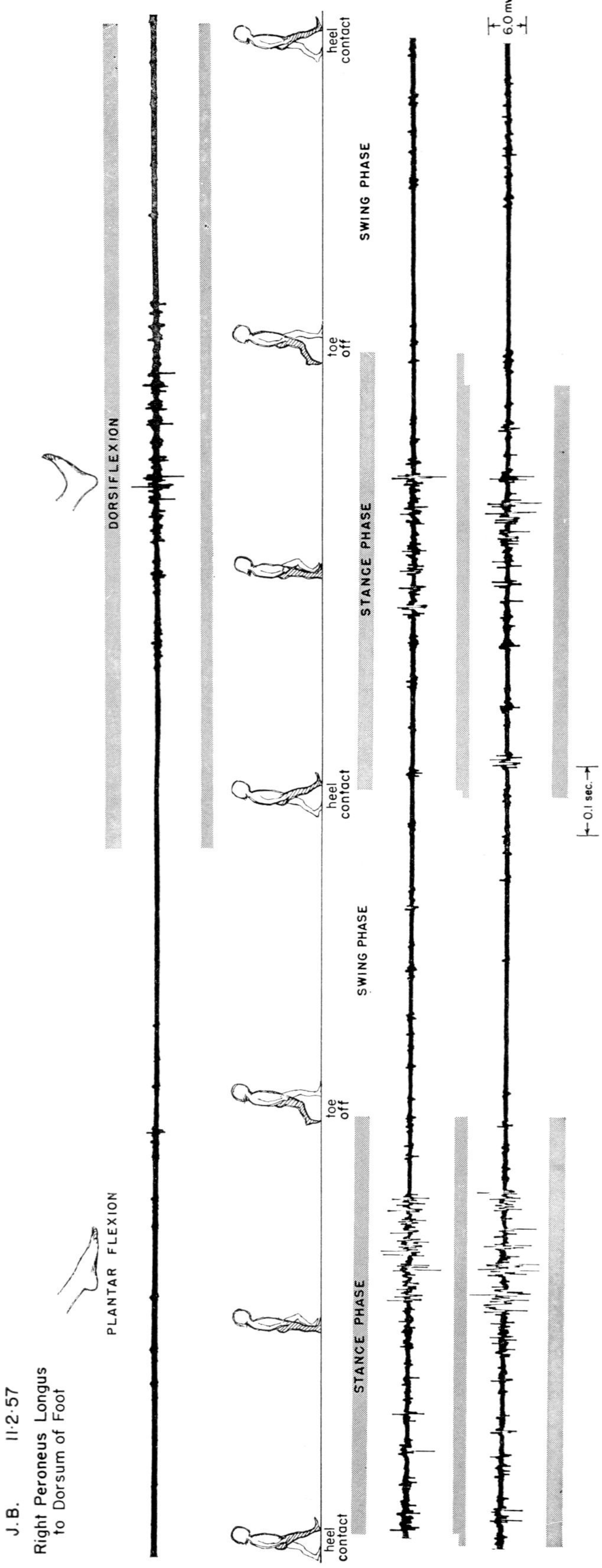

Figure 99. Peroneus longus transfer to dorsum of foot. The muscle's original stance-phase behavior has been retained. (From *Journal of Bone and Joint Surgery*, March, 1959.)

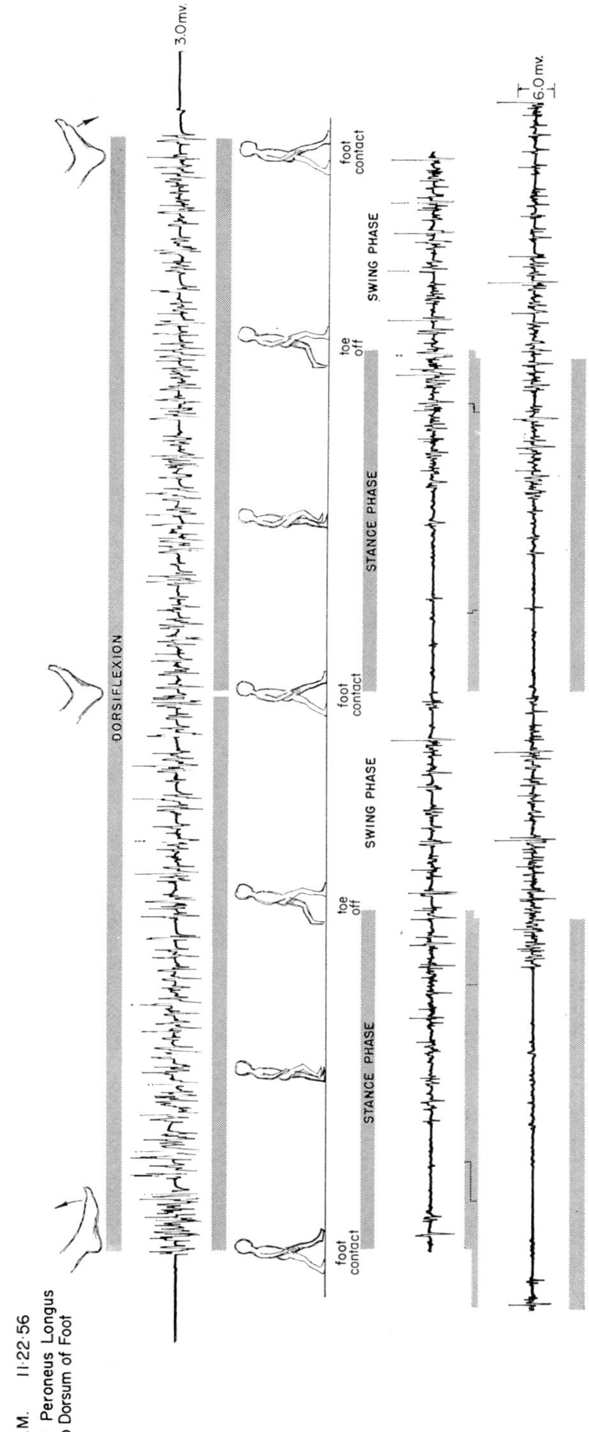

Figure 100. Peroneus longus transfer to the dorsum of foot. There is considerable conversion of the muscle's activity to the swing phase. (From *Journal of Bone and Joint Surgery*, March, 1959.)

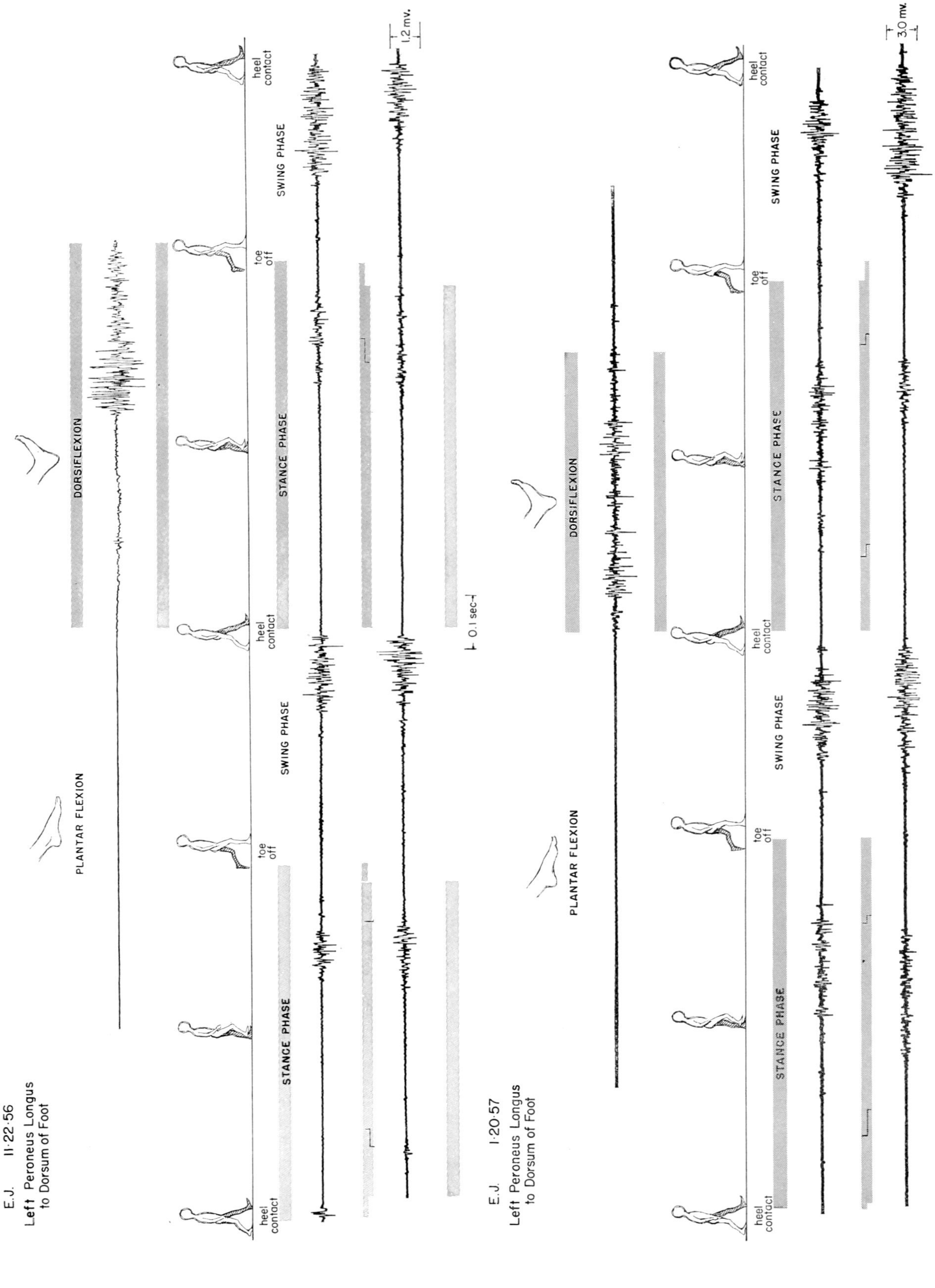

Figure 101. Peroneus longus to dorsum of foot. Two tests of the peroneus longus muscle which has been transferred to the dorsum of the foot. This transfer was done because of a traumatic severance of the anterior tibial muscle a few years previously. Almost identical patterns are obtained from both tests although it is noted that the oscilloscope sensitivity is not the same. The peroneus longus, now serving as a dorsiflexor, has a considerable amount of useful swing-phase activity, although some stance-phase activity persists. (From *Journal of Bone and Joint Surgery*, March, 1959.)

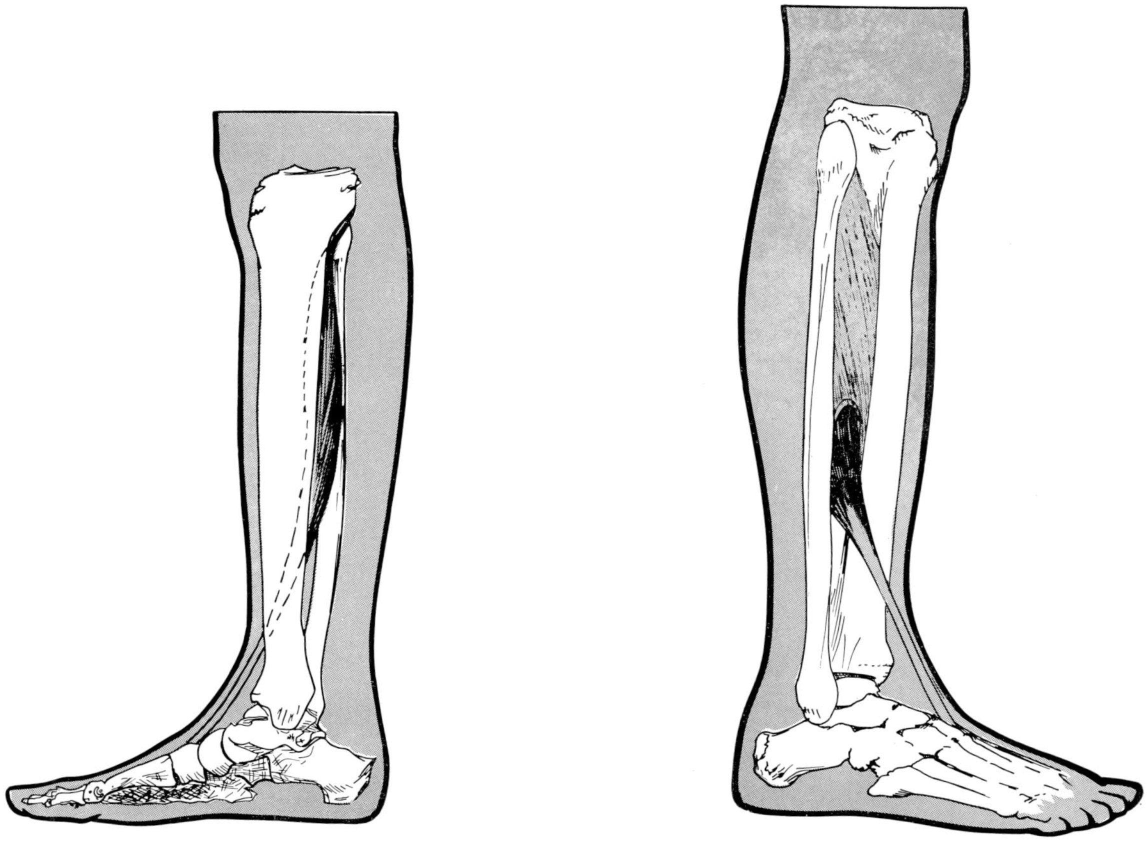

Figures 102 (*left*), 103 (*right*). The posterior tibial tendon transferred through the interosseous membrane to the dorsum of the foot.

to the surface electrodes, its action potentials can be demonstrated only by properly inserted internal wire electrodes.

The study described in this chapter consisted of individual electromyographic tests conducted on 170 muscles in which tendon transfer had been done. Approximately 150 patients were studied. All except two of the patients chosen for the study had muscle weakness resulting from poliomyelitis. Often it was possible to make a preoperative, as well as postoperative, record of the electrical activity of the muscle. A comprehensive series of postoperative tests were often made at regular intervals. Normally, only a preoperative and postoperative analysis was done. Often, the subject was not tested preoperatively, since many of the operations preceded this study. Preoperative electrical studies, when performed, assisted materially in the evaluation of the patient under consideration for surgery.

The patients studied are classified in Table III according to the type of tendon transfer performed. The patients chosen were those who walked well without appliances and with a continuous regular gait, rapidly enough to produce the rhythm of phasic activity. The muscles studied met the usual criteria of a good tendon transfer: (a) deformity was improved* and (b) voluntary control of the muscle was evident.

Most of the studies were made on normal and good muscles. It is often difficult to penetrate poor muscles with the electrodes and to stimulate them electrically in order to verify the position of the electrodes. The action potentials from weak muscles may require excessively high amplification. This produces

* Analyses were sometimes made on transfers considered clinical failures. Some of this group demonstrated scant electrical activity. These subjects were not included in this series.

TABLE III
CLASSIFICATION OF TRANSFERS STUDIED

Group	*Phasic*	*Nonphasic*
Extensor to dorsum	Extensor hallucis longus Extensor digitorum longus	
Flexor to dorsum		Peroneus longus Peroneus brevis Tibialis posterior Medial gastrocnemius
Extensor and flexor to dorsum	Extensor hallucis longus Extensor digitorum longus	Peroneus longus Peroneus brevis Tibialis posterior
Flexor to calcaneus	Peroneus longus Peroneus brevis Tibalis posterior Flexor hallucis longus Flexor digitorum longus	
Extensor (tibialis anterior) to calcaneus		Tibialis anterior
Flexor and extensor to calcaneus	Peroneus longus Peroneus brevis Tibialis posterior Flexor hallucis longus Flexor digitorum longus	Tibialis anterior
Extensor to dorsum and flexor to calcaneus	Extensor hallucis longus Extensor digitorum longus Peroneus longus Flexor hallucis longus	
Hamstrings to patella		Biceps femoris short head Biceps femoris long head Semitendinosus Semimembranosus
Hip flexor to greater trochanter		Iliopsoas to greater trochanter

technical difficulties, such as the increased amplification of electrical interference. Occasionally, excessively high amplification produced faint recordings of adjacent, more vigorous muscles. These difficulties were not encountered in good or normal muscles where lower amplification could be used. In these subjects, it was possible to stimulate the muscle adequately and definitely verify the position of the electrodes. Many of the records were checked by repeating the postoperative test (Fig. 101). Dual-wire internal electrodes were used (Fig. 6). The electrodes were connected by shielded dual cable to the differential preamplifier of a cathode-ray oscilloscope (Fig. 3). The oscilloscope tracing was photographed with a continuous recording oscilloscope camera. The vertical signal output of the cathode-ray oscilloscope was connected to the amplifier of the sound motion picture camera. Thus, simultaneous synchronized records were made at the same speed by both recording instruments.

In many of the charts, the phasic activity of the opposite extremity is shown. This makes it possible to compare the lengths of the stance and swing phases of each lower extremity. This is a good index of the amount of limp.

Most of the recordings were made at a frequency range of 60,000 Hz to 80 Hz. The upper frequency limit of 60,000 could readily be lowered to 10,000. This produced no apparent difference in the recordings. One recording from each muscle was usually also made, using a lower frequency limit of 8 Hz. This lower limit was not used in most of the graphs reproduced here because of resulting irregularity of the base line. However, no significant discrep-

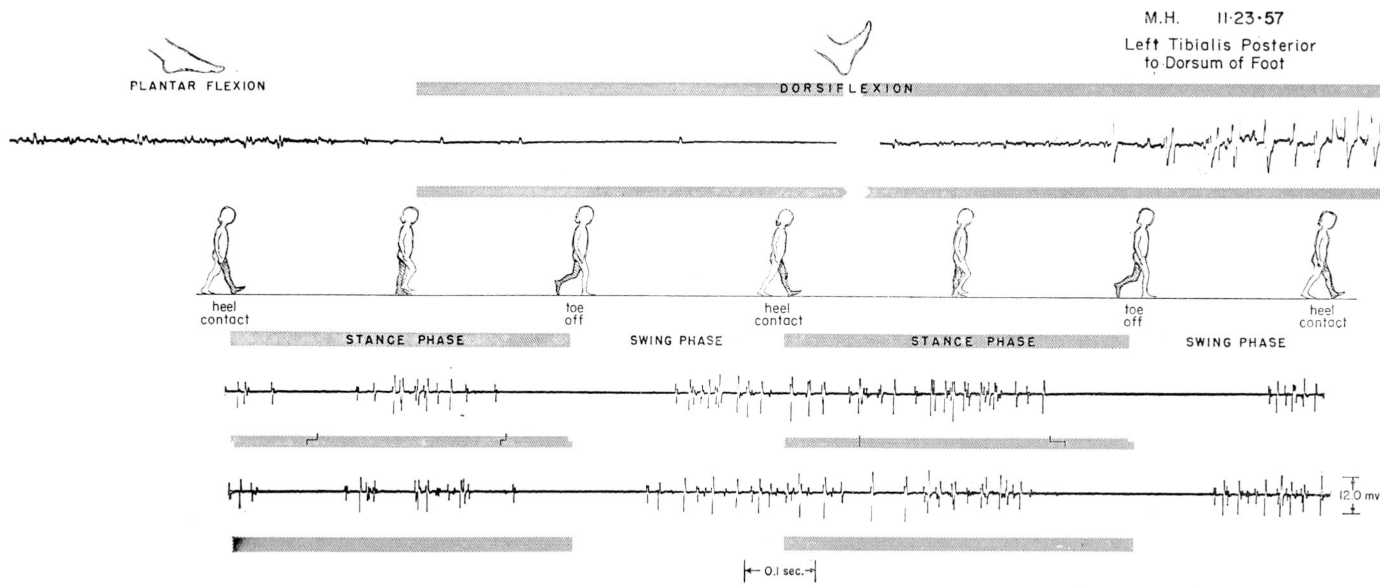

Figure 104. Posterior tibial to dorsum of foot. Following transfer of the tibialis posterior to the dorsum of the foot, the muscle for the most part retains its stance-phase activity. There is, however, some useful late swing-phase activity. (From *Journal of Bone and Joint Surgery,* March, 1959.)

ancy between the recordings made at 60,000 to 80 Hz and those made over a range of 60,000 to 8 Hz was noted (Fig. 105). Normal musculature required a calibrated oscilloscope sensitivity setting of 0.5 to 1.0 mv/cm. The post-poliomyelitis muscle usually required a setting of 1.0 to 2.0 mv/cm. The oscilloscope sensitivity settings used during the test are recorded on each figure shown. The same setting was used during each comprehensive patient study in order to make the various activities such as level walking, walking on stairs and voluntary isotonic contraction while in a sitting position, reflect comparative amounts of electrical activity produced. Since the reduction factor of the oscilloscope camera is six, the oscilloscope sensitivity setting is multiplied by this number in order to express the calibration of each record. A rejection ratio of as high as 50,000 to 1 could be achieved. The preamplifier input impedance was 10 megohms paralleled by 50$\mu\mu$f.

By setting the oscilloscope sensitivity so that peaks of electrical activity during walking, ascending stairs and voluntary contraction are just contained on the oscilloscope screen, one is able to place the amount of electrical activity of each action in its proper relationship. For example, Figure 91 reveals the relatively minor role of the hamstring muscles during walking compared with their contribution to flexion of the knee, and the relatively low level of quadriceps activity during walking compared with extension of the knee during activities other than level walking.

Accurate placement of the electrodes within the belly of the muscle to be tested was ascertained by the use of the electrical stimulation through the electrodes before and after each test. No data were accepted unless isolated contraction of the stimulated muscle could be observed. This procedure eliminates the utter confusion that results from simultaneous recording of adjacent muscles not otherwise concerned in the experiment.

Oscillograms for continuous walking cycles can be readily plotted from the data obtained. The occurrence of the events of walking are recorded by analysis of the individual frames of the 16 mm film and transferred to the oscillogram on the 35 mm paper of the oscillograph camera. The action potential during walking is thus precisely synchronized with the phases of the walking cycle. The same oscillation, displayed by the oscilloscope and recorded by the oscilloscope camera as the subject walks, is simultaneously recorded on the sound track of the sound camera. A great deal of material may be reviewed in a short time with the use of sound film, much more than is covered in the charts presented in this section where only a single voluntary contraction and four consecutive walking cycles are usually presented for each test.[34,39]

The procedure readily yields the following information:

1. The phasic activity of the muscle for repeated walking cycles.
2. The duration of each contraction.
3. The electrical intensity and rate of motor-unit activity.

After the injection of local anesthesia into the skin overlying the area to be examined, wire electrodes are inserted into the muscle (Fig. 6). The subject is then asked to walk in his usual manner. No instruction which might alter the subject's customary gait is given. After tracings of normal walking are made, the effect of specific instruction may be recorded. For instance, if the peroneus longus, transferred to the dorsum of the foot, appears to be silent, irregular or ineffective, data may be taken after the patient is asked to concentrate on the use of the muscle as a dorsiflexor as he walks. Tracings from a muscle of this type are shown in Figures 97 and 98.

Evidence of the reliability of the method is found in its ability to reproduce on repeated tests the same particular pattern of electrical activity for a given muscle. In two normal individuals, when the electrodes were placed in the upper one third of the adductor magnus, the muscle produced a characteristic burst of activity during the swing phase followed by a lesser burst of activity during the stance phase

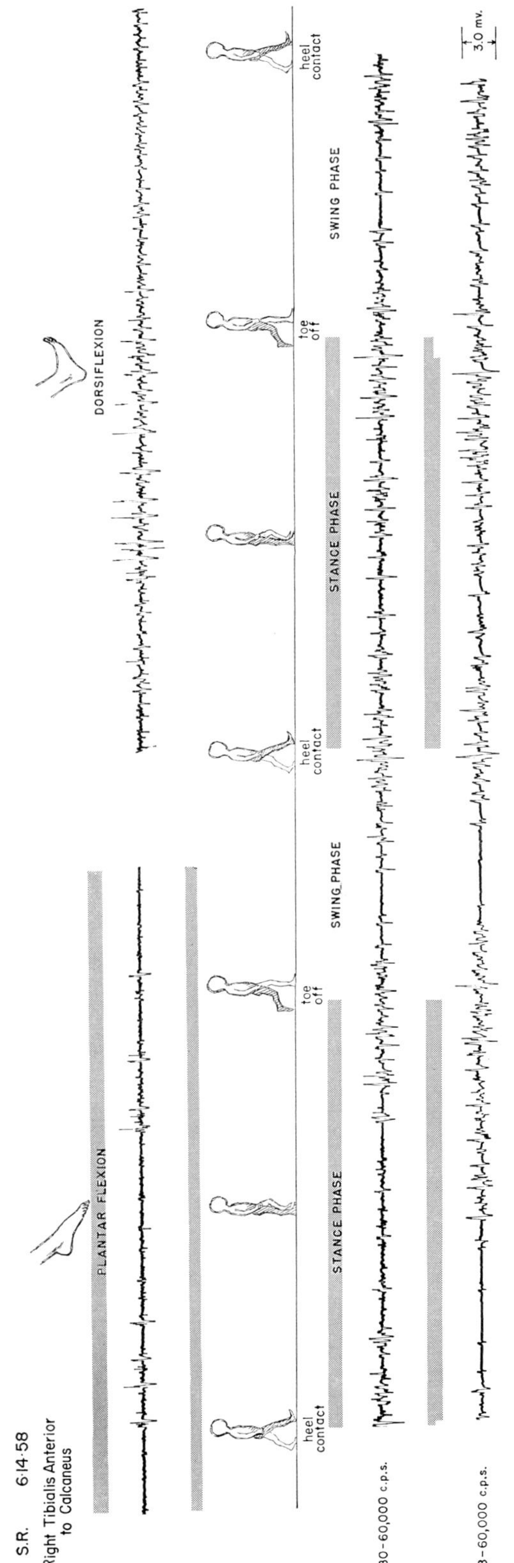

Figure 105. Anterior tibial transfer to calcaneus. There is almost continuous activity during walking. As in Figure 115, the muscle is more active on dorsiflexion than on plantar flexion. (From *Journal of Bone and Joint Surgery*, March, 1959.)

of walking (Fig. 91). Tracings obtained from the contralateral muscle of the same subjects were almost identical. The two bursts of activity are thought to represent the sciatic and obturator innervated portions of the muscle. After the peroneus longus muscle of one patient had been transferred to the dorsum of the foot, it demonstrated a unique pattern with bursts of activity during the middle portions of both swing and stance. The muscle reproduced a remarkably similar pattern of activity at a subsequent test (Fig. 101). This is evidence that the muscle possesses an inherent pattern of behavior represented precisely by the dual internal electrode technique. This complicated pattern is present throughout the muscle and can be reproduced at subsequent tests.

Phasic transfers, such as extensor transfers to the dorsum of the foot, appear to regain their preoperative swing-phase activity. Often their original amplitude and rate of action potentials were regained in as short a time as seven to eight weeks (Figs. 106, 107, 108). An index to the degree of recovery and effectiveness of the muscle with tendon transfer is found in its postoperative tracing when its electrical activity and rate of action potentials are compared with the preoperative tracing. Studies at repeated intervals postoperatively were assembled in some cases to show the customary progressive return of electrical activity to be expected (Fig. 107). Other phasic transfers, such as peroneal muscles, or flexor muscles to the calcaneus, similarly retain their preoperative phasic activity (Figs. 109, 110, 112). It is noted that in the normal subject (Fig. 90), the peroneal muscles are scarcely used during normal walking. Following transfer into the calcaneus (Fig. 112), a great deal of stance phase activity may be acquired. The muscle may undergo marked hypertrophy.

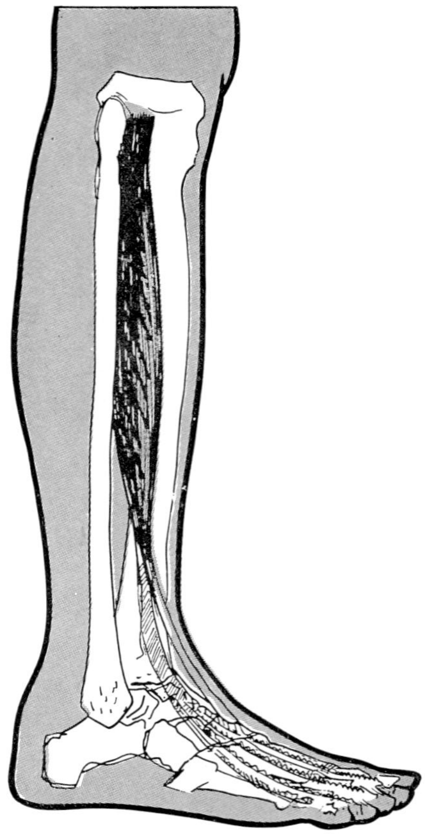

Figure 106. The extensor digitorum longus tendons transferred to the dorsum of the foot.

In general, nonphasic transfers such as hamstrings to patella, tibialis anterior to heel, and flexors to the dorsum of the foot demonstrated a strong tendency to retain their preoperative pattern of phasic activity, failing to make the desired conversion to the phase of the muscle for which they were substituted (Figs. 99, 105, 113).

Among the nonphasic transfers, conversion was demonstrated in a number of instances (Figs. 95, 100, 112, 115, 116, 118).

TRANSFERS

Anterior Tibial to Calcaneus

In the 16 patients studied in whom the tibialis anterior tendon was transferred to the calcaneus (Fig. 114), 6 transfers showed a significant amount of acquired stance-phase electrical activity (Figs. 112, 115, 116). Where the tendon was transferred to the calcaneus as a single procedure, phasic conversion appeared to be a great deal more likely than if the tendon were transferred simultaneously with others. Apparently, training the muscle of the non-

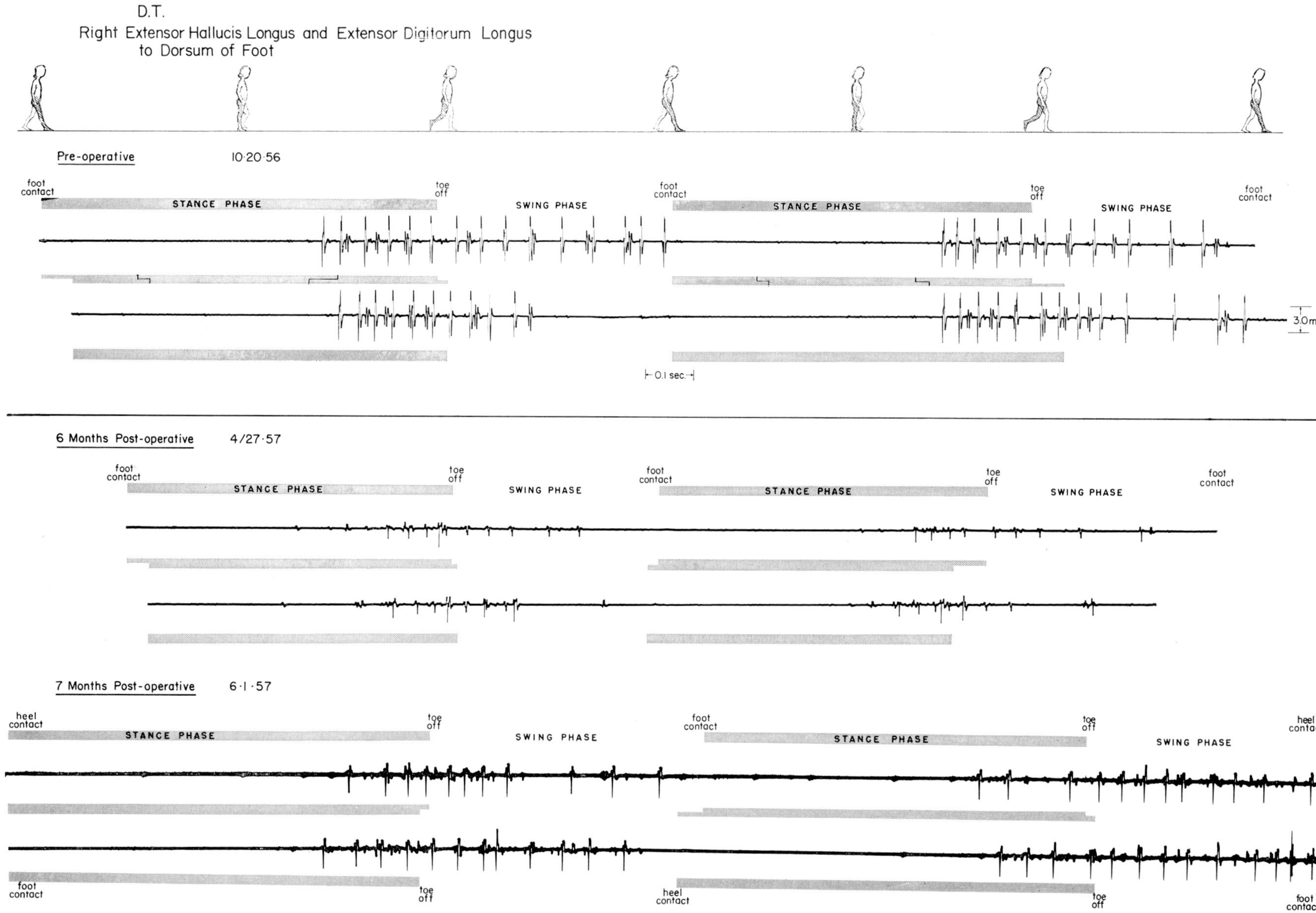

Figure 107. Preoperative and postoperative studies taken from the extensor hallucis longus and extensor digitorum longus muscles. Intermittent motor unit activity is demonstrated, with the muscles having some weakness following poliomyelitis. Six months postoperatively, electrical activity is considerably diminished. Seven months following transfer, the muscle has almost regained its preoperative electrical intensity. The preoperative pattern of swing-phase activity has been retained. (From *Journal of Bone and Joint Surgery*, March, 1959.)

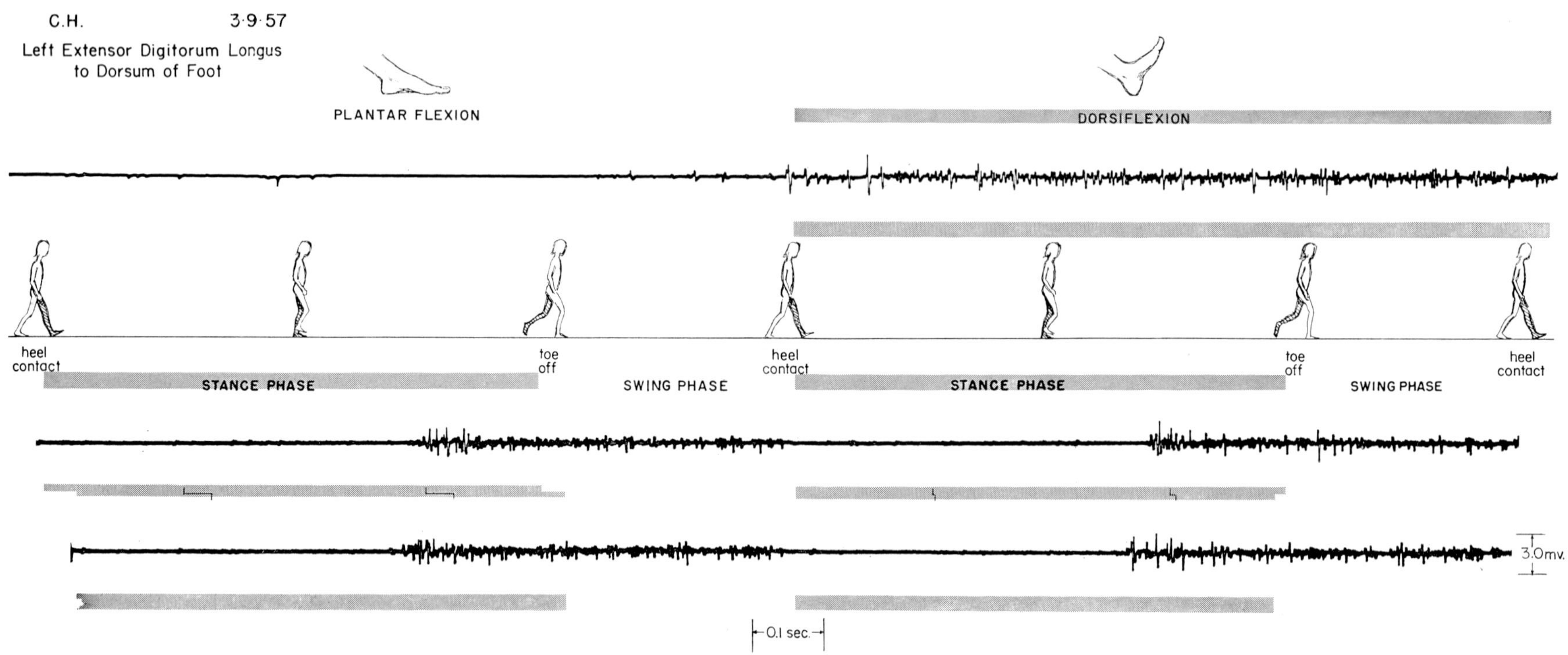

Figure 108. Extensor digitorum longus to dorsum of foot. The long extensors transferred to the dorsum of the foot demonstrate the characteristic swing-phase behavior of the anterior tibial group. (From *Journal of Bone and Joint Surgery,* March, 1959.)

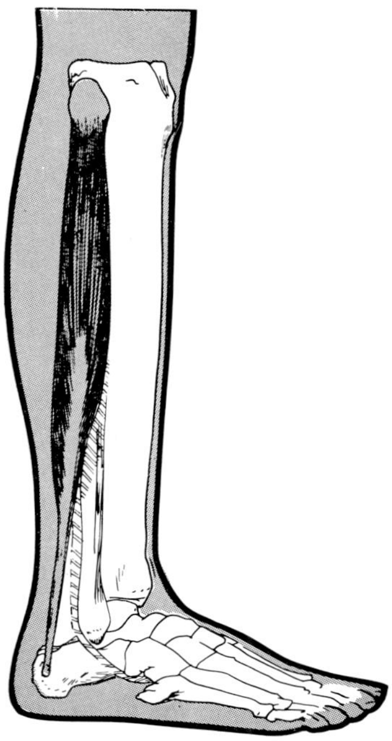
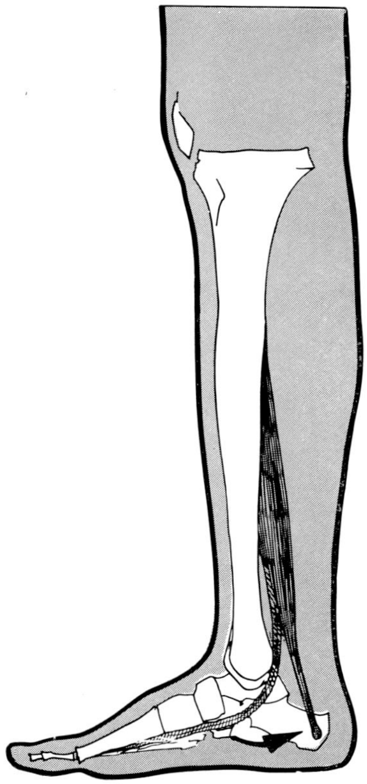
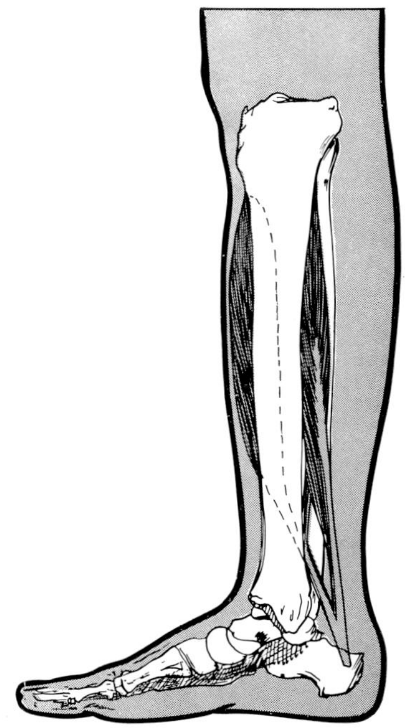

Figure 109 (*left*). Transfer of the peroneus longus to the calcaneus.
Figure 110 (*center*). Transfer of the flexor hallucis longus to the calcaneus.
Figure 111 (*right*). Transfer of the tibialis anterior to the calcaneus.

phasic type is more effective when it can be applied to a single muscle, its tendon having been transferred individually to its new insertion. Phasic conversion of the tibialis anterior transferred to the calcaneus was not evident before an eight-month postoperative period had elapsed.

The postoperative test of one patient is shown where continuous anterior tibial muscle activity resulted. The tendon of the anterior tibial muscle was transferred to the calcaneus one year prior to the test. Weekly physiotherapy was given for six months (Fig. 105).

Peroneus Longus to the Dorsum

Two of the thirty-two transferred muscles studied showed considerable involuntary swing-phase activity (Figs. 100 and 120). Another transfer of this type demonstrated bursts of activity during swing and stance (Fig. 101). Other peroneus longus transfers to the dorsum of the foot showed some activity throughout the entire walking cycle.

The most frequent result of peroneus longus transfer to the dorsum of the foot was retention of the original stance-phase pattern (Fig. 99). If foot-drop is overcome by a transfer with such inappropriate phasic activity, this can be attributed to the shortness of the transferred muscle. This shortness results from the length chosen at surgery when the tendon is placed in its new insertion. If such a transfer is opposed in its new insertion by a strong triceps surae, it finally becomes stretched out and fails to prevent foot-drop. Good temporary results are often reported for such transfers where the nonphasic transfer is tightly inserted into its new site and its new antagonist, the triceps surae, lengthened.

The nonphasic, anteriorly transferred peroneus attempts to dorsiflex the foot at the same

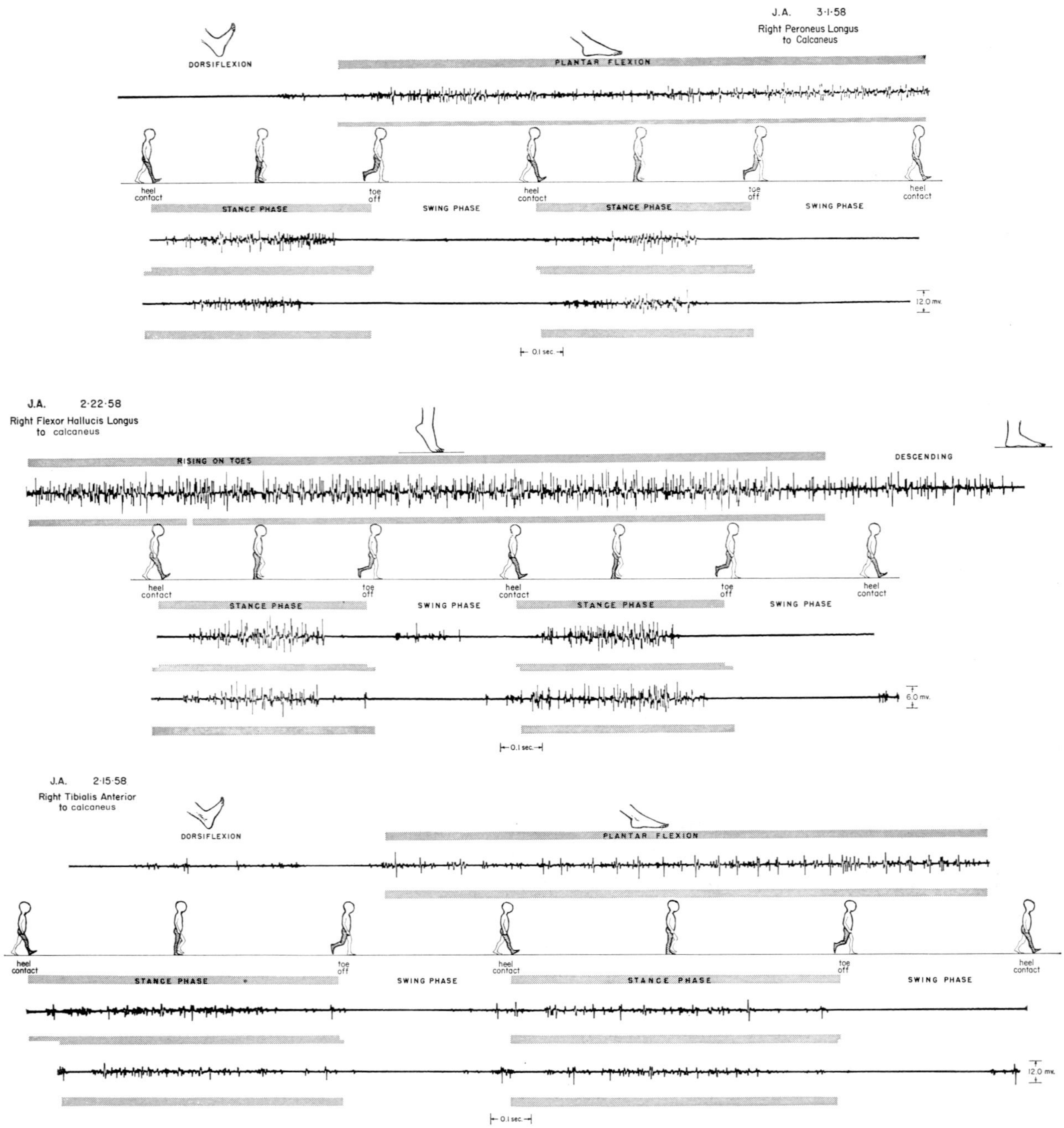

Figure 112. The peroneus longus and flexor hallucis longus muscles transferred into the calcaneus readily demonstrate their desired stance-phase behavior. The anterior tibial has undergone conversion to the stance phase. It was transferred as a separate procedure. (From *Journal of Bone and Joint Surgery*, March, 1959.)

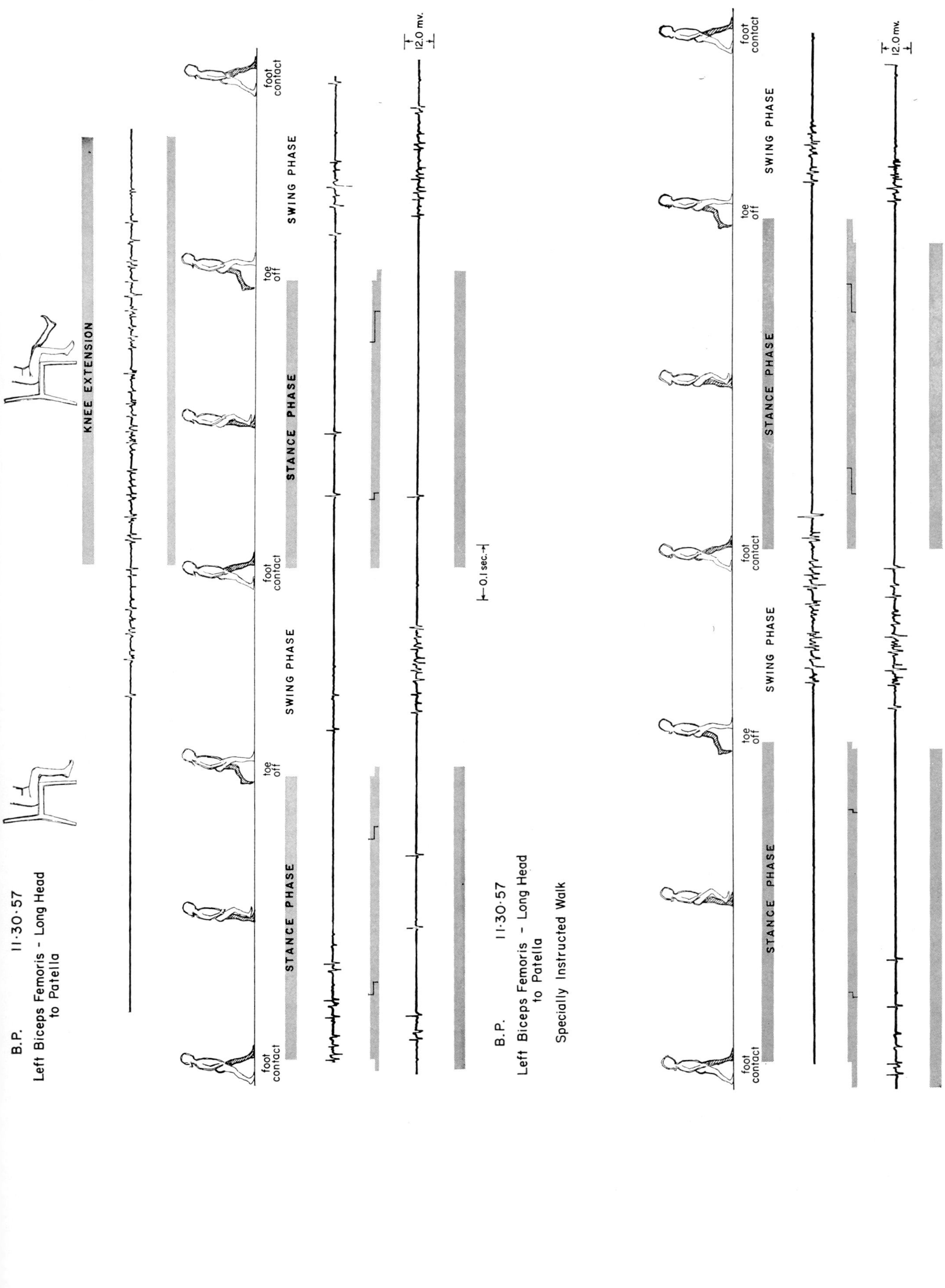

Figure 113. Biceps femoris transferred to patella. The long head of the biceps demonstrates swing-phase behavior. Temporary conversion does not follow instruction. (From *Journal of Bone and Joint Surgery,* March, 1959.)

time (stance phase) that the triceps surae is plantar flexing it. The peroneus longus then has not only a nonphasic action but also the untenable role of antagonist to the calf. The transfer eventually becomes stretched out or even pulled out of its new insertion as it continuously attempts to dorsiflex the foot against the force of the calf musculature.

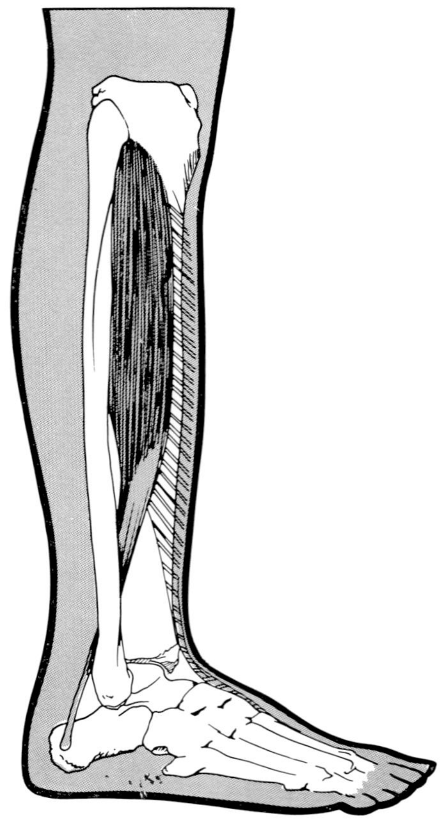

Figure 114. The anterior tibial transfer to the calcaneus.

Peroneus Brevis to the Dorsum of the Foot

Fourteen of the sixteen patients studied demonstrated a remarkable conversion from stance to the swing phase of walking. Spontaneous conversion was observed as early as three months postoperatively (Figs. 117 and 118). This single muscle was not usually of sufficient strength to serve as the sole dorsiflexor. However, after transfer, a considerable amount of hypertrophy was often observed. The unfortunate complication, dorsal bunion, does not follow transfer of the peroneus brevis, a complication frequently seen after transfer of the peroneus longus.

Hamstrings to Patella

Study of twenty patients with biceps transfers to the patella showed that their characteristic swing-phase pattern was usually retained (Fig. 113). In two patients where swing-phase activity persisted during walking, spontaneous stance-phase activity developed when the subject ascended stairs. Two patients with biceps transfer to the patella, however, demonstrated complete conversion to the stance phase (Figs. 93, 94, 95). It is noted that preoperatively, one patient walked with the involved knee in hyperextension during the stance phase (Fig. 95). The transferred biceps femoris, with its newly acquired stance-phase activity, made it possible for the subject to walk with the knee slightly flexed. This is thought to be the function of the normal quadriceps mechanism during level walking. It was found that valid tests and precise stimulation of the transferred biceps femoris were possible only when the muscle was penetrated with the testing electrodes well proximal to the blending of the transfer with the weakened quadriceps muscle. Otherwise, reaction to electrical stimulation on the part of the transfer could not be distinguished from that of the remaining quadriceps. Furthermore, some electrical activity from the remaining weakened quadriceps could influence the recording if the electrodes were inserted into this area. Thus, a site closer to the muscle's origin at the ischial tuberosity rather than the region near its new patellar insertion was chosen for insertion of the electrodes. This made possible a test of the long head of the biceps to the exclusion of the short head. It was not ordinarily possible to make a known insertion into the short head alone. It was possible to test both heads simultaneously. In one case (Fig. 95), the two heads of the transferred biceps femoris could be differentiated by electrical stimulation, making individual recordings possible.

Figure 122 shows the tracings from a semi-

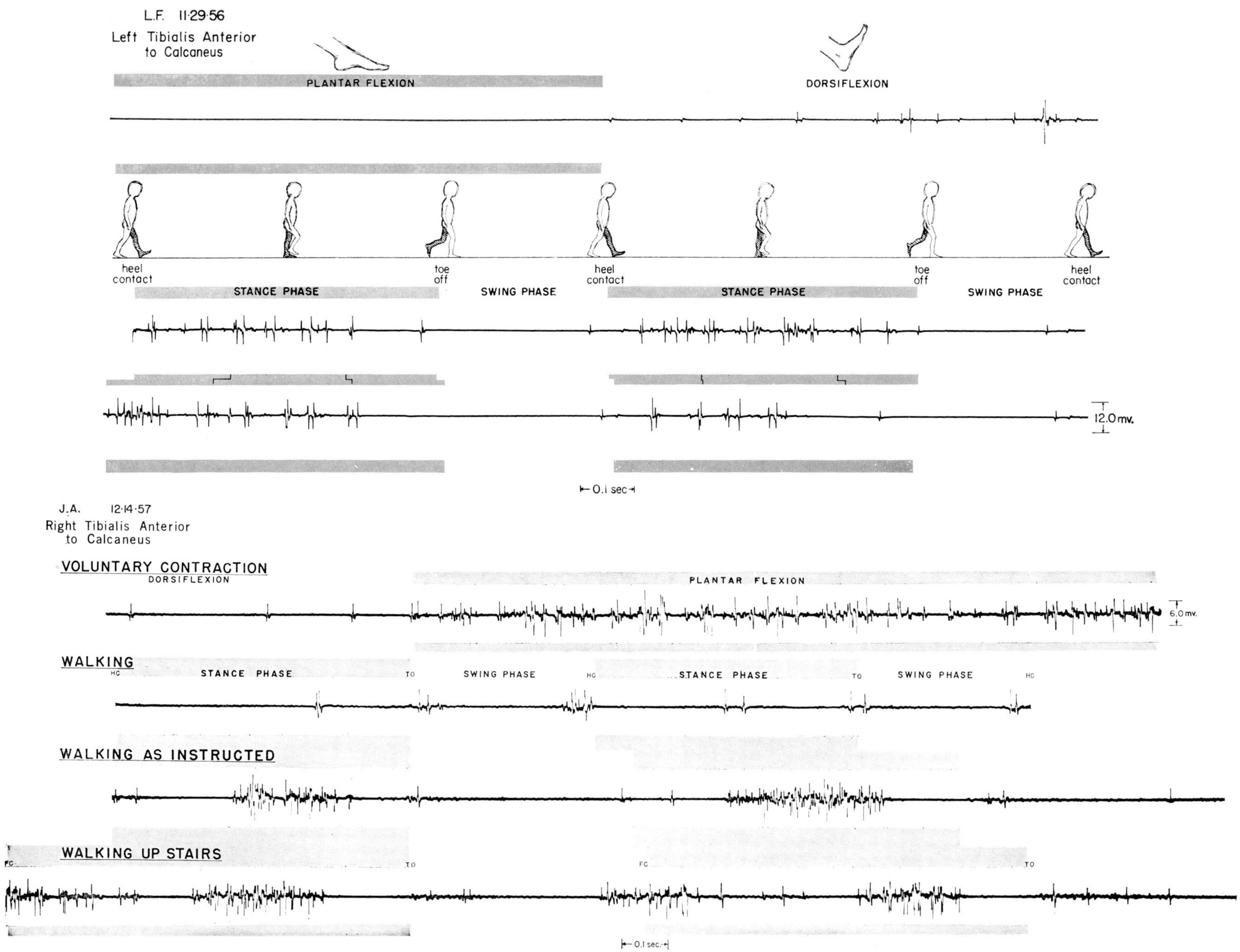

Figures 115, 115A. The anterior tibial transfer to the calcaneus. There is voluntary activity during the stance phase of walking.

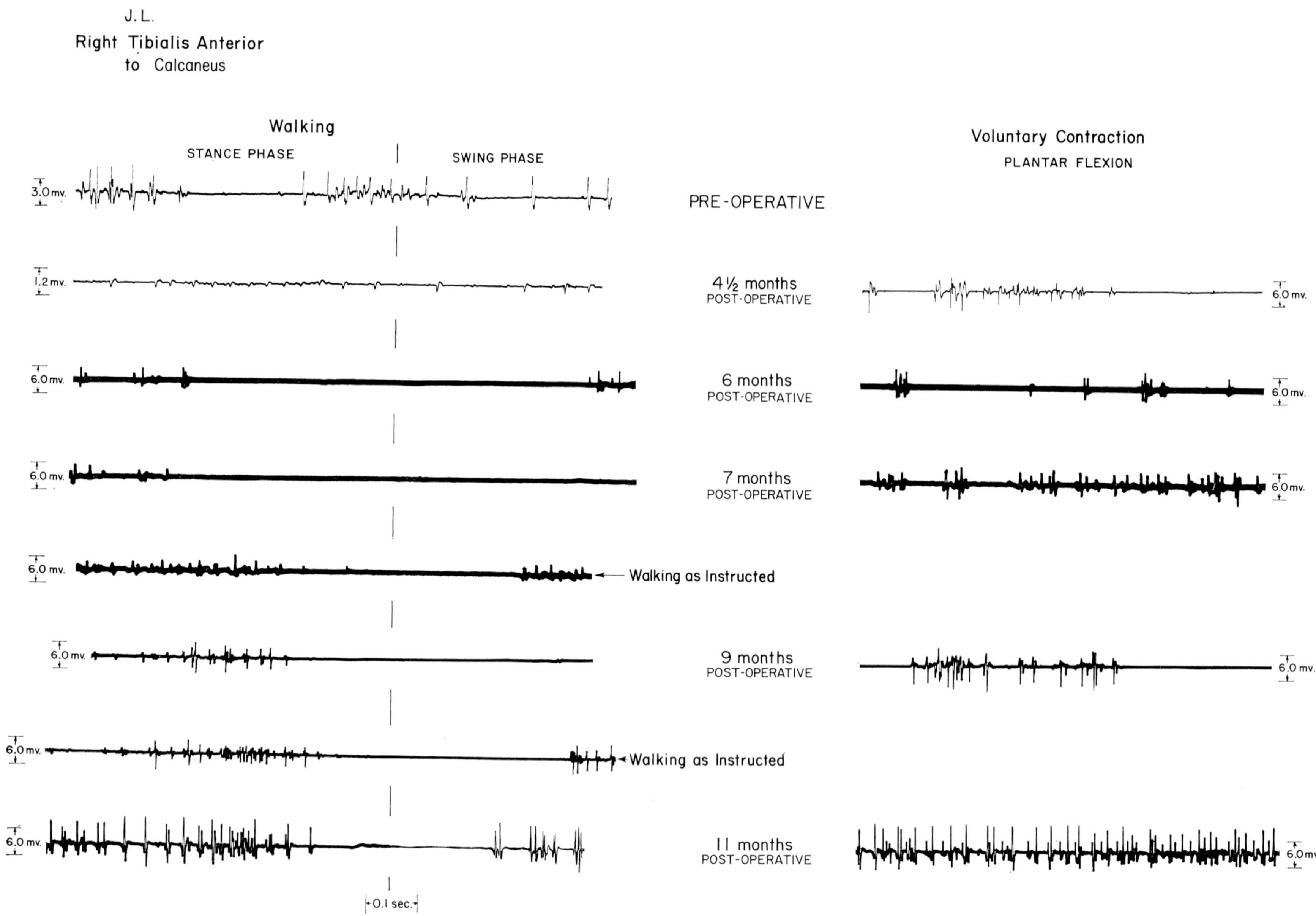

Figure 116. Preoperatively, there was almost no calf strength. Preoperative studies on the anterior tibial muscle are followed by periodic studies after transfer to the calcaneus. Gradual recovery and conversion of the muscle's activity to the stance phase are noted. The extent of recovery appears complete. (From *Journal of Bone and Joint Surgery,* March, 1959.)

tendinosus muscle transferred to the patella. No preoperative studies were made on this subject. The patient, in whom the biceps femoris was also transferred to the patella, developed complete voluntary extension of the knee. The tracings show that, postoperatively, the semitendinosus is a stance-phase muscle. The biceps femoris transfer in this patient failed to undergo phasic conversion. Possibly it was not necessary for the semitendinosus to make a complete phasic change. Reference to the records of normal thigh muscles (Fig. 91) reveals that the activity of the normal semitendinosus may extend into the stance phase. Preoperative testing of this muscle might have revealed that such was the case. Preoperative testing of such muscles in the paralytic lower extremity may be of great value in selecting the muscle most adaptable for transfer if a choice is available. Preoperative and postoperative tests were done on the biceps femoris and semitendinosus muscles (Fig. 121) of another subject (Figs. 123 and 124). These muscles were virtually silent preoperatively during normal walking. Their postoperative behavior was irregular. The transfers were of considerable benefit to the patient, however, in eliminating persistent flexion of the hip and knee during walking.

In one case, where there was severe bilateral lower extremity weakness, the semimembranosus was the only muscle available for transfer to the patella. The branches of the sciatic nerve (two levels) were carefully preserved, and the transfer became very useful to the patient.

Discussion

There is great value in understanding the phasic activity of normal muscles. In the paralytic subject, it is valuable to establish the phasic activity of the muscles which might be under consideration for transfer. Where a choice of muscles available for transfer exists, knowledge of the phasic activity of those muscles is of obvious importance in planning the procedure. For example, if hamstring muscles are to be brought forward to replace or reinforce the quadriceps, a transfer having some

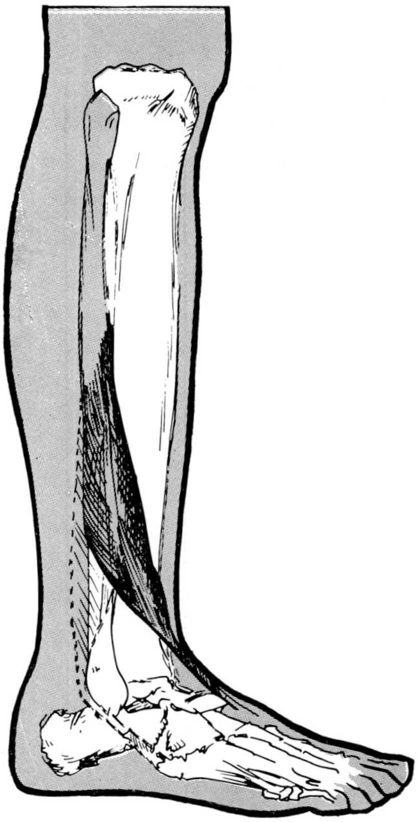

Figure 117. The peroneus brevis transfer to the dorsum of foot.

stance-phase activity may be the muscle of choice. Furthermore, if one has a choice between certain nonphasic transfers, such as the peroneus brevis or peroneus longus, for transfer to the dorsum of the foot, it is recalled that one muscle (peroneus brevis) possesses far more likelihood of conversion to the swing phase than the other. Evaluation of the amplitude and amount of electrical activity of the muscle under consideration for transfer is also a valuable adjunct to clinical testing of the muscle.

Where a choice is available between a phasic and nonphasic transfer, such as the long extensor and the peroneus longus to the dorsum of the foot, one can be confident that the long extensor muscles will retain their swing-phase activity, while the likelihood of conversion of the stance-phase peroneus longus to the swing phase is remote. It may be true, but to no avail, that the peroneus longus transferred to the dorsum of the foot can voluntarily produce dorsi-

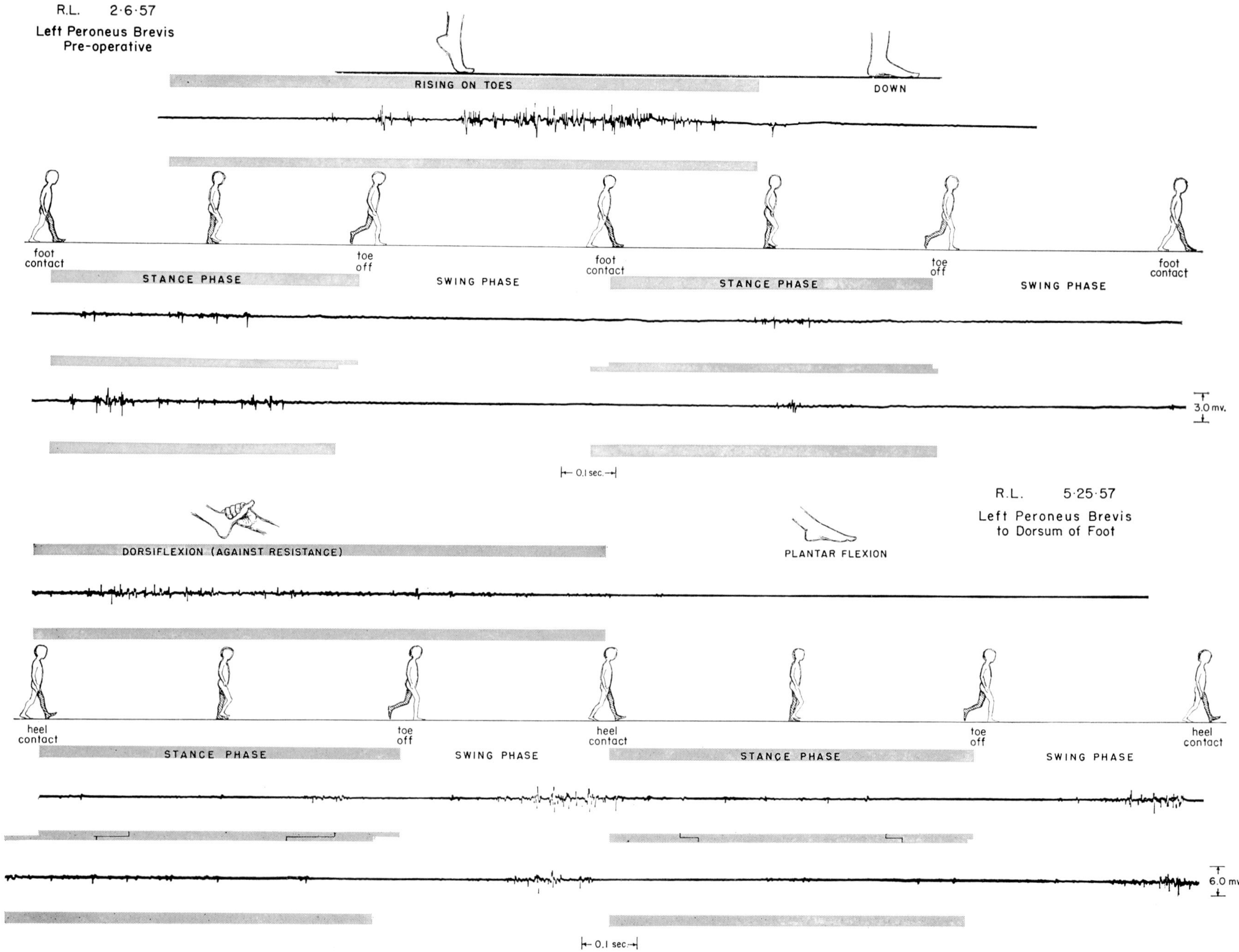

Figure 118. Preoperatively, the peroneus brevis muscle demonstrates the usual stance-phase activity. The muscle is not used extensively during normal walking. Three months postoperatively, there is decisive conversion to the swing phase of walking. No instruction was given before the recording was made. (From *Journal of Bone and Joint Surgery*, March, 1959.)

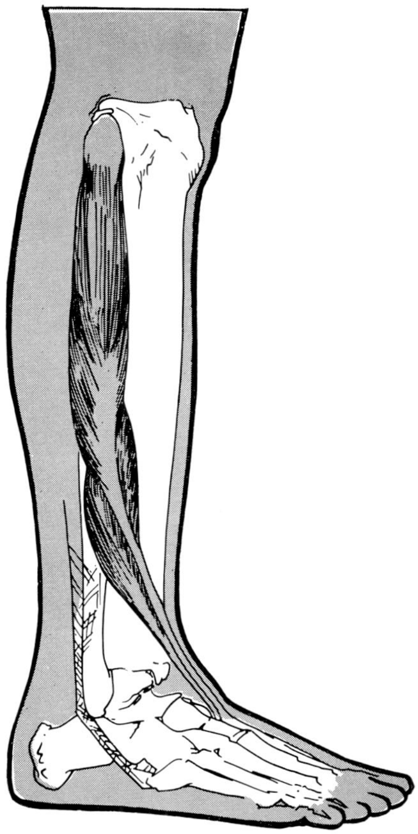

Figure 119. Transfer of peroneus brevis and peroneus longus tendons to dorsum of foot.

flexion. The phasic transfers, such as extensors to the dorsum of the foot or flexors to the heel readily regain most of their electrical activity soon after surgery. It is not uncommon to find considerably increased electrical activity (in amplitude and duration) in the phasic transfer (Figs. 112 and 125). Marked hypertrophy has been observed in the peroneus longus and flexor hallicus longus after transfer to the calcaneus, where it is desirable to add to the strength of the triceps surae (Fig. 112). In the same subject, the anterior tibial muscle transferred to the calcaneus has undergone phasic conversion. The level of electrical activity of the muscle, however, is quite low.

Where phasic transfers are combined with nonphasic transfers at one procedure, considerable difficulty is encountered in training the nonphasic muscle. If the tibialis anterior is transferred to the heel along with flexor muscles, phasic conversion of the tibialis anterior muscle is very unlikely. Occasionally, the peroneus longus is transferred to the dorsum along with an extensor (Figs. 126 and 127). In such combinations, we are aware of no instance where phasic conversion of the peroneus longus has taken place. Apparently nonphasic transfers, if transferred individually, can be trained far more successfully. The nonphasic transfer, if accompanied by phasic muscles to the same site, appears to fail.

In one instance, conversion of the peroneus longus and peroneus brevis to the swing phase was found (Figs. 119 and 120). The tendons of these muscles were transferred forward ten years prior to the test demonstrated in the figure.

The difference in the fate of the transferred peroneal muscles to the dorsum of the foot is of considerable interest. As Schaeffer states,[154] the ankle joint is found only in mammals. In lower forms, the peroneal muscles are anterior muscles and serve as extensors. As the calcaneus migrated beneath the talus, the peroneus longus tendon and probably later, the brevis, were displaced to the posterior aspect of the lateral malleolus. This migration apparently is more recent in the case of the peroneus brevis, as it can still be found in some amphibians to be anterior to the lateral malleolus.

The extensor and peroneal rudiments in the human embryo[10,109] are a single muscle mass until the sixth week. The peroneal mass then separates from the extensor as the individual peroneal rudiments become distinct. The peroneus longus tendon is at first lateral to the rudimentary lateral malleolus, passing posterior to it when the embryo reaches the size of 20 mm. Posterior transposition of the tendon of the developing peroneus brevis must await that of the peroneus longus if it is to retain its position anterior to that structure.

In several of the subjects tested in whom both of the peroneals were transferred anteriorly, the peroneus longus failed to adapt itself to the swing phase, the peroneus brevis making a complete phasic conversion, often without training (Fig. 125). Years later, obvious hypertrophy was observed in the transferred pero-

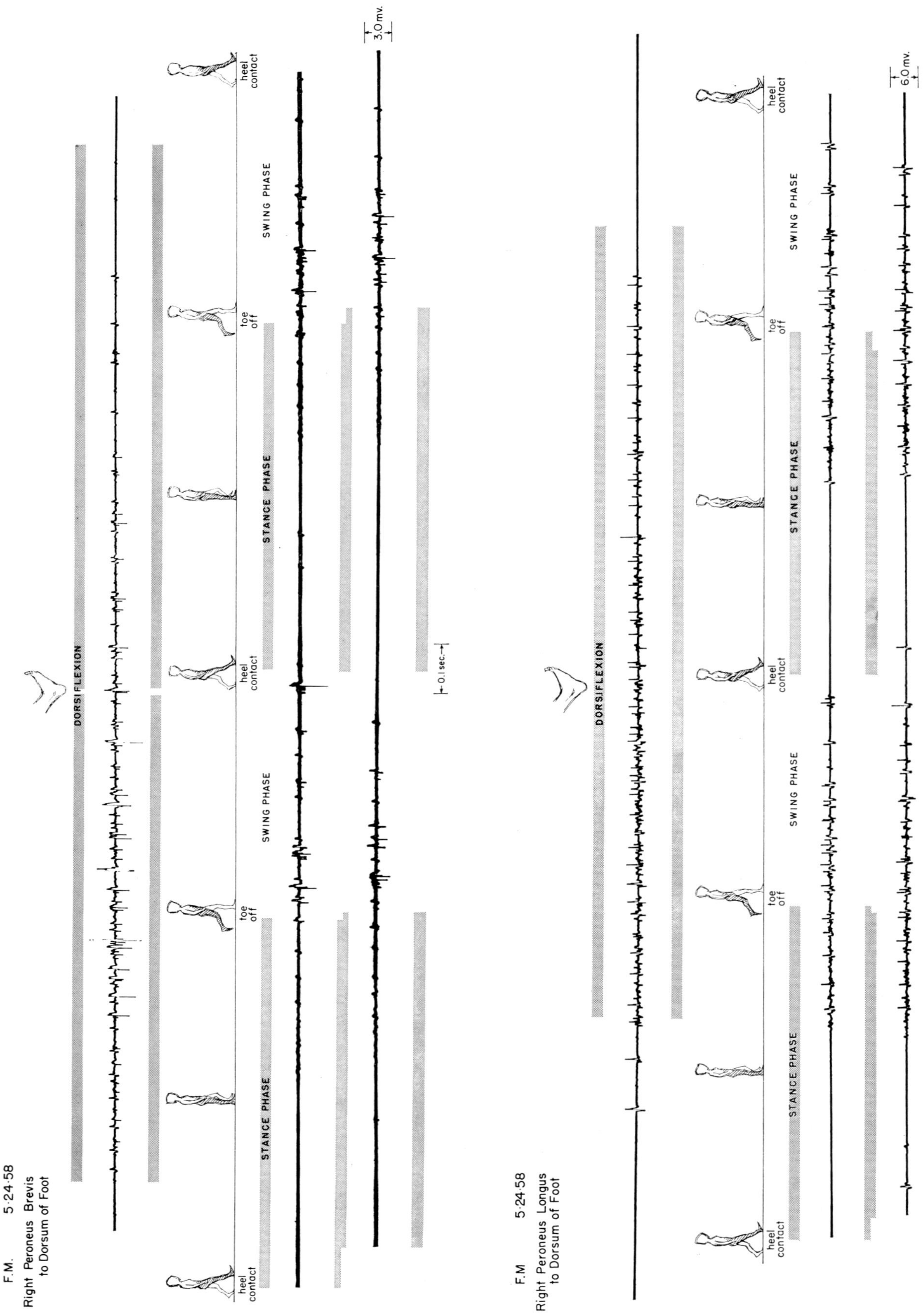

Figure 120. Peroneus brevis, peroneus longus to dorsum of foot. The transfers were done ten years prior to the tests. Conversion to the swing phase is demonstrated by both muscles. (From *Journal of Bone and Joint Surgery*, March, 1959.)

neus brevis muscle, with the peroneus longus demonstrating atrophy. In one patient, however, in whom the peroneal tendons were transferred forward, both muscles demonstrated complete conversion to the swing phase without postoperative training (Fig. 120).

Transfer of the peroneus brevis (Fig. 117), performed as a single procedure, may result in very prompt conversion of the phasic activity of the muscle from stance to swing. In one instance (Fig. 118), spontaneous conversion took place in three months. No training or reeducation of the muscle had yet been given.

Training in the form of reeducation directed toward the use of the muscle during the proper phase of the walking cycle appeared to be of definite value following tendon transfer. In particular, the tibialis anterior, when transferred to the calcaneus, can undergo phasic conversion only after several months of intensive physical therapy. Even after such training, as demonstrated in one case (Fig. 105), only continuous activity throughout the walking cycle resulted. The only subjects with hamstring transfer to the patella followed by phasic conversion were extensively trained by the physical therapist. Here the patient was taught to use the transfer just before placing the extremity on the walking surface. The two patients with transfers of the peroneus longus to the dorsum, where a considerable amount of phasic conversion took place (Figs. 100 and 120), received little or no instruction. Many others were extensively trained for several years but produced no swing-phase involuntary activity. The rare conversion of the peroneus longus transferred to the dorsum appears to be a spontaneous phenomenon.

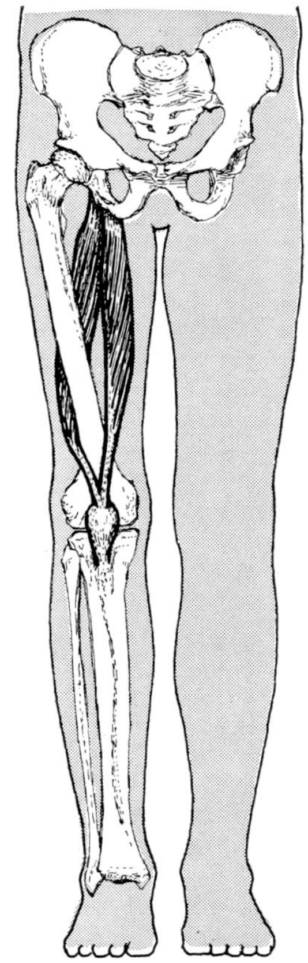

Figure 121. The biceps femoris and semitendinosus transfers to the patella.

Summary

The sound motion picture camera, its sound tract in synchrony with the cathode-ray oscilloscope, using internal individual muscle electrodes, represents an excellent and practical method for recording the phasic activity of the muscles of the lower extremity. The value of the study of phasic activity in the normal and the preoperative and postoperative paralytic patient is apparent.

Phasic transfers, in general, retain their preoperative phasic activity. They also appear to regain their preoperative duration of contraction and electrical intensity. Many nonphasic transfers retain their preoperative pattern of phasic activity, failing to duplicate the desired action of the muscle for which they are substituted. Certain nonphasic transfers are capable of phasic conversion.

The phasic or physiological transfer, such as flexor hallicus longus or peroneus longus to the calcaneus, may result in hypertrophy and markedly increased electrical activity in the transferred muscle (Fig. 112).

The scant electrical activity found in the normal peroneus longus is shown in Figure 90. The vigor of the muscle usually acquired after transfer to the calcaneus is apparent in Figure 112.

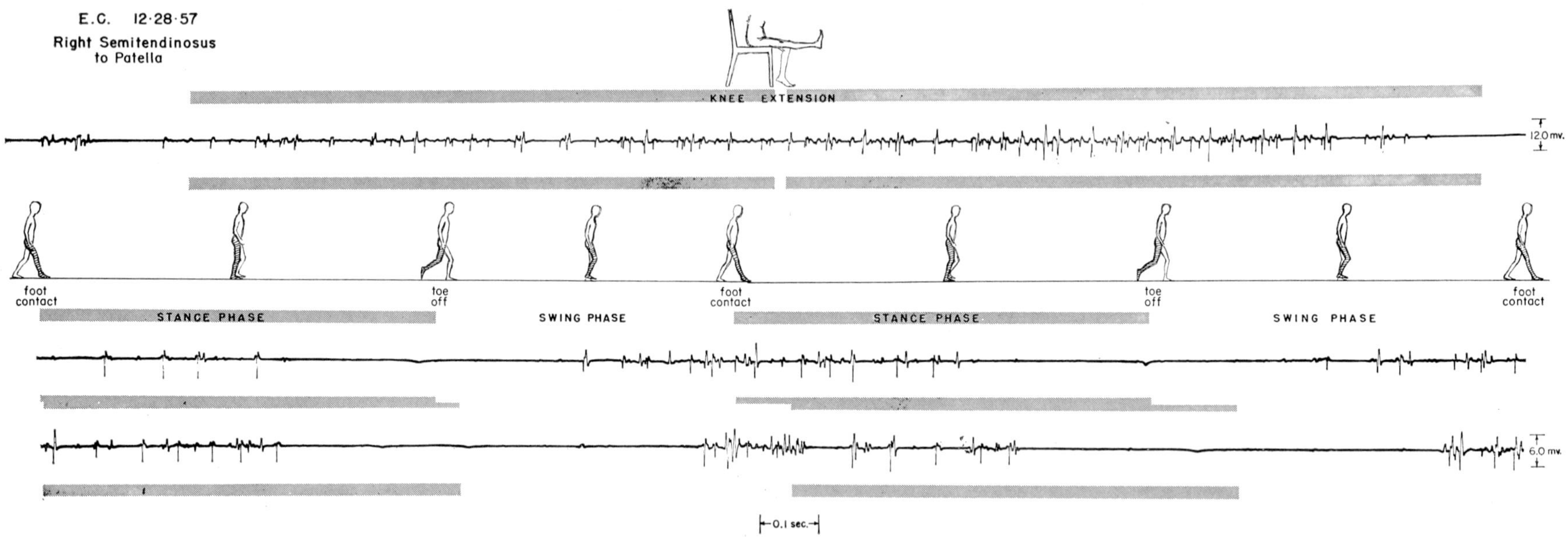

Figure 122. The patient is six months postoperative following transfer of the right semitendinosus and biceps femoris tendons into the patella. His quadriceps muscle was rated as zero. He now has complete voluntary extension, as indicated by the figure. The semitendinosus muscle demonstrated a remarkable amount of stance-phase activity, its behavior resembling the action of the quadriceps muscle. (From *Journal of Bone and Joint Surgery*, March, 1959.)

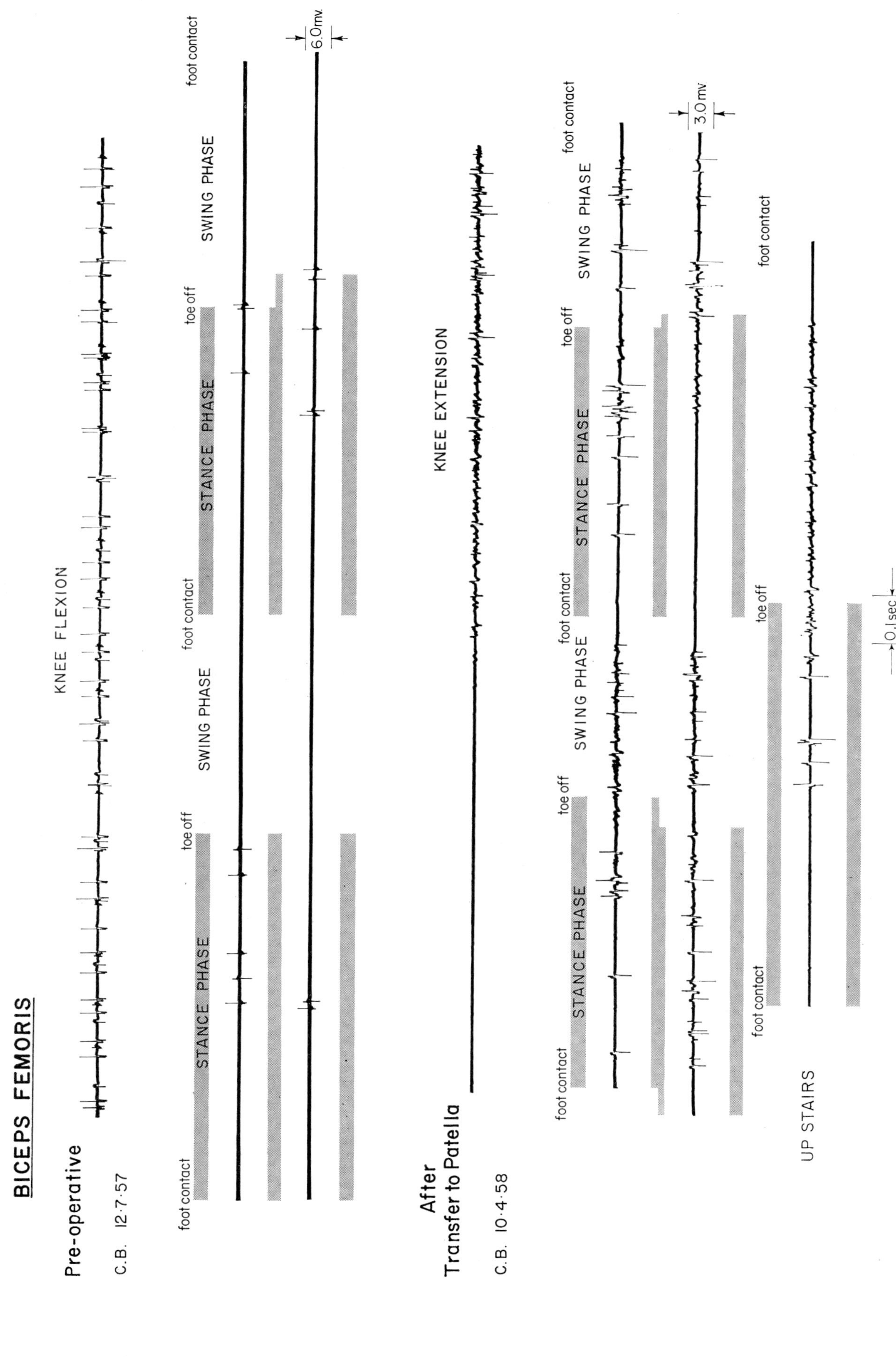

Figure 123. Preoperative and postoperative studies on a patient with biceps femoris and semitendinosus tendons transferred to the patella. Dicussion in text.

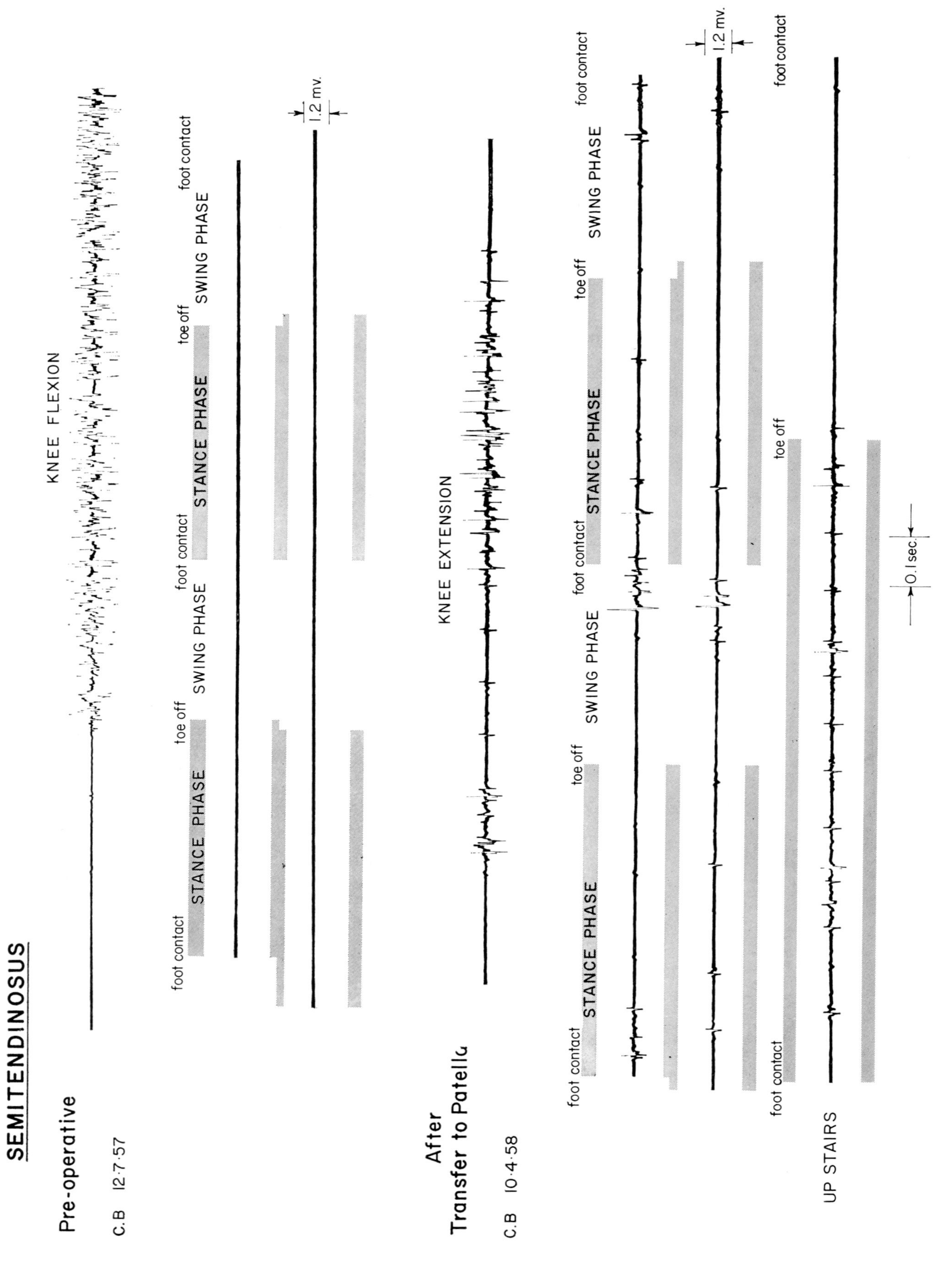

Figure 124. The semitendinosus transfer in the same case shown in previous figure.

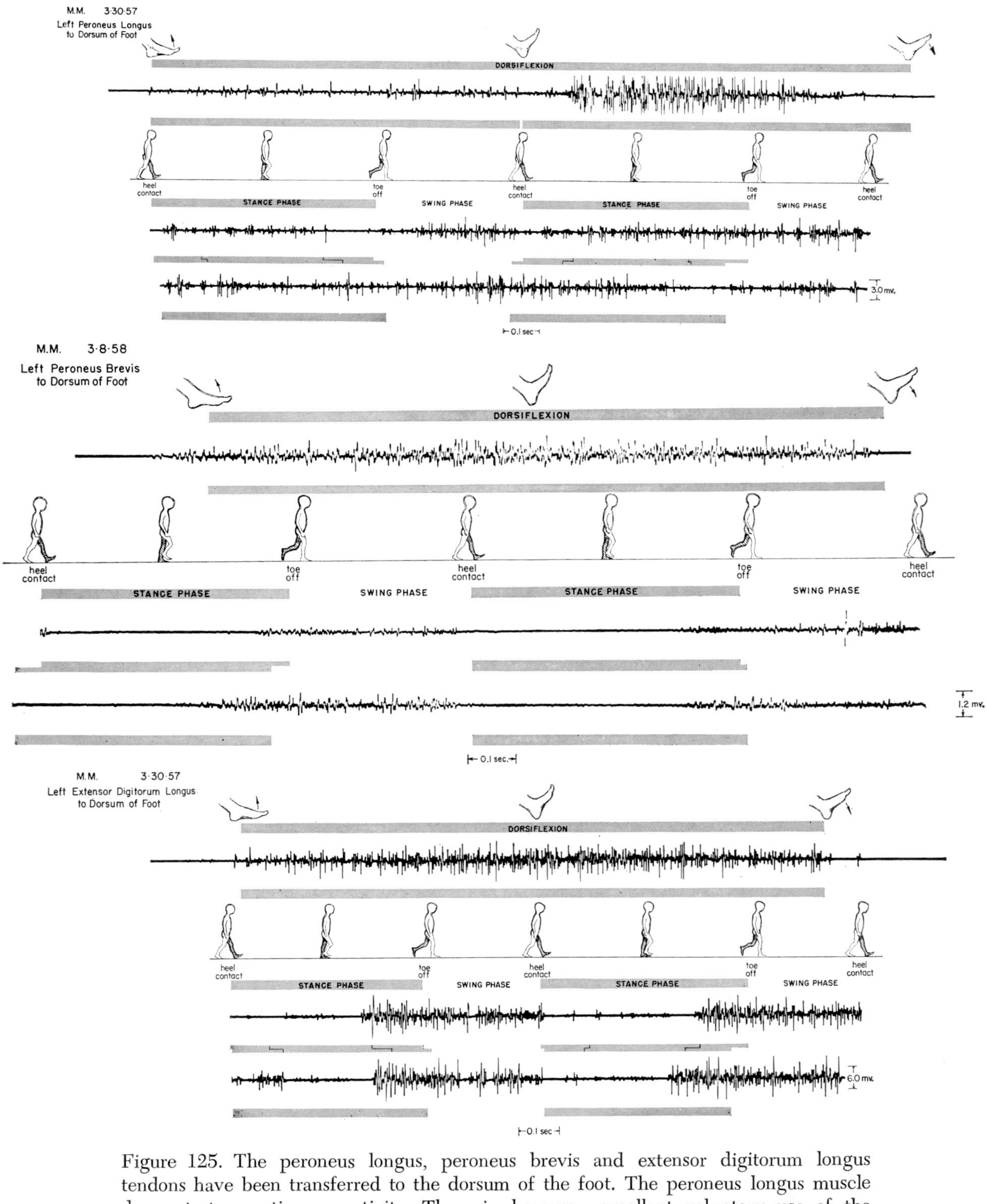

Figure 125. The peroneus longus, peroneus brevis and extensor digitorum longus tendons have been transferred to the dorsum of the foot. The peroneus longus muscle demonstrates continuous activity. There is, however, excellent voluntary use of the muscle as a dorsiflexor. The peroneus brevis muscle has made a complete conversion to the swing phase of walking. The extensor digitorum longus also demonstrates the anticipated excellent swing-phase activity. (From *Journal of Bone and Joint Surgery*, March, 1959.)

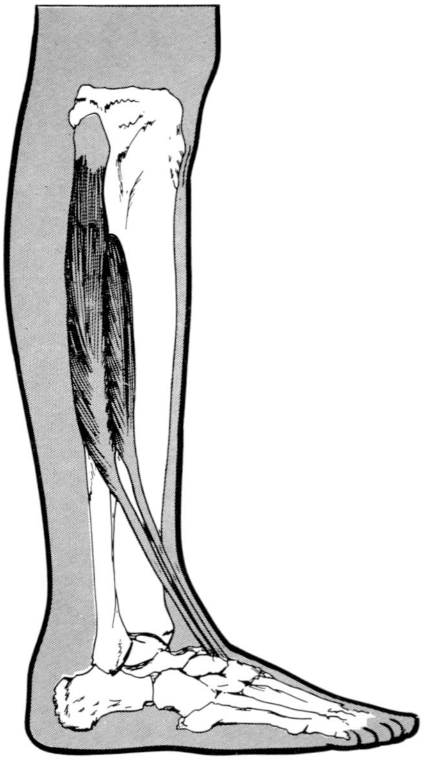

Figure 126. Transfer of the peroneus longus and extensor hallucis longus tendons to the dorsum of the foot.

Conclusions

The method described is a simple procedure by which reconstructive tendon transfers in the lower extremity can be better evaluated. The test can be conducted with a limited amount of equipment and may be relied upon to produce an excellent display of the electrical activity of the muscle. The orthopaedic surgeon engaged in the rehabilitation of the paralytic patient can profit by electromyographic evaluation before and after tendon transfer.

In the normal subject, electromyographic studies may be of considerable basic value. For example, it is often observed that some of the hamstring muscles are almost inactive during walking (Fig. 91). This lends further justification to the procedure of transposition of one or more of the medial hamstring tendons if it becomes desirable to reinforce an injured medial collateral ligament.

The phasic transfer, in general, is superior to the nonphasic transfer. Results of the phasic transfer are predictable. In some instances, however, nonphasic transfers can undergo phasic conversion.

It is possible that certain instances of prompt, spontaneous phasic conversion may be attributed to the use of ancestral pathways once used in forebears where the muscle had the function of the transfer in its new position.

ANALYSIS OF THE FUNCTION OF THE ILIOPSOAS AND SOME OF THE MUSCLES OF THE THIGH

The size of the iliopsoas muscle in man indicates that it merits careful consideration and analysis. It has been demonstrated[35] that in addition to being active in flexion, the muscle also works in abduction. Although abduction by the iliopsoas alone may not take place in paralytics where the gluteal musculature is weakened, it may nevertheless assist in abduction in conjunction with other muscles. For example, after the gluteal musculature has succeeded in abducting the thigh to a sufficient extent that the tendon of the iliopsoas passes somewhat lateral to the center of the head of the femur, it is readily understood how the muscle may then take advantage of this new position and assist in further abduction (Fig. 37). Observation of lateral displacement of the distal portion of the iliopsoas mechanism is readily demonstrated on abducting the lower extremity of the cadaver with the muscles anterior to the hip exposed. The attachment of a few fibres to the hip joint capsule and the adjacent position of the heads of the rectus femoris muscle do not prevent this change in the line of action of the iliopsoas muscle. Nature apparently calls upon all available muscles to assist in movement, seldom restricting one muscle's activity to single simple types of well-defined motion.

The possibility of a rotatory function of the iliopsoas in addition to adduction has been studied by Basmajian.[14] In this work, no sig-

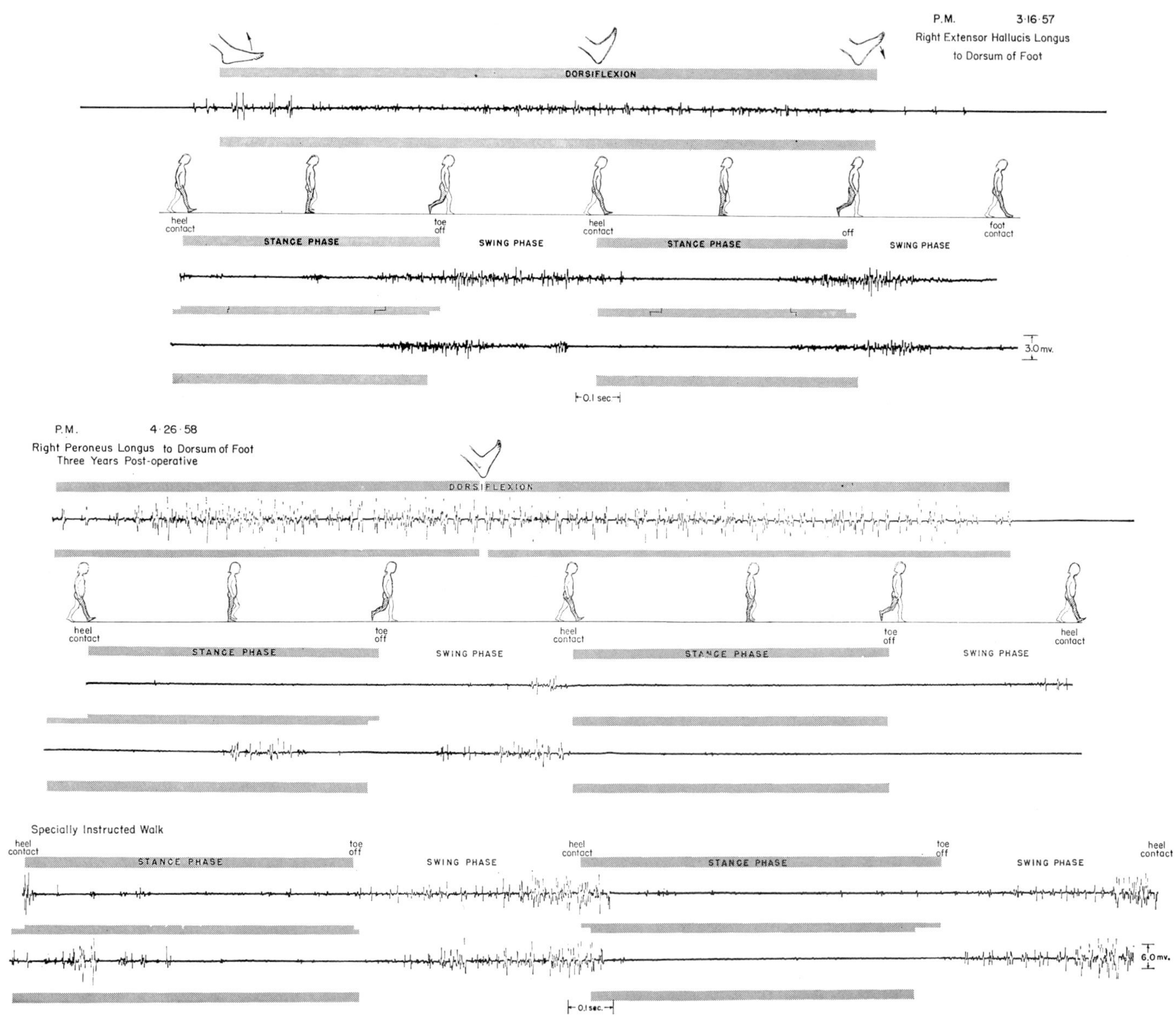

Figure 127. Peroneus longus to dorsum of foot. Extensor hallucis longus to dorsum of foot. The peroneus longus muscle demonstrates spasmodic swing-phase activity. The usefulness of the muscle increases markedly when the patient concentrates on its use following instruction. In contrast, the extensor hallucis longus muscle has retained its spontaneous swing-phase activity. (From *Journal of Bone and Joint Surgery*, March, 1959.)

nificant activity of the muscle was reported on rotation.

Four cases in which iliopsoas transfer to the greater trochanter[135] was done have been studied. In these studies, only patients with good ambulatory ability were included, since principles of phasic activity often do not apply in the severe paralytic. It was noted, however, that preoperatively in some of these patients there was activity in the iliopsoas muscle during walking. This was not necessarily a regular pattern of behavior and differed from the normal where we have been unable to find electrical activity in this muscle during normal walking. In two of the cases where the muscle was transferred, it was routed through a large perforation made in the ilium (Fig. 128). In both of these cases, fairly exhaustive postoperative studies were also performed (Fig. 129). In the other two cases, the tendon was transferred to the greater trochanter via a groove deepened in the anterior portion of the ilium, making a perforation of the ilium unnecessary (Fig. 130). This route might be preferable where external rotation of the lower extremity, possibly due to weak opposition to the gluteus maximus, must be overcome. The transfer by the anterior route assists in internal rotation of the hip.

If phasic conversion is claimed, one must actually demonstrate preoperative phasic activity as well as postoperative activity; otherwise, although the transferred muscle may act in a strange phase, one cannot say without reservation that any change has taken place following surgery. This is particularly true in patients with poliomyelitis deformities where the body may claim any muscle that may be useful and use it during any beneficial, if abnormally timed, period. One cannot say that because the iliopsoas muscle, for instance, is a swing-phase muscle in the normal patient that it is also a swing-phase muscle in the preoperative poliomyelitic patient. One patient (Fig. 129) preoperatively demonstrated considerable late stance-phase activity in the iliopsoas muscle, in contrast to the normal where there is practically no activity during walking (Fig. 91). In the normal, iliopsoas activity, if there is any, is swing phase in character.

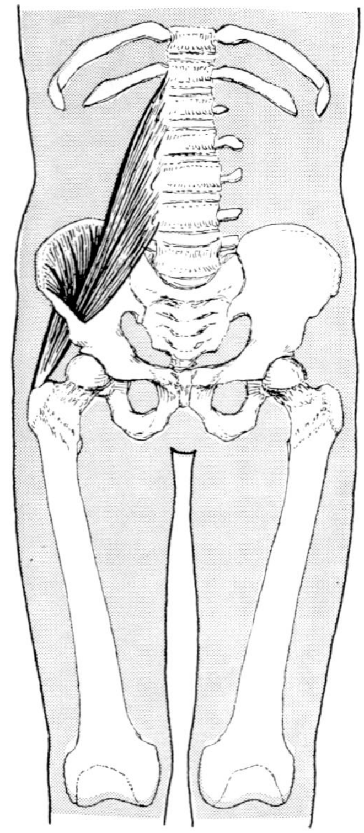

Figure 128. Transfer of the iliopsoas to the greater trochanter through a perforation in the ilium.

Exhaustive 16-mm sound motion picture films have been made based on the activity of the iliopsoas muscle. Here, body activity is photographed with the electromyogram synchronously recorded on the sound track of the film. When this material is projected, one can readily correlate the activity of the muscle with body activity. It is practical to study only the iliacus portion of the muscle, since the approach to the psoas major is difficult and hazardous. Preoperatively, and in most of the normal subjects, only the iliacus portion is studied. The psoas major, however, has been studied a few times in the normal. This examination was done by a posterior sympathetic block approach. It was possible to clearly demonstrate that the activity of the psoas major was quite similar to that of the iliacus portion of the iliopsoas muscle. Postoperatively, both portions of the muscle are easily accessible for study. However, after the muscle has been thus transferred, one cannot

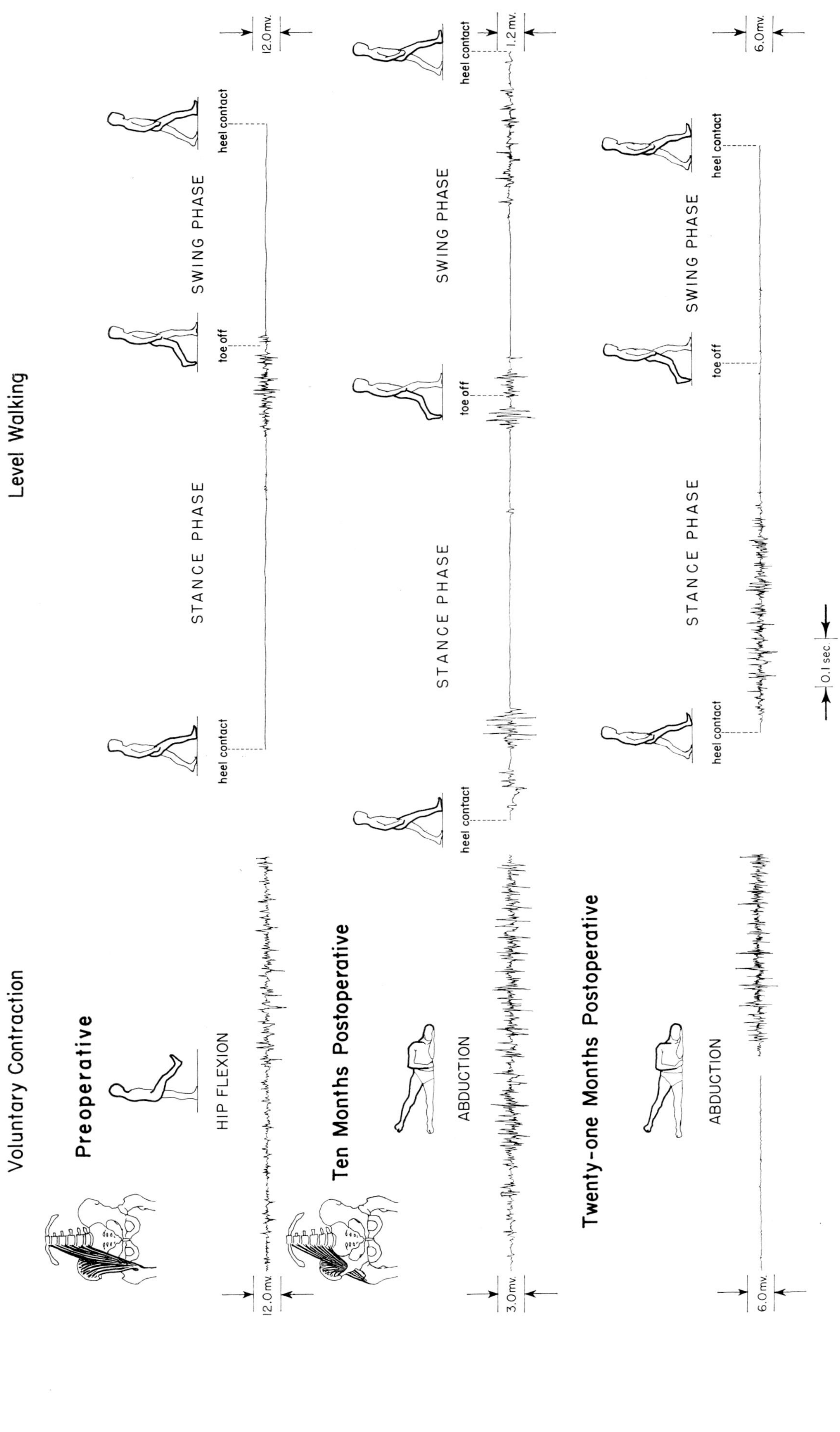

Figure 129. Preoperative and postoperative studies on the iliopsoas muscle. Transfer to the greater trochanter was done through a perforation in the ilium.

be certain which portion of the muscle the electrodes (anteriorly inserted) may have penetrated. They may have been inserted into the psoas major or the iliacus, and even electrical stimulation does not clearly differentiate the two portions after transfer to the greater trochanter.

The technique is also excellent for evaluation of any real improvement derived from transfer of the insertion of the iliopsoas muscle from the lesser trochanteric region of the femur to the greater trochanter. It has been our policy to observe any possible changes in the lumbar spine. In some of the cases, preoperative and postoperative x-ray films of the lumbar spine were made. A good many months postoperatively, it was not possible to detect any abnormal curvature resulting from the transfer. If the muscle is to serve as an abductor in its new position, a function which requires considerable force, then it must also have a very stable origin. One might question whether the lumbar spine, particularly in the frontal plane, might not be subject to considerable stress, this resulting in deformity of the lumbar spine. Preoperative and postoperative films of the lumbar spine are therefore a very desirable addition to the study of the effect of this transfer.

In the two cases where preoperative and postoperative studies were possible, the type of transfer done was the type requiring a perforation in the ilium. In these cases, no abnormal curvatures in the lumbar spine were produced. In the two cases where no preoperative electrical studies were possible, the transfer was done by the anterior route.

The 16 mm sound film with synchronized sound track (the EMG) and image proved to be a superior method of evaluation of such a procedure. One has the opportunity to observe the patient's capabilities preoperatively while evaluating the electrical activity of any single muscle involved. Finally, it is possible to compare the preoperative study to the postoperative study. It was felt that such a system of study resulted in the most accurate and most practical method of evaluation available for such a procedure.

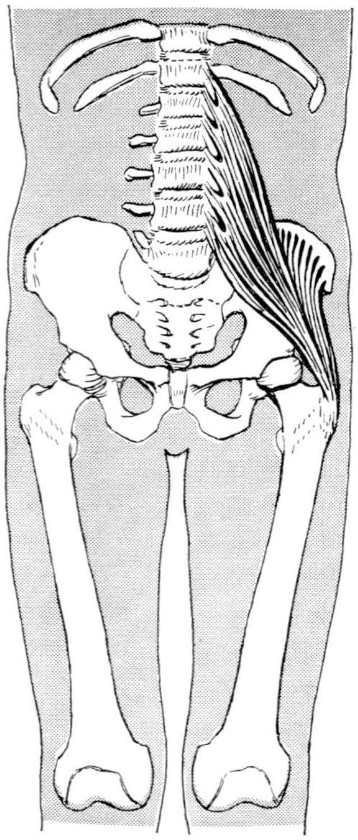

Figure 130. Transfer of the ilipsoas to the greater trochanter through an anterior groove in the ilium.

As stated above, our preoperative studies were confined to the iliacus muscle. Our postoperative studies on the transfer were not specific regarding the identity of the two portions. Since our studies on the normal subject have shown virtually identical behavior of the psoas major and the iliacus, it is concluded that our postoperative studies are valid in the sense that they can be compared to the preoperative studies which were, of necessity, confined to the iliacus muscle.

Actual records of the action potentials are shown in Figure 129, since amplitude, rate, oscilloscope sensitivity or amplification, and the characteristics of the action potential with its fine variations in form indicate many important aspects of the muscle's action, particularly regarding its vigor and the amount of original vitality which is ultimately regained after the transfer. Very scant electrical activity, the weak signal from a weak muscle, may mean little, regardless of the time of contraction. Also, conversion of an extensively affected muscle could often be of only academic significance.

In all of our studies, careful electrical stimulation of the portions of the iliopsoas mechanism was carried out. This was done before and after each test in order to avoid the possibility of confusion due to penetration of adjacent muscles with the electrodes and to rule out the possibility of migration of the electrodes during the test. The necessity of internal placement of electrodes in the study of this muscle is obvious, since there is no portion where surface electrodes may be accurately applied to such a deep muscle. Since migration of electrodes is possible, no records are considered valid unless unquestionable and correct response to electrical stimulation following, as well as preceding, the examination is observed.

One can gain information of unquestioned value regarding the desirability of this transfer. If this technique is followed, one may record the following data:

1. Preoperative and postoperative 16 mm sound moving pictures with synchronized sound track of the electromyogram.
2. Continuous synchronized oscillograms from the oscilloscope.
3. Preoperative and postoperative films of the lumbar spine.

Although one can gain considerable information by comparing the amount of electrical activity as demonstrated on the sound track of the sound motion picture film, the oscillograms made synchronously during the activity provide further information regarding the characteristics of the action potential, such as its amplitude, rate and duration of activity. In this case, at least, the simpler coaxial electrode is not feasible. Satisfactory electrical stimulation for identity of the muscle is not possible with this applicator, and it is felt that its rigid structure inhibits normal activity of the muscle.

The general concept of phasic activity may not apply in the case of the badly crippled child. It is well known that such an individual with chronic poliomyelitis, for instance, will make use of practically any available muscle that may assist him. Phasic activity, in general, applies to the normal or only moderately crippled patient. This brings out the necessity of preoperative as well as postoperative studies in order to assess the value of tendon transfers.

There can be no doubt that preoperative electrical studies add to our knowledge of the physiological state of the muscle and may be of value in designing the procedure best adaptable to the patient's problem. From preoperative studies, we can also establish the existing phasic activity and general quality of the muscle to be transferred. Even normal subjects may vary in phasic activity (Fig. 91). For example, the semitendinosus muscle often follows the general pattern of the swing-phase hamstrings. However, many normal individuals demonstrate rather extensive stance-phase activity. The phasic activity of the iliopsoas muscle in particular may differ in the polio patient from the normal and this has been found in several of our subjects (Figs. 91, 129). As previously stated, the iliopsoas is virtually silent during locomotion in the normal subject.

The function of the iliopsoas muscle in the normal has been studied exhaustively using the 16 mm sound film with synchronized sound track in several normal individuals. We have paid particular attention to

1. Activity during flexion.
2. Activity during abduction.
3. Internal rotation.
4. External rotation.
5. Phasic activity during locomotion.

The oscillogram, as anticipated, demonstrated abundant activity during flexion. There is, however, considerable activity on abduction. This activity on abduction usually begins after the gluteus mechanism has succeeded in abducting the extremity until the tendon of the iliopsoas muscle has moved lateral to the center of the head of the femur. Therefore, on abduction, the iliopsoas may be suppressed until it is in a position to be useful. Repeated oscillograms and 16-mm sound films have demonstrated only scant activity on internal and external rotation. During normal walking, the muscle is surprisingly silent except for the first step. The picture changes entirely as the subject ascends the stairs when the muscle becomes useful and is used extensively to flex the thigh.

Complete conversion of the iliopsoas muscle from swing to stance was observed in all of the subjects. In Figure 129, preoperative and two postoperative studies on one patient are demonstrated. In this case, the transfer was done through a perforation in the ilium (Fig. 128). Twenty-one months postoperative, the muscle had become a well-timed stance-phase muscle with a gluteus medius type of action. Note that on voluntary abduction (Fig. 129), the muscle was at some times silent and at others vigorously active.

THE ADDUCTORS

Another system of muscles of considerable size which critically influences the movement of the lower extremity is the adductor group. In order best to understand the complicated action of this system of musculature, one must consider the axis of motion of the lower extremity and the innervation of the adductor group.

The axis of the lower extremity is complex and to a great extent explains the rather strange behavior of the adductor system. As demonstrated in Figures 38, 39 and 40, the axis of the lower extremity extends from the center of the head of the femur to the point of contact of the lower extremity with the walking surface. If the lower extremity is not in contact with the walking surface, this axis passes from the center of the head of the femur to the center of gravity of the dependent lower extremity. The great bulk of the adductor system, particularly the adductor magnus, though inserting posteriorly on the linea aspera, except for the hamstring portion, passes anterior under most circumstances to the axis of the lower extremity. It is quite possible that a minor part of the adductor magnus, in some positions, passes posterior to the axis and contributes to external rotation. The adductor longus, however, arising more anteriorly along the inferior ramus of the pubis close to the symphysis and passing downward, must be exclusively an internal rotator. As discussed in Chapter III, these muscles, the adductor longus and adductor magnus, when stimulated with the patient standing or supine, produce weak external rotation. When the patient is asked voluntarily to internally and externally rotate the lower extremity, however, it is clear that abundant electrical activity is produced by internal rotation alone.

The adductor longus is simpler to analyze than the more complex adductor magnus. It has a small area of origin. It is innervated exclusively by the obturator nerve. During locomotion, the phasic activity of this muscle is exclusively stance (Fig. 91).

The adductor magnus, on the other hand, has both obturator and sciatic innervation. Due presumably to its obturator innervation, during locomotion it demonstrates a burst of stance-phase activity. There is, however, a consistent burst of swing-phase activity. This must be due to its sciatic portion, which might be compared to a hamstring muscle. This portion, inserting on the medial femoral epicondyle, is in fact philogenetically related to the hamstrings. It is thought to have been continuous in ancestral forms with the medial collateral ligament which once had no attachment at the medial femoral epicondyle, attaching only distal to the knee joint.

This analysis of the action of the large adductor system may be of some importance in explaining the scissors gait. It may also be of value in assessing the possible results of adductor tenotomy and/or obturator neurectomy. Perhaps we can better predict what the loss of the muscle through operative procedures might be if we understand fully its normal action.

The adductors appear to be the most important internal rotators of the hip.

BICEPS FEMORIS

There is a difference in the action of the long and short heads of the biceps femoris, as demonstrated easily by the sound motion picture with synchronized sound track and oscillograms. The long head has been found to be silent in normal individuals during knee flexion. It is found to be very active on external rotation of the knee. The short head, on the other hand, is electrically active only on flexion. Both heads are active on forceful flexion or external rotation of the knee with the knee flexed.

The phasic activity of the two heads differs. Electrical activity of the two heads is scant in the normal subject during walking and is sometimes virtually absent.

A good example of the ability of the biceps femoris to externally rotate the lower extremity is often seen in the postpoliomyelitis patient where there is paralysis of the medial hamstrings in the presence of a strong biceps femoris muscle.

QUADRICEPS

Four-channel studies have been done on this muscle group. This consists of simultaneous recording by the Visicorder (galvanometer) of the four major components of the quadriceps mechanism. Virtual identity of behavior of these muscles on extension was demonstrated. There is some difference, however, in the case of the rectus femoris. Here the length of the muscle is influenced by the position of the patient. If the patient is sitting, the muscle is shorter than if the patient is supine. Records of the temporal behavior of each of the components of the quadriceps is virtually identical, as the knee is extended when the subject is sitting. In the case of the rectus femoris, however, the bulk of electrical activity is recorded during the latter part of extension, since the muscle's length is less with the subject in the sitting position. This particular study was done by single-channel recordings. Many contractions, however, were observed and typical contractions recorded. Multichannel work which has been done is not shown here. These simultaneous records of the four major components of the quadriceps muscle similarly showed the delayed activity of the rectus femoris with the subject in the sitting position. The subject was then placed in a supine position. The rectus femoris adopted a pattern of behavior quite similar to the vasti.

The phasic activity of each component of the quadriceps muscle is demonstrated in Figure 92.

It is interesting and of great value to relate angular displacements to the electrical activity that the angle at the knee in the sagittal plane of certain muscles. In Figure 91A it is observed that the angle at the knee in the sagittal plane is close to 180 degrees at the onset of the stance phase of walking. A considerable degree of flexion takes place during the first part of the stance phase of walking, since it is undesirable to raise the center of gravity of the body more than necessary. This increases work. Therefore the knee bends slightly and at just this time the quadriceps demonstrates electrical activity. This is virtually the only time that the quadriceps demonstrates electrical activity during normal walking. Flexion of the knee during the stance phase of walking is a superb example of the ability of the human subject to conserve energy wherever possible.[153]

Chapter VI

THE MUSCLES OF THE FOREARM AND HAND

Methods for recording from several muscles simultaneously were described in Chapter II. In Figure 16, a method is shown for recording from four muscles. The recordings were synchronized with a cinematic record of the motion of the body part involved. The sound track of the electrical activity of one muscle at a time was recorded. The activity of the four muscles was printed in synchrony on the recording medium of the multichannel galvanometer (Visicorder). Angular displacement of the parts of the upper extremity was also displayed on the same medium (see Figs. 17 and 18, which represent an assembly of all of these data).

It was considered that a simultaneous synchronized display of the oscillograms of several muscles accompanied by a direct visual display of the activity would be ideal. It was considered essential that the method be a single process producing high fidelity, automatically synchronized recordings of (a) a complete, uninterrupted series of action potentials of several —as many as six or eight—related muscles and (b) the motion of the body or body part with which these muscles are concerned.

This chapter concerns such a direct, single process and some of the multichannel data recorded from a number of muscle groups of the upper extremity. The method is readily adapted for the study of the function of the muscles of the shoulder, hip or knee, or for locomotion studies. The method has also proved ideal for recording the electrical activity of six interrelated muscles of a single digit as a series of prescribed motions is performed. The resulting display makes possible rapid and convenient analysis of the six oscillograms and the motion of the digit, all recorded in automatic synchrony. Thus it became possible to obtain recordings of many combinations of extrinsic and intrinsic muscles of the hand and the muscles involved in wrist motion and stabilization. Since the tests could be carried out rapidly, with little discomfort or inconvenience to the subject, a great many records were obtained, as shown in Figures 133 to 156. The material was also studied extensively as a moving picture. For this, a variable-speed projector was used, with which it was possible to view the activity and associated oscillograms at speeds as slow as 1 to 2 frames/sec.

METHOD

A recording instrument that will make a complete record of the muscle action potentials in synchrony with the motion of the body part is the dual-lens motion picture camera with rotary prism-shutter. This shutter makes possible moving pictures without interruption of the motion of the film. The *objective lens* is used to record movement. Eight rotating prisms direct consecutive images to their proper position on the film, which is in constant motion (Fig. 132). The rotary prism-shutter employs normally a 1/2.5 focal plane shutter disk. It is convenient to change this ratio of exposure time to 1/10 or 1/20 by changing the segmented shutter disk. This results in a sharper photographic image and is necessary for camera operation at slow speed. The instrument, when operated at very slow speeds, such as 4, 5 or 6 frames per

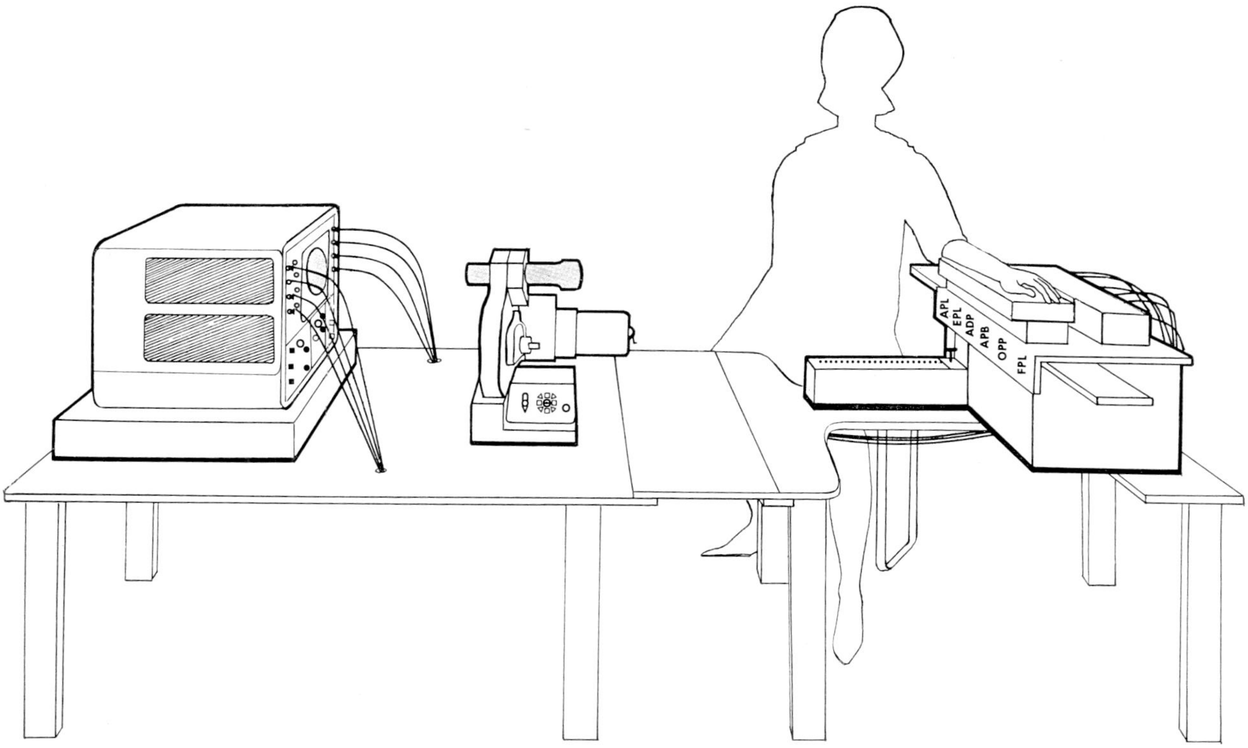

Figure 131. Subject in position for recording, showing optical bench, dual-lens camera and multichannel oscilloscope.

second, compresses the data into dimensions such that movements and action potentials may be displayed in their entirety in chart form, as described below.

The second lens, the *oscilloscope lens,* facing the opposite direction, has no shutter. Taking advantage of the constant motion of the 16-mm film, the *oscillo*-optical system directs the vertical oscillations of as many as six or eight oscilloscope channels from the oscilloscope screen to the same frame that is also being exposed by the *objective lens.* The constant motion of the film provides the horizontal or time dimension for the oscillogram ordinarily provided by the sweep system of the oscilloscope. The oscilloscope sweep system is turned off; only vertical oscillations are observed on the oscilloscope screen during muscular activity.

Action potentials and movement are thus recorded in synchrony. The film motion is horizontal, perpendicular to the vertical oscillations displayed on the oscilloscope screen. These are produced by the action potentials of the muscles under study and appear on the final film as oscillograms with a time base equal to the film speed.

For graphic demonstrations of single movements, one may take advantage of the slow camera speeds. For example, at 6 frames/sec, all of the electrical activity occurring during a complete motion may be recorded and conveniently grouped, due to the reduced time base. Such records are more easily compared to one another than those spread out over a time base four times greater (24 frames/sec). The images of the moving part, much fewer in number, are kept sharp by changing the focal plane shutter disk from one of 1/2.5 exposure ratio to one of 1/10 or 1/20. A direct print may be made from records of complete movements recorded at 6 frames/sec. These prints are of convenient size. Complete movements are often contained on as little as four or five inches of 16 mm film. The series of action potentials, grouped closely together, approach highly accurate, totalized series that are readily com-

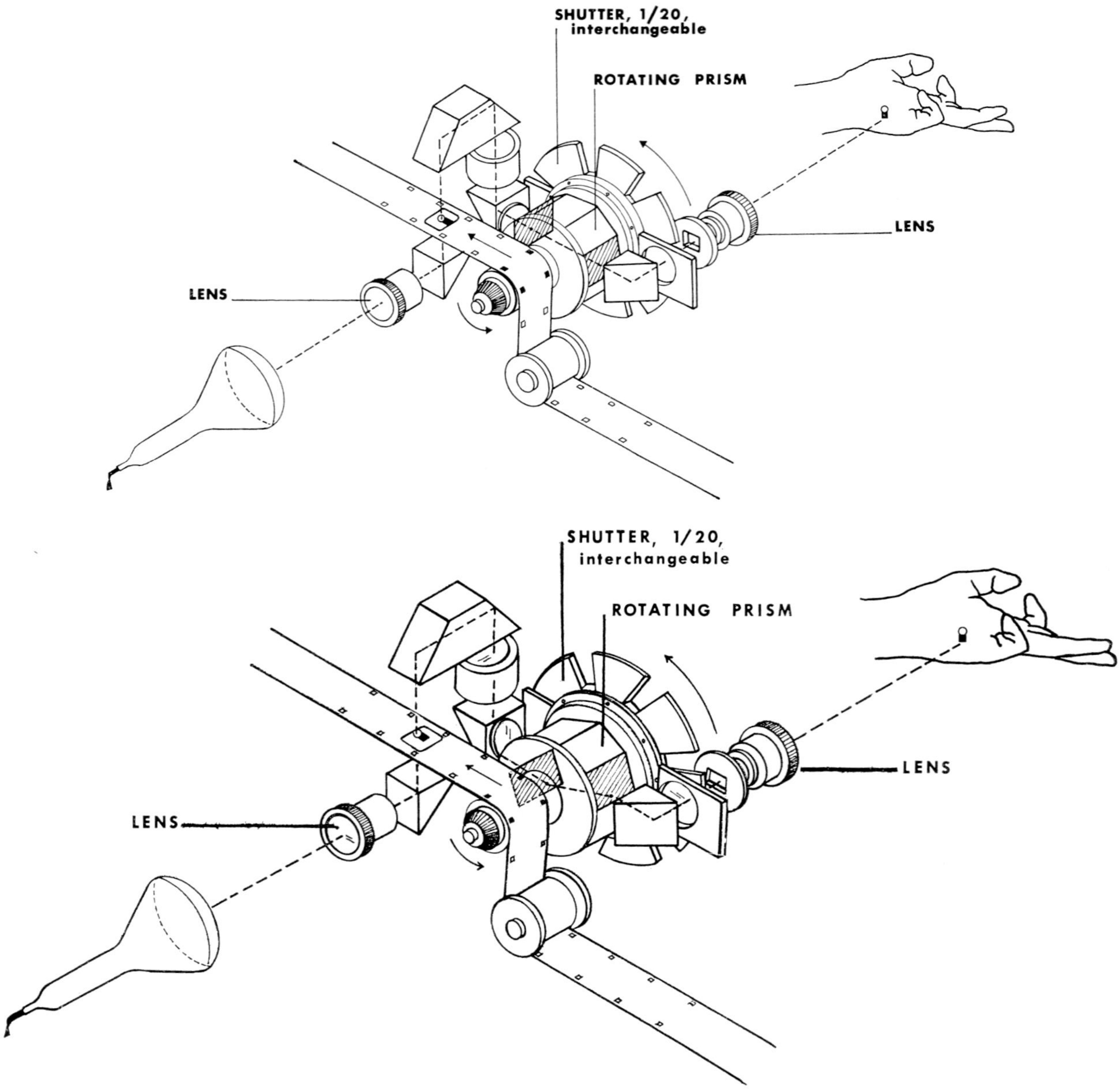

Figure132. Mechanisms of dual-lens, rotating prism-shutter camera with uninterrupted transportation of film (Hycam).

pared for relative amount of activity, time of onset and termination of electrical activity.

The method outlined above produces in a single operation moving pictures and chart records of movements and oscillograms of several muscles in synchrony. All of the data, in one process, are placed on a single recording medium. The moving picture taken at conventional speed of operation is ready for complete analysis. Material taken at very slow speeds, already in chart form, approaches in convenience of evaluation the "treated" expressions of the oscillogram: the averaged curve, the action potential total count or the analog curve of the action potential count. Processes of averaging, integrating, or counting greatly complicate the recording procedure, particularly when as many as six channels are to be so treated. Furthermore, each of these procedures, in changing the data into a different, simpler form, introduces its own compromises. Their products are final forms of data in which something is lost in the process of simplification.

Although no form of the record of muscle electrical activity is a direct indicator of tension except under specified conditions,[36,41] one may legitimately compare amplitude and action potential counts of various contractions to those produced during maximal contractions. When it has been demonstrated that the maximal contractions are comparable, the amplitudes of individual channels may then be compared among themselves. Differences among various series of action potentials are readily apparent on records made at slow speeds, e.g. 5 frames/sec. These useful comparisons are meaningful only if the data are consistently recorded according to several specified conditions:

1. Orientation of electrodes (longitudinal).
2. Plastic-coated copper wire electrodes of diameter 75μ.
3. Constant distance between electrodes—1 cm.
4. Recording frequency range; in these studies a range of 40 Hz to 10 kHz was used.
5. Each channel calibrated and of equal sensitivity.
6. Known constant speed of photographic film during recording.
7. Interference ("noise") at acceptable level.
8. Linear differential input amplifiers.
9. Proof of position of electrodes in individual muscle, muscle component or muscle digitation by electrical stimulation.
10. Comparisons among channels confined to the data of each experiment.

With electrodes properly placed in each muscle, maximal effort of contraction should reach the same level of electrical activity in each channel or for each muscle. In reviewing the recordings, one may readily judge the percentage of the muscle's motor units that have been activated by comparing the level of activity to the known maximum, as established for each muscle prior to the test. Known reservations regarding the relation of total electrical output to tension are kept in mind. We must also remember in comparing the electrical output of various muscles that to a limited extent, for isometric contractions, cross-sectional area is proportional to the ability of the muscles to develop tension.[144] Thus the level of electrical output of each muscle should be corrected by its relative cross section, since in all cases, the electrode tips conduct signals from a small volume, the extent of which is well within the confines of the borders of the muscle. Data are available for making approximate correction for each muscle's cross-sectional area.[85] Although application of the correction for difference in cross-sectional area from muscle to muscle would be misleading, awareness of its existence serves to prevent direct comparisons of channels in terms of tension.

The totalized chemical events, the aggregate of transmembrane passage of ions, are reflected directly by the analog of the muscle's electrical output. This may be a satisfactory, if not superior, indication of the muscle's participation.

Summaries of data in charts obtained in studies of thumb, middle and index fingers, and of components of m. flexor digitorum sublimis are presented. Final summaries are presented in Tables IV, V and VI. Additional data are reported showing the effect of motor nerve block of m. lumbricalis No. 1 (Figs. 144, 145, 146, 155). The activity of some of the wrist extensors and flexors is shown in Figure 156.

RESULTS

Forty-eight series of studies were completed. The usual series consisted of a six-channel, or six-muscle, recording. The series were distributed as follows:

Muscles of the thumb	6 studies
Muscles of the middle finger	6 studies
Muscles of the index finger	10 studies
Components of m. flexor digitorum sublimis	10 studies
Muscles of the forearm	6 studies
Muscles with tendon transfer	10 studies
Total multichannel studies	48 studies

The above series consisted of the following groups of muscles (abbreviations used are given.) Charts summarizing typical records of data obtained in studies of the above-described muscle groups were prepared and may be conveniently referred to as follows:

THUMB

FPL Flexor pollicis longus
OPP Opponens pollicis
APB Abductor pollicis brevis
ADD Adductor pollicis
EPL Extensor pollicis longus
APL Abductor pollicis longus
Alternates included on two occasions:
FPB Flexor pollicis brevis
EPB Extensor pollicis brevis

INDEX FINGER

EDC Extensor digitorum communis, index finger digitation
EIP Extensor indicis proprius
DI-1 1st dorsal interosseous
LUMB 1st lumbrical
FDP Flexor digitorum profundus, index finger digitation
FDS Flexor digitorum sublimis, index finger component (referred to as S2 in study of flexor digitorum sublimis, below)
Alternate included on one occasion:
PL Palmaris longus

MIDDLE FINGER

LUMB 2nd lumbrical
DI-2 2nd dorsal interosseous
DI-3 3rd dorsal interosseous
EDC Extensor digitorum communis, middle finger digitation
FDP Flexor digitorum profundus, middle finger digitation
FDS Flexor digitorum sublimis, middle finger component

COMPONENTS OF M. FLEXOR DIGITORUM SUBLIMIS

Superficial portion
SM Flexor digitorum sublimis, middle finger component
SR Flexor digitorum sublimis, ring finger component
(See Fig. 150.)

PROX 2, 5 Proximal belly of trigastric portion common to index finger and little finger
S 2 Distal belly of trigastric portion, to index finger
S 5 Distal belly of trigastric portion, to little finger

TENDON TRANSFERS

Muscles with tendon transfer (wrist, thumb) Figures 154D, 154E, 154F

MUSCLES OF THE FOREARM

ECRL Extensor carpi radialis longus
ECRB Extensor carpi radialis brevis
ECU Extensor carpi ulnaris
FCR Flexor carpi radialis
FCU Flexor carpi ulnaris

The charts representing the actions of the thumb, middle and index fingers are preceded by a drawing of the individual muscles involved and placement of the electrodes (Figs. 133 (thumb), 147 (middle finger) and 136, 140 (index finger).

The most satisfactory method for evaluating the data is repetitive study of the projected 16 mm film. The charts presented contain representative displays of the data. From the charts, one may readily evaluate the comparative participation of each muscle involved. A means of evaluation of all of the data was also employed. The electrical activity of each muscle during the motions prescribed was evaluated according to a simple scale: F, full activity; M, medium activity; O, zero activity (see Tables IV, V, VI).

Corrections for differences in cross-sectional area were not made. Tables similar to Tables IV, V and VI were also made directly from the displays in Figures 134 through 143, and 148 and 149, using the same basis of evaluation. The results were virtually the same. The figures or the tables may therefore be used interchangeably—the "sample" data of the figures may be considered authenticated.

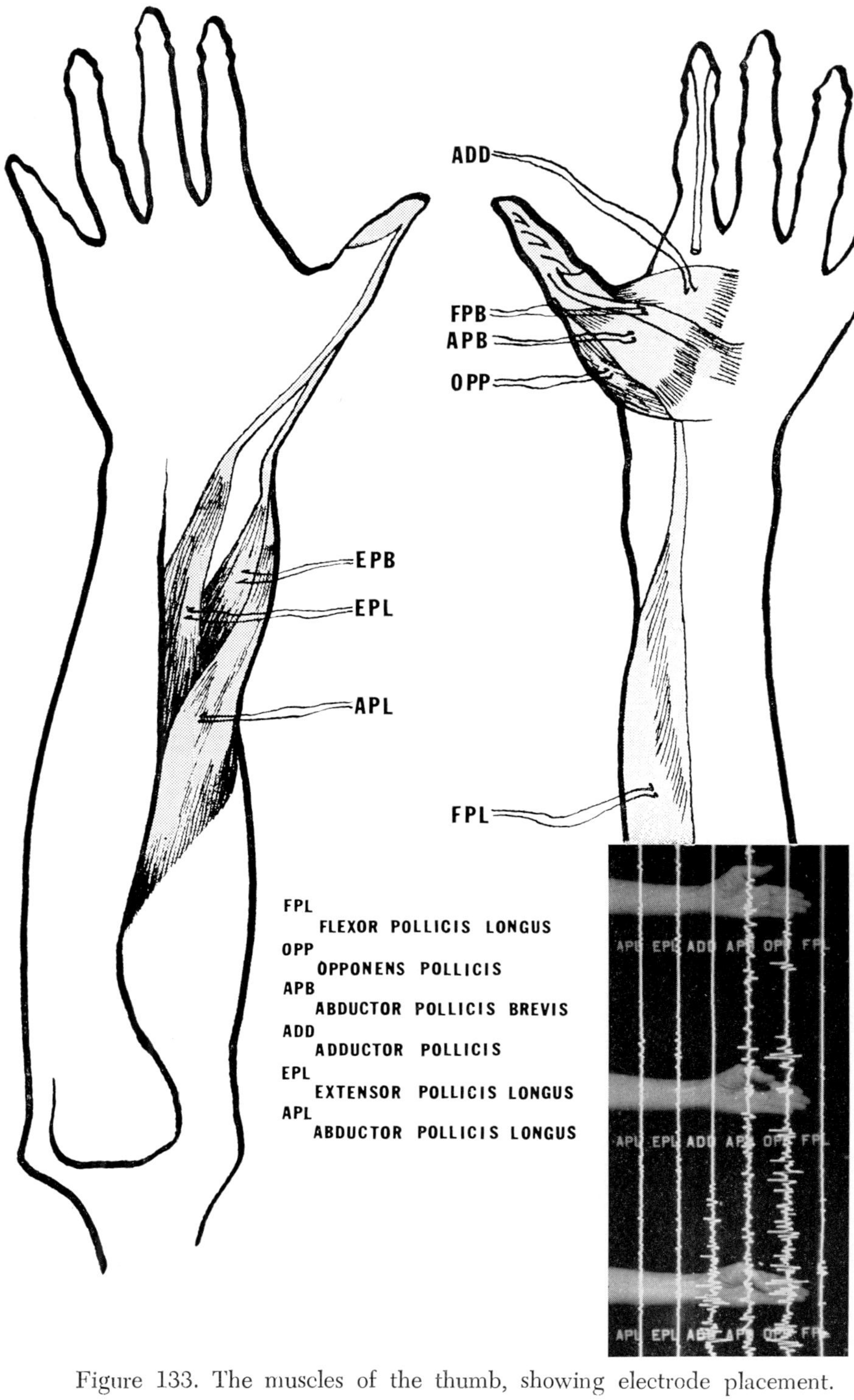

Figure 133. The muscles of the thumb, showing electrode placement.

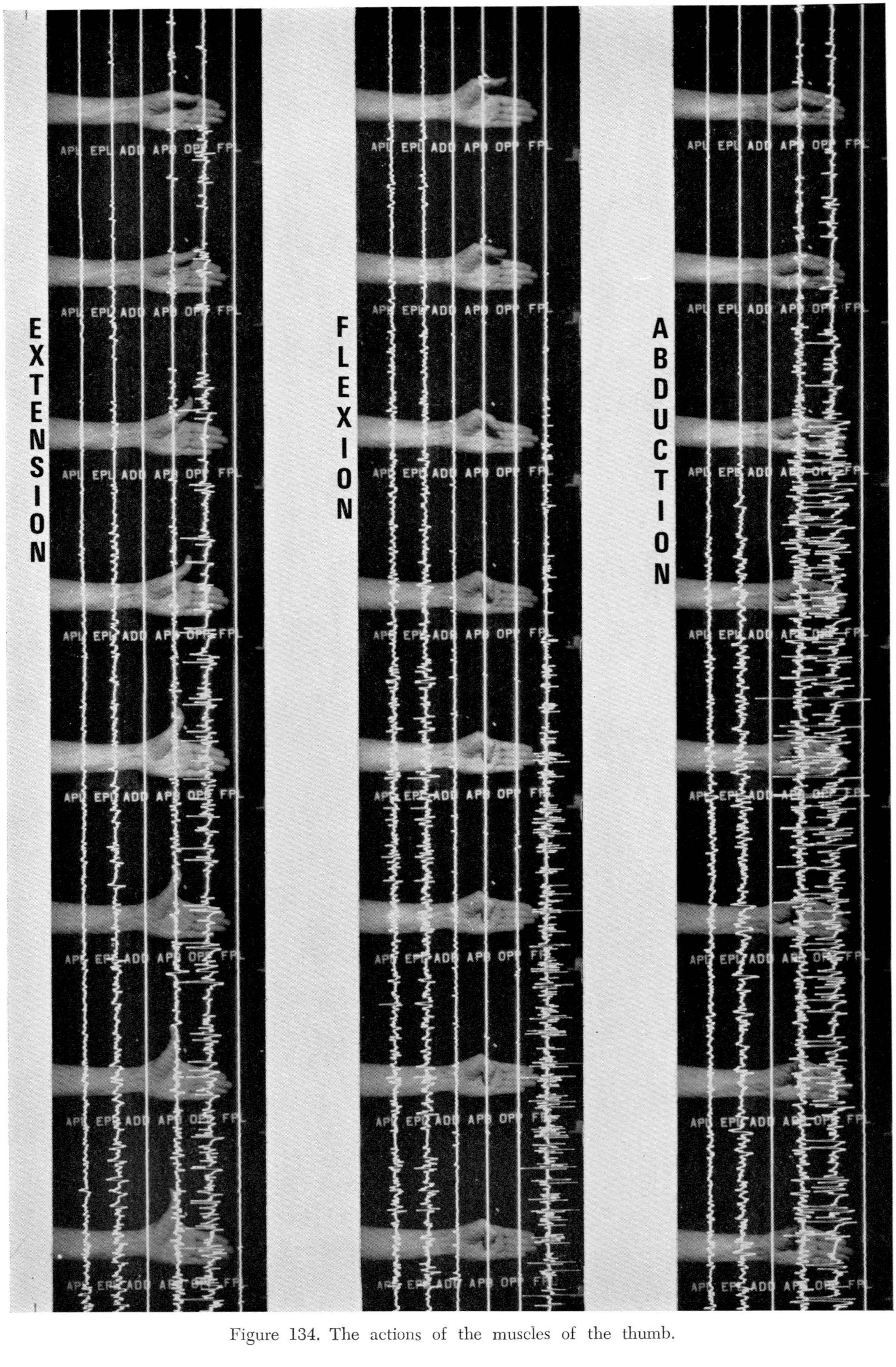

Figure 134. The actions of the muscles of the thumb.

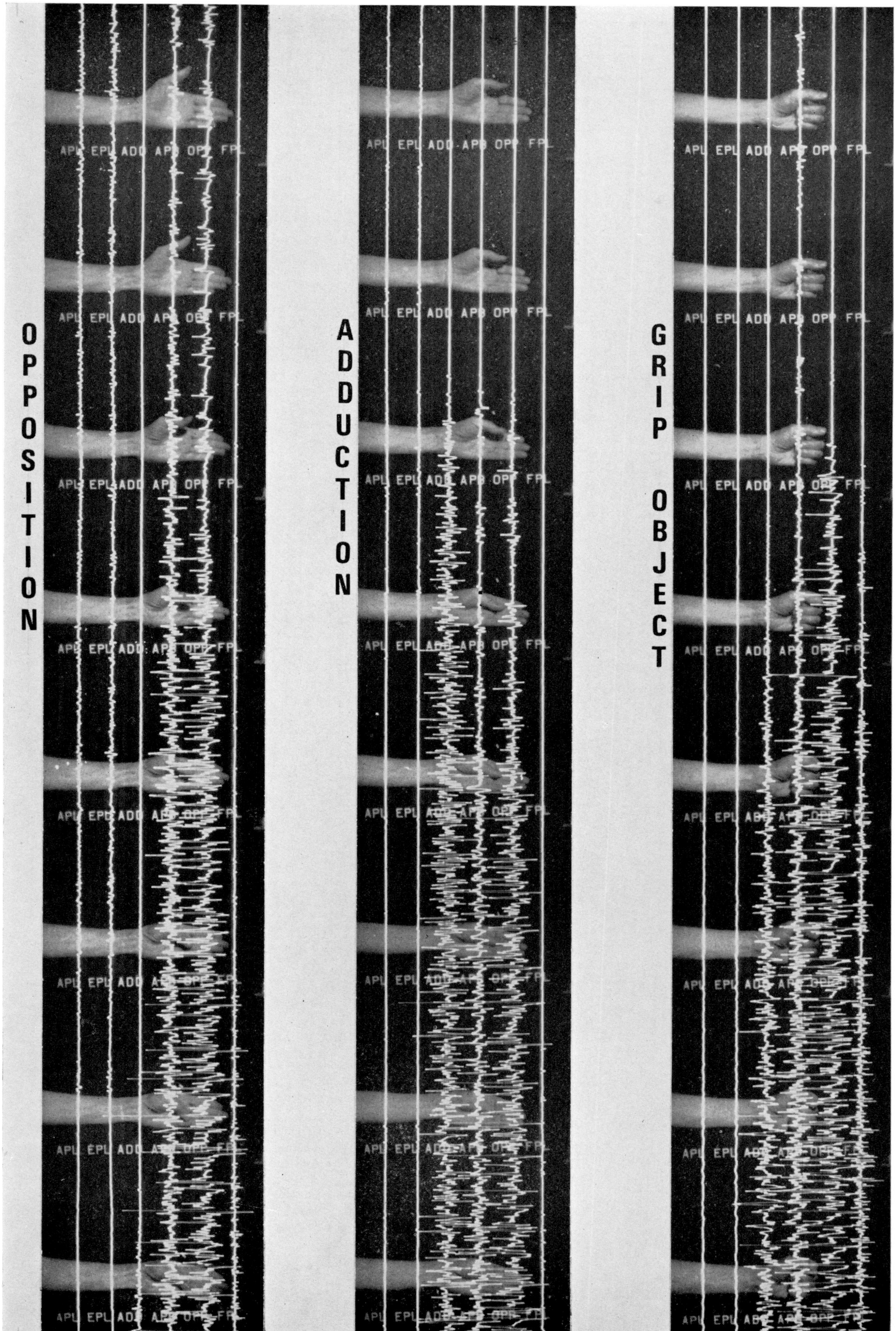

Figure 135. The actions of the muscles of the thumb.

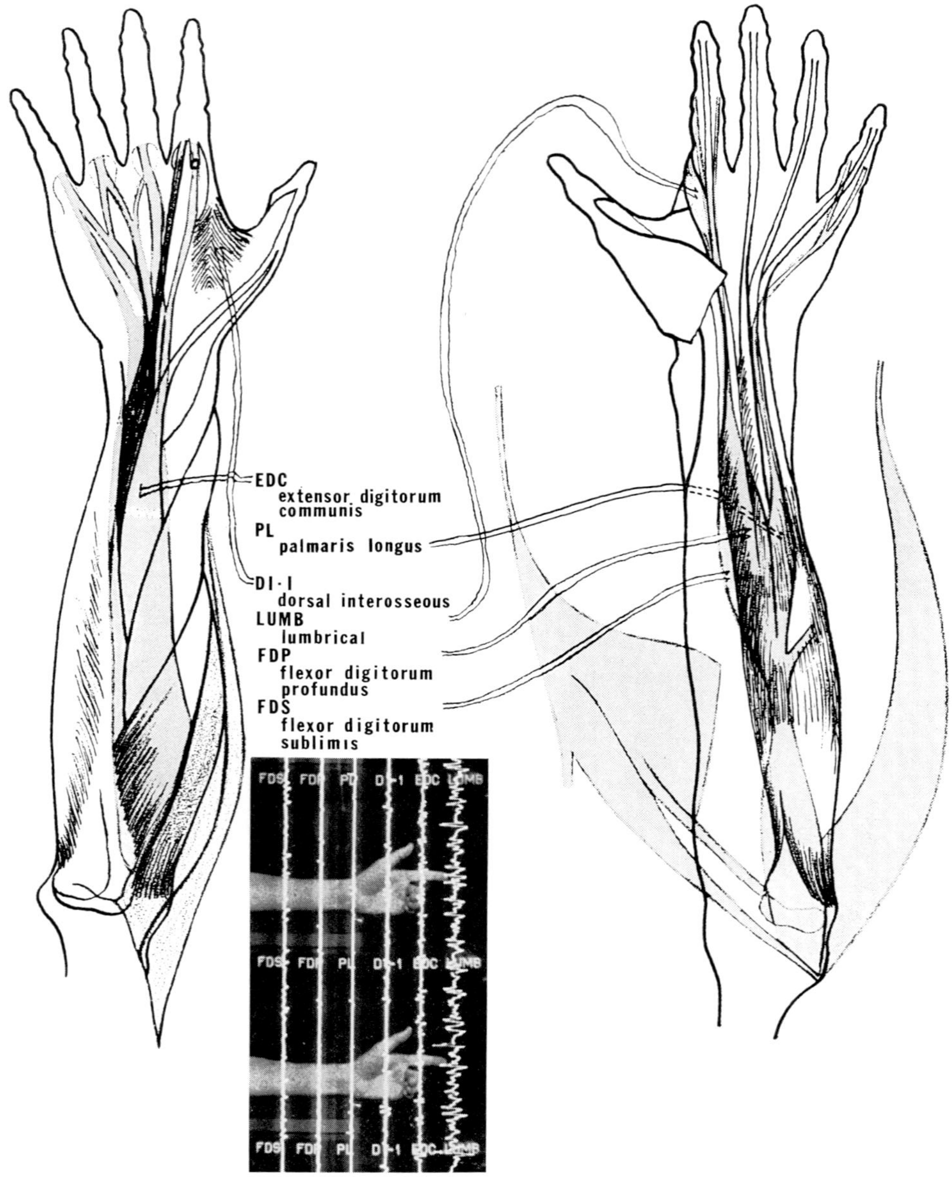

Figure 136. The muscles of the index finger, showing electrode placement; Subject 1.

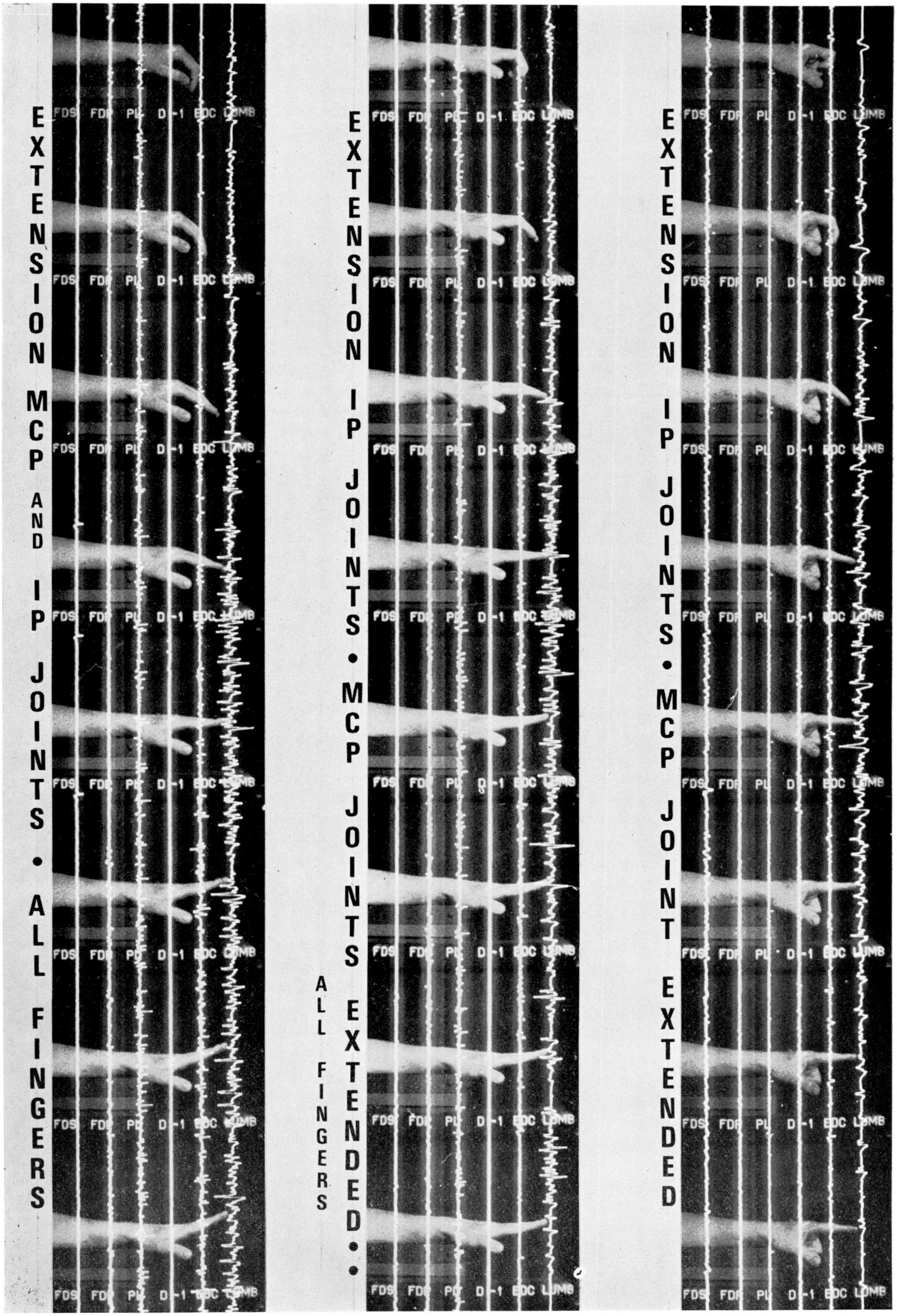

Figure 137. The actions of the muscles of the index finger; Subject 1.

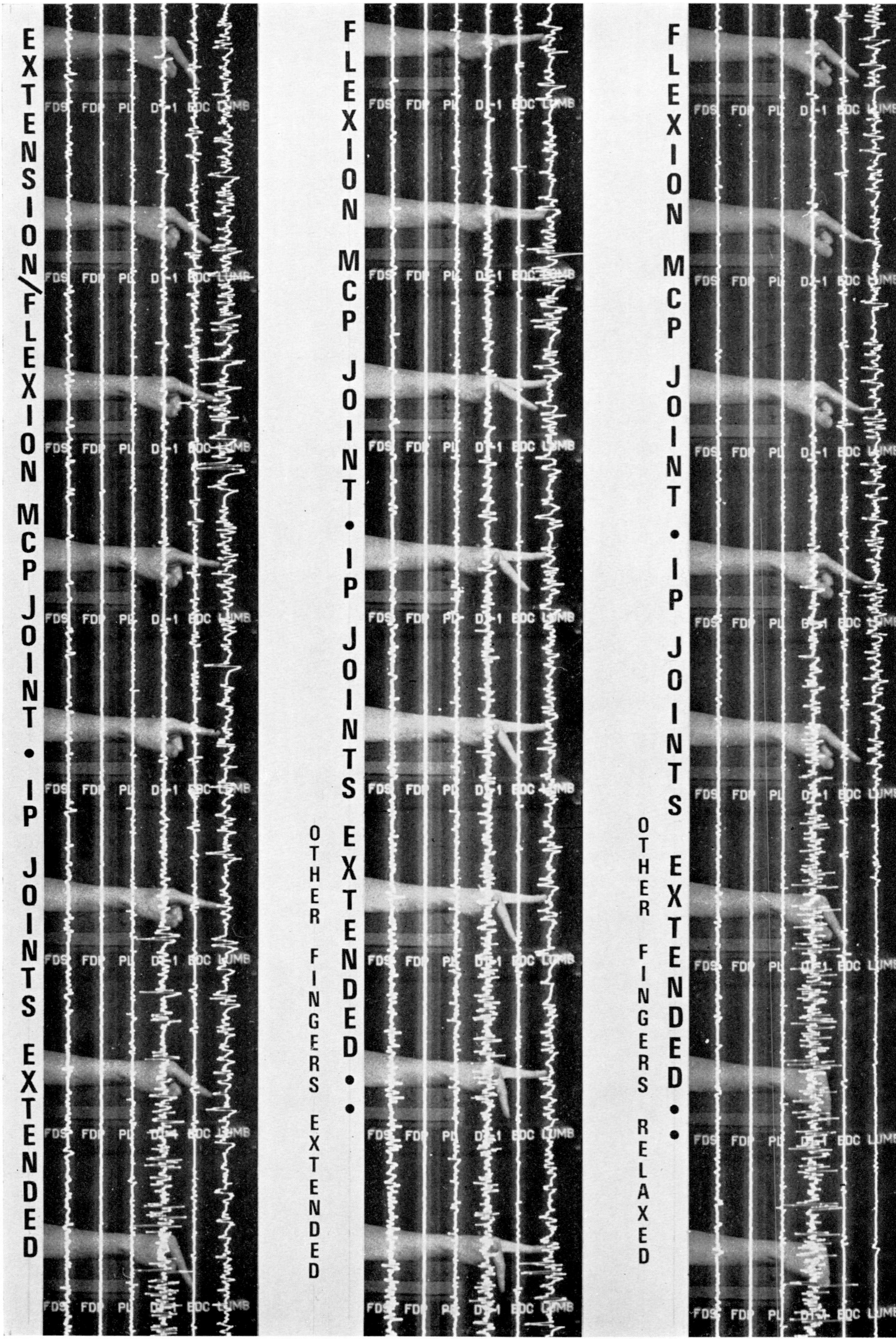

Figure 138. The actions of the muscles of the index finger; Subject 1.

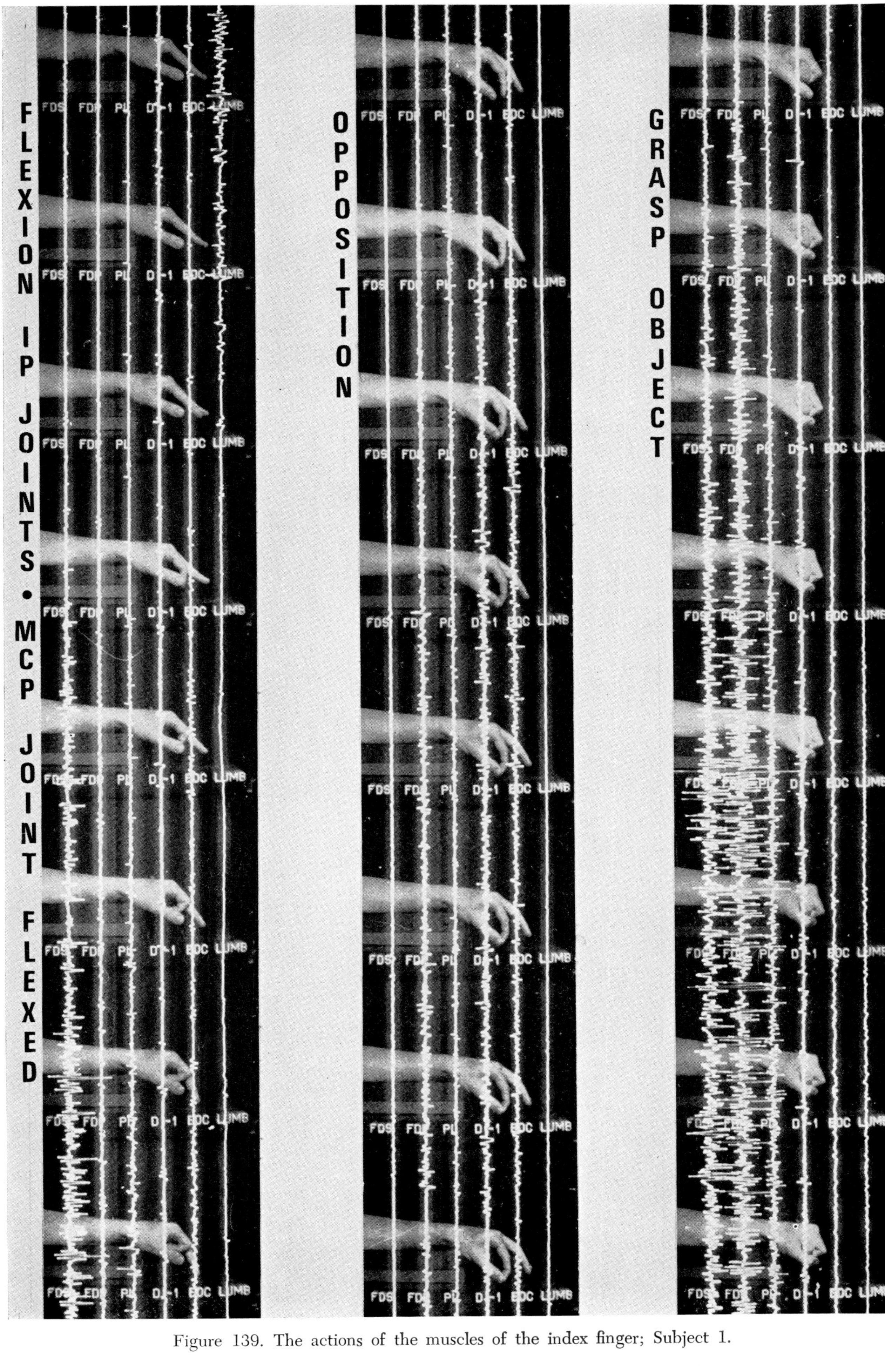

Figure 139. The actions of the muscles of the index finger; Subject 1.

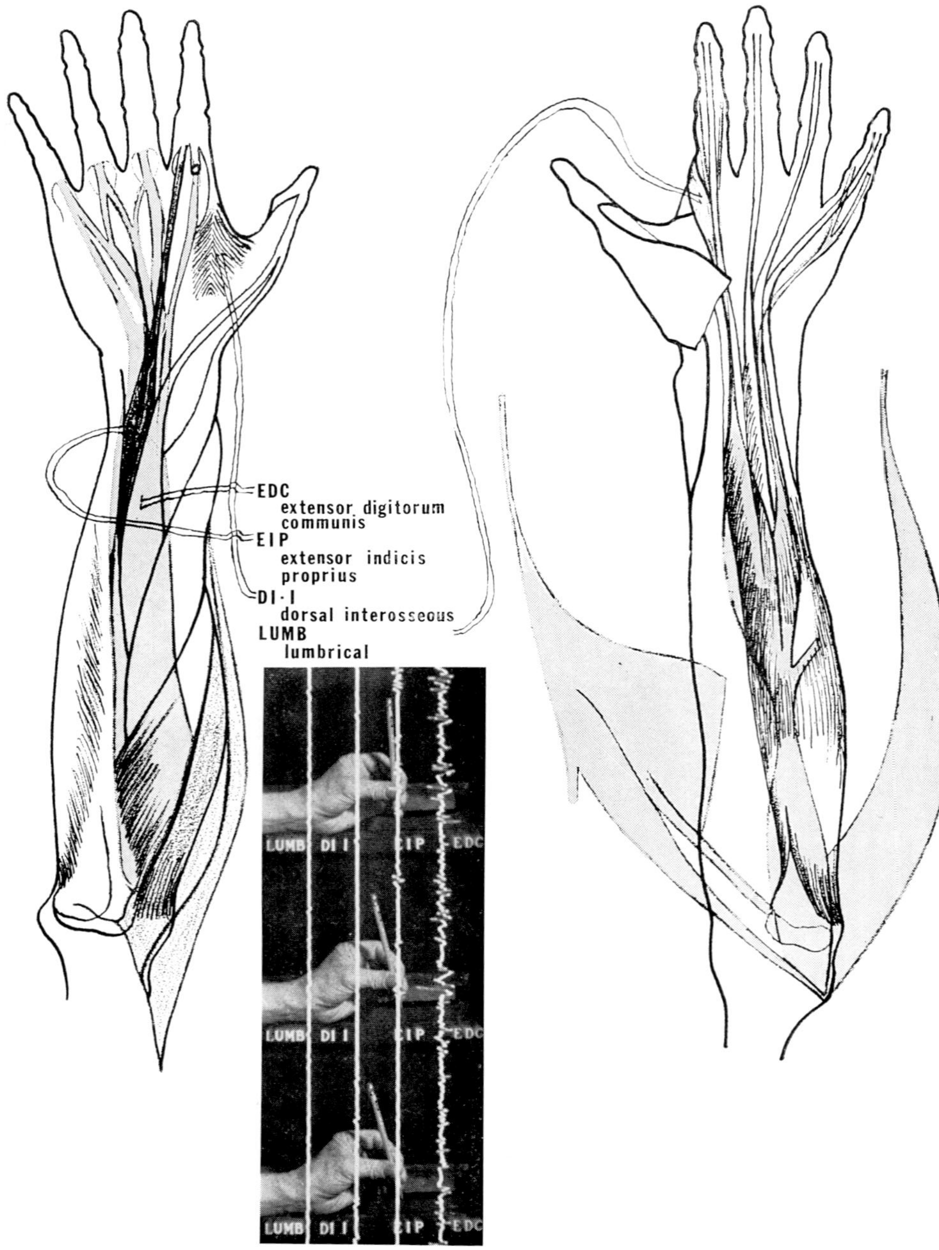

Figure 140. The index finger (Subject 2), showing electrode placement.

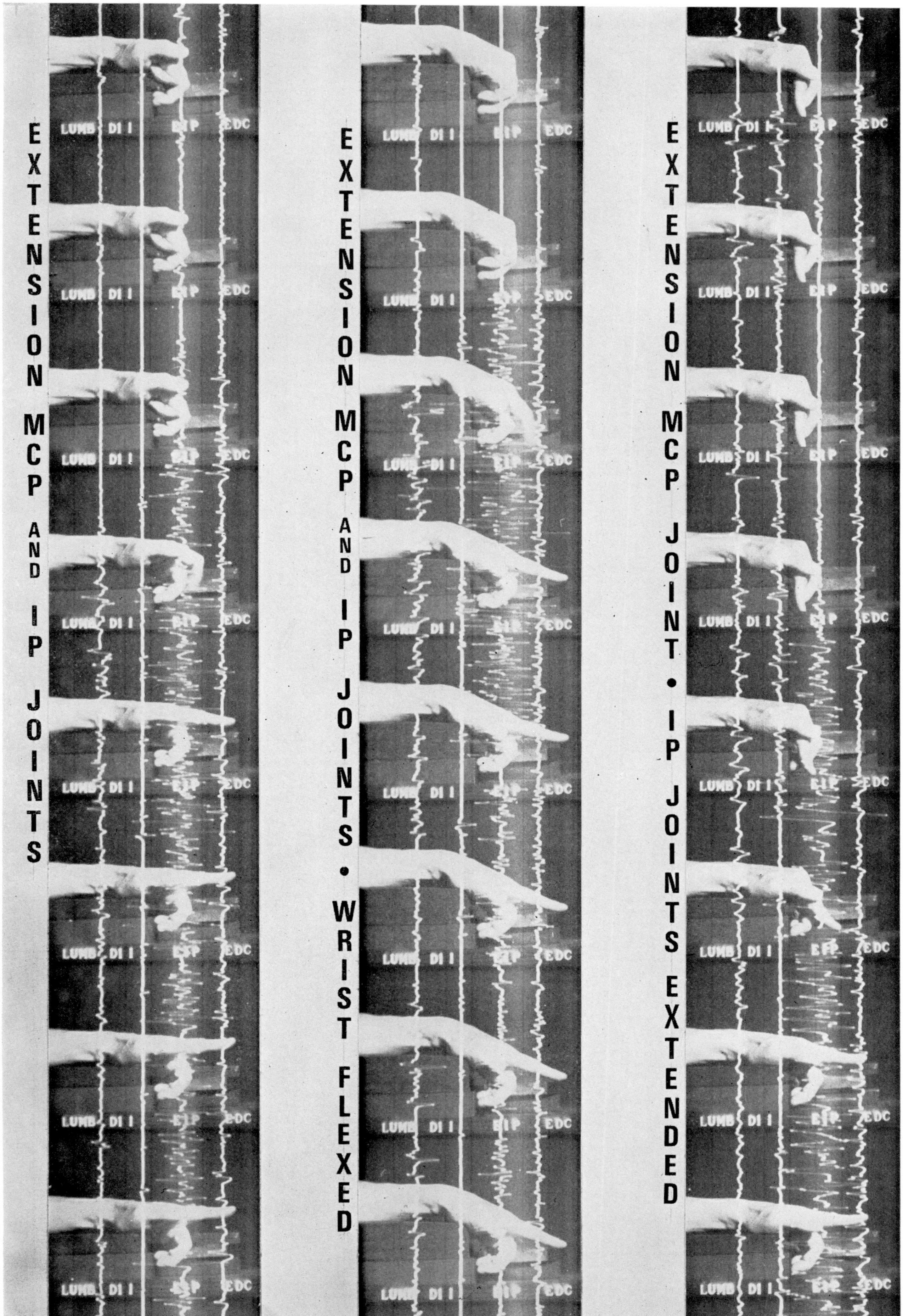

Figure 141. The actions of the muscles of the index finger; Subject 2.

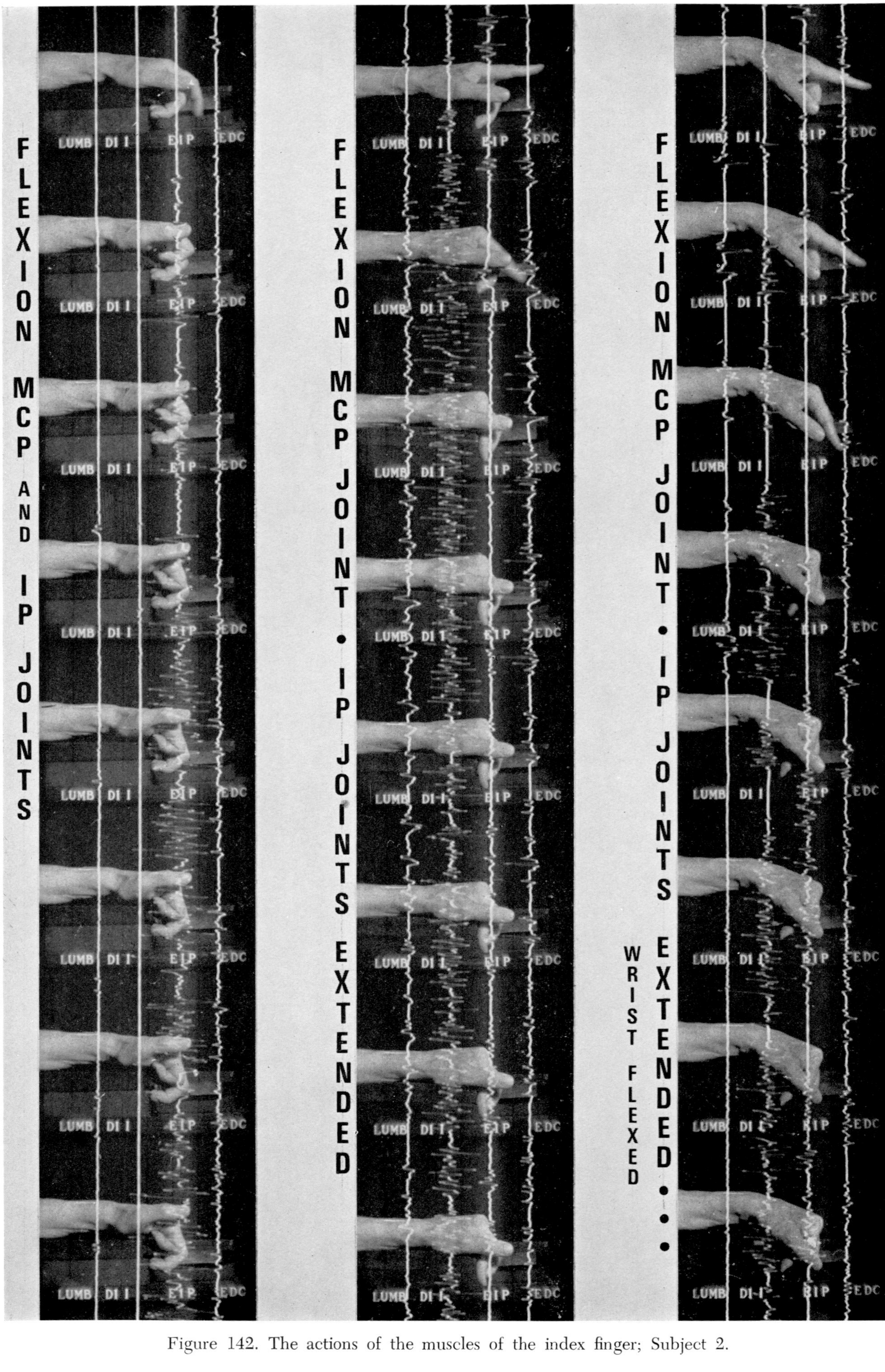

Figure 142. The actions of the muscles of the index finger; Subject 2.

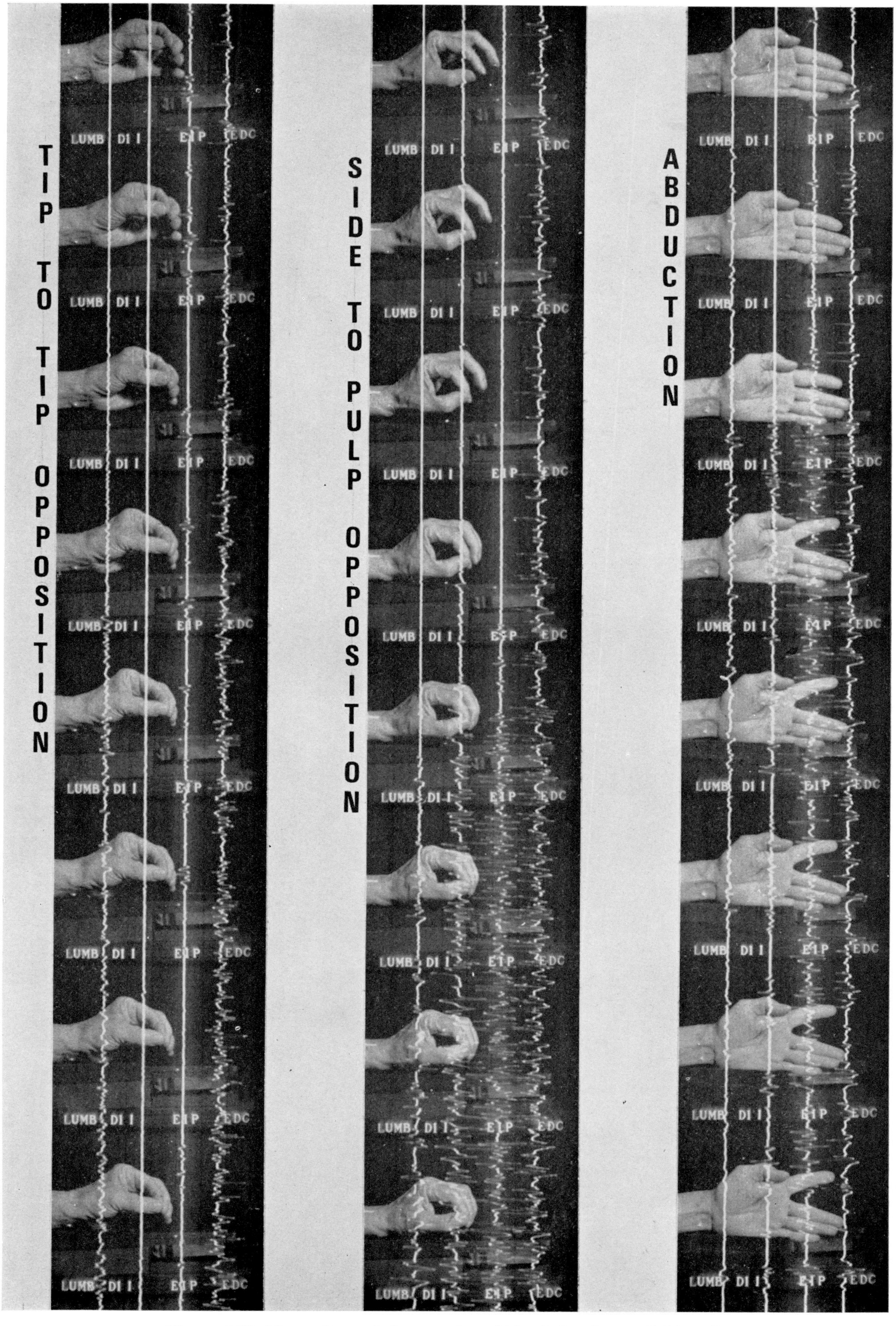

Figure 143. The actions of the muscles of the index finger; Subject 2.

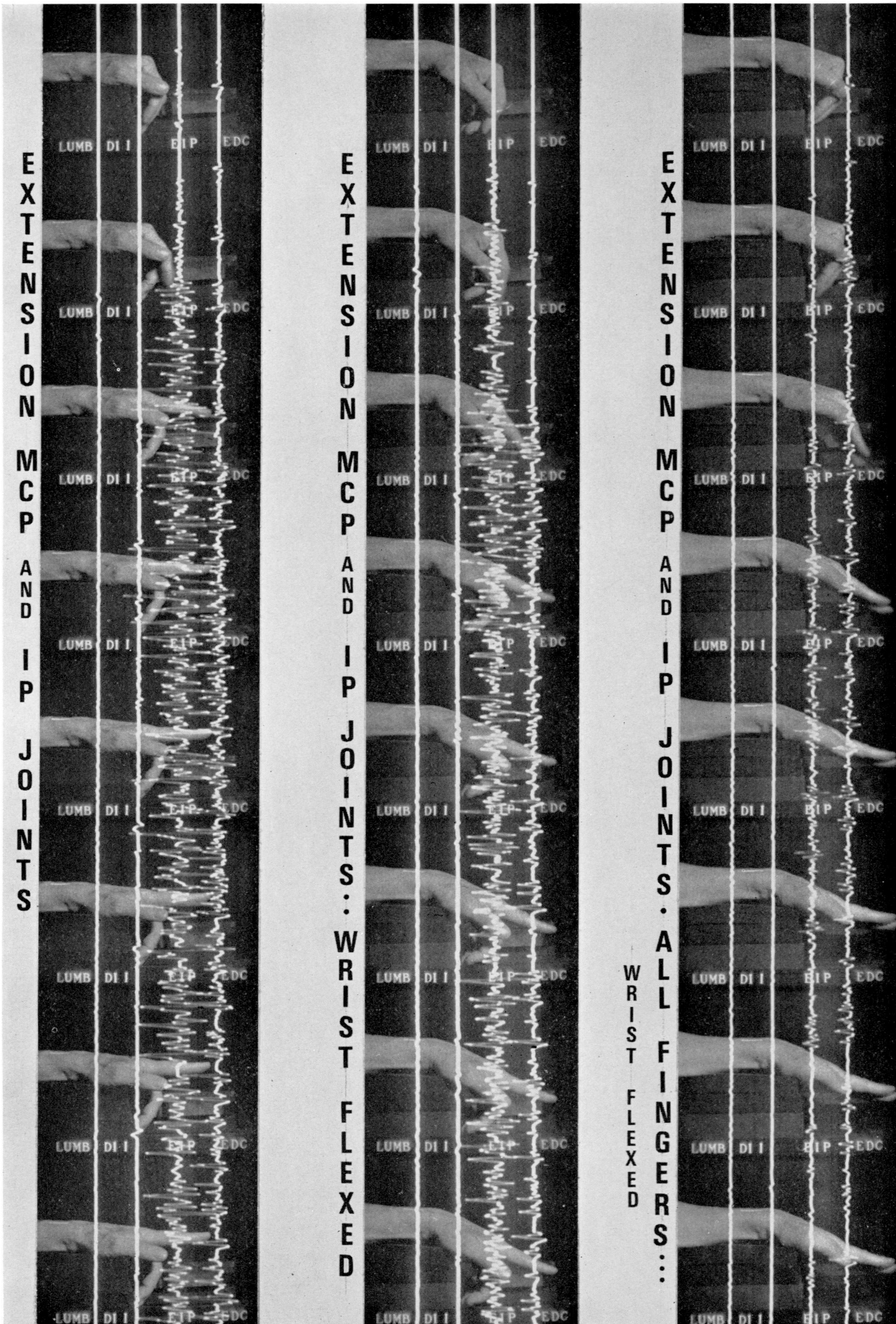

Figure 144. Same study, after xylocaine motor nerve block of m. lumbricalis No. 1.

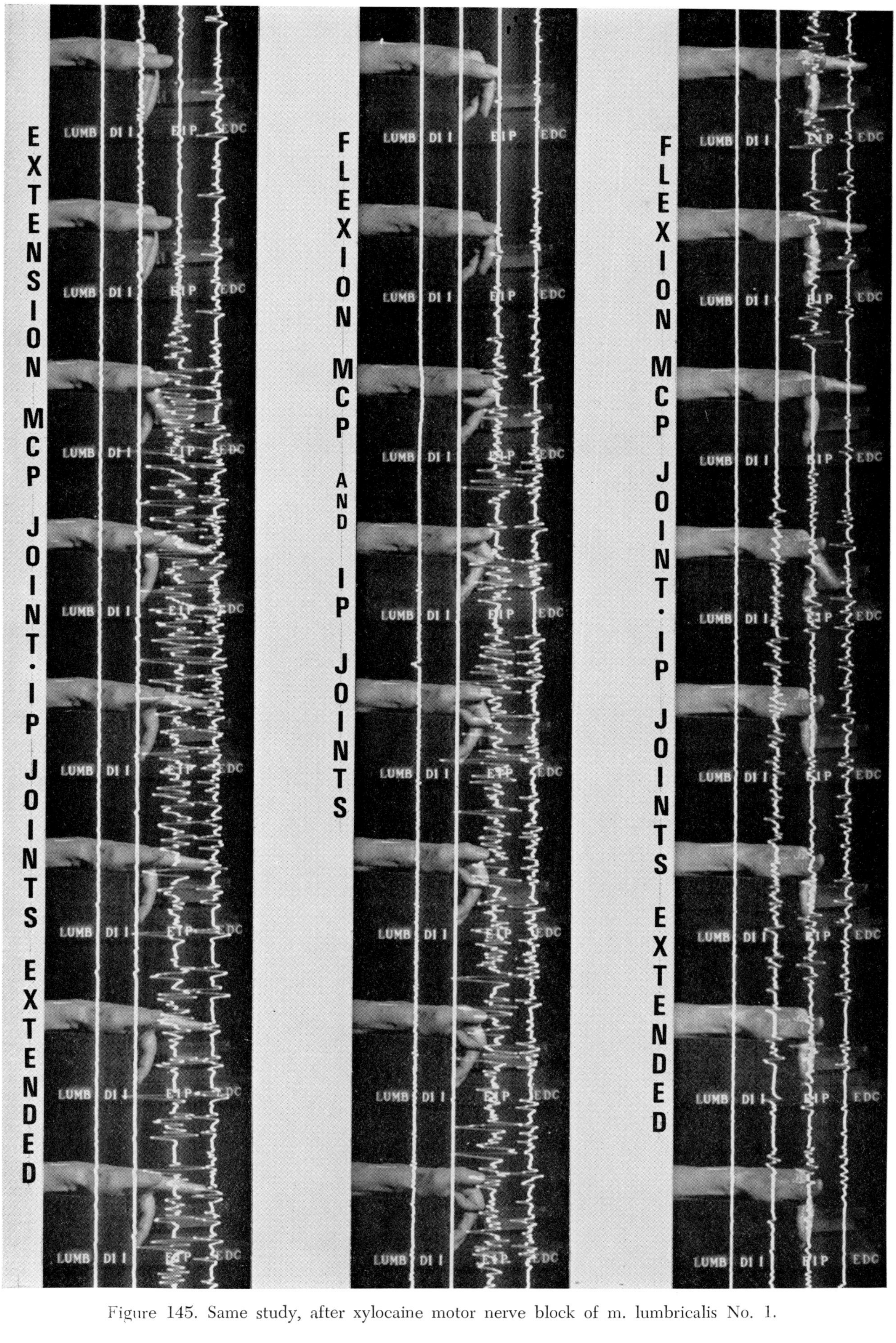

Figure 145. Same study, after xylocaine motor nerve block of m. lumbricalis No. 1.

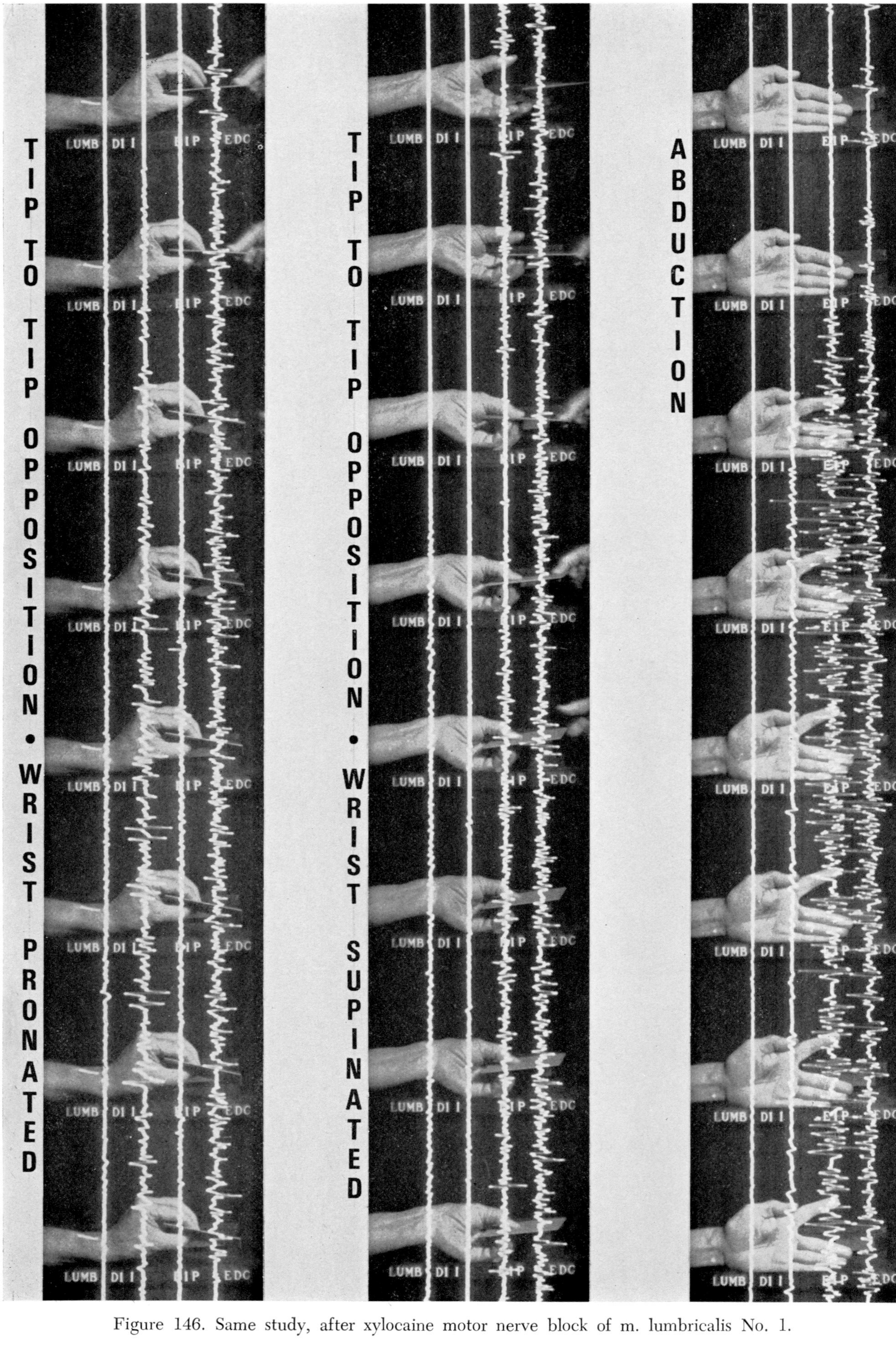

Figure 146. Same study, after xylocaine motor nerve block of m. lumbricalis No. 1.

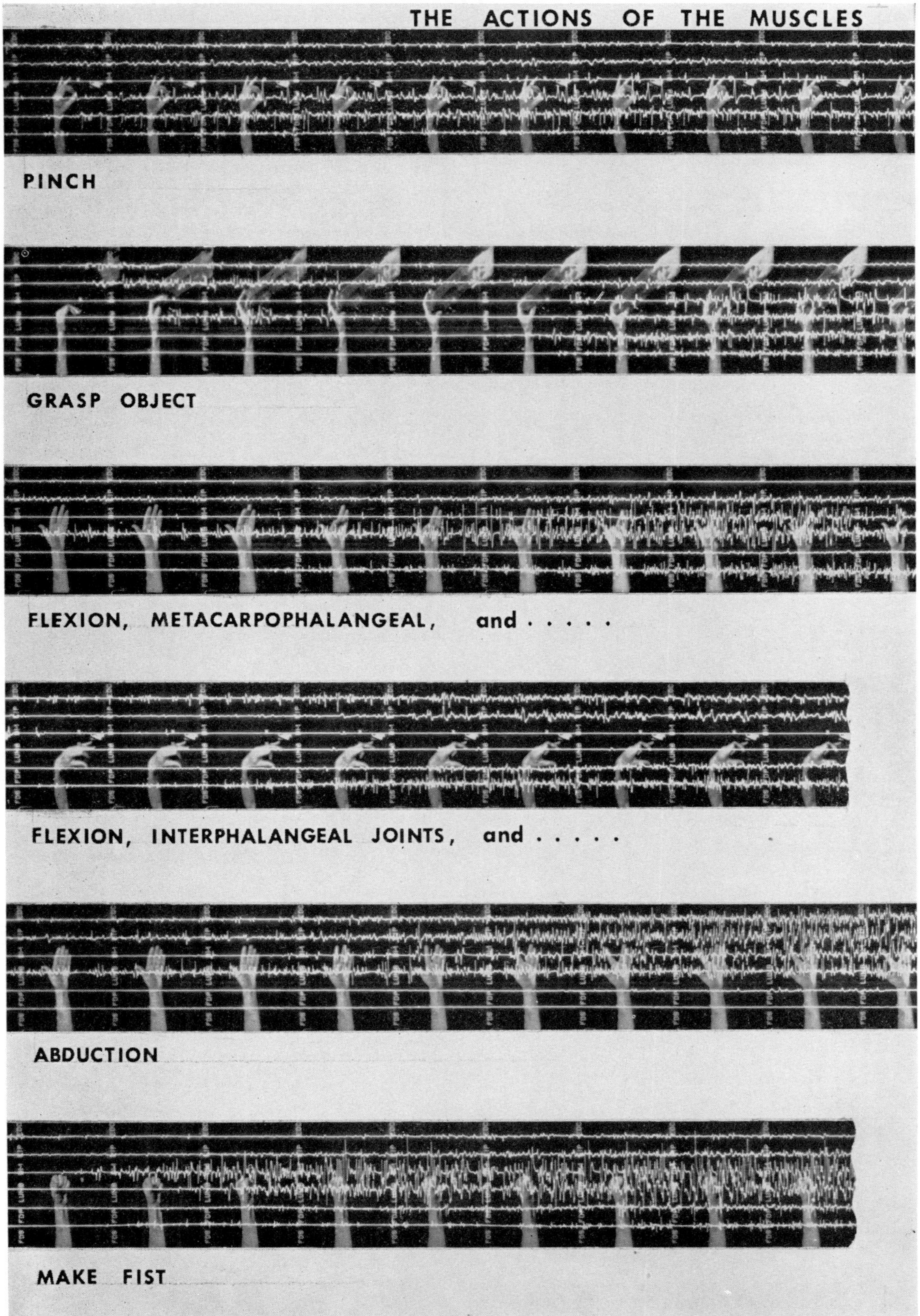

Figure 146A. The index finger; Subject 3.

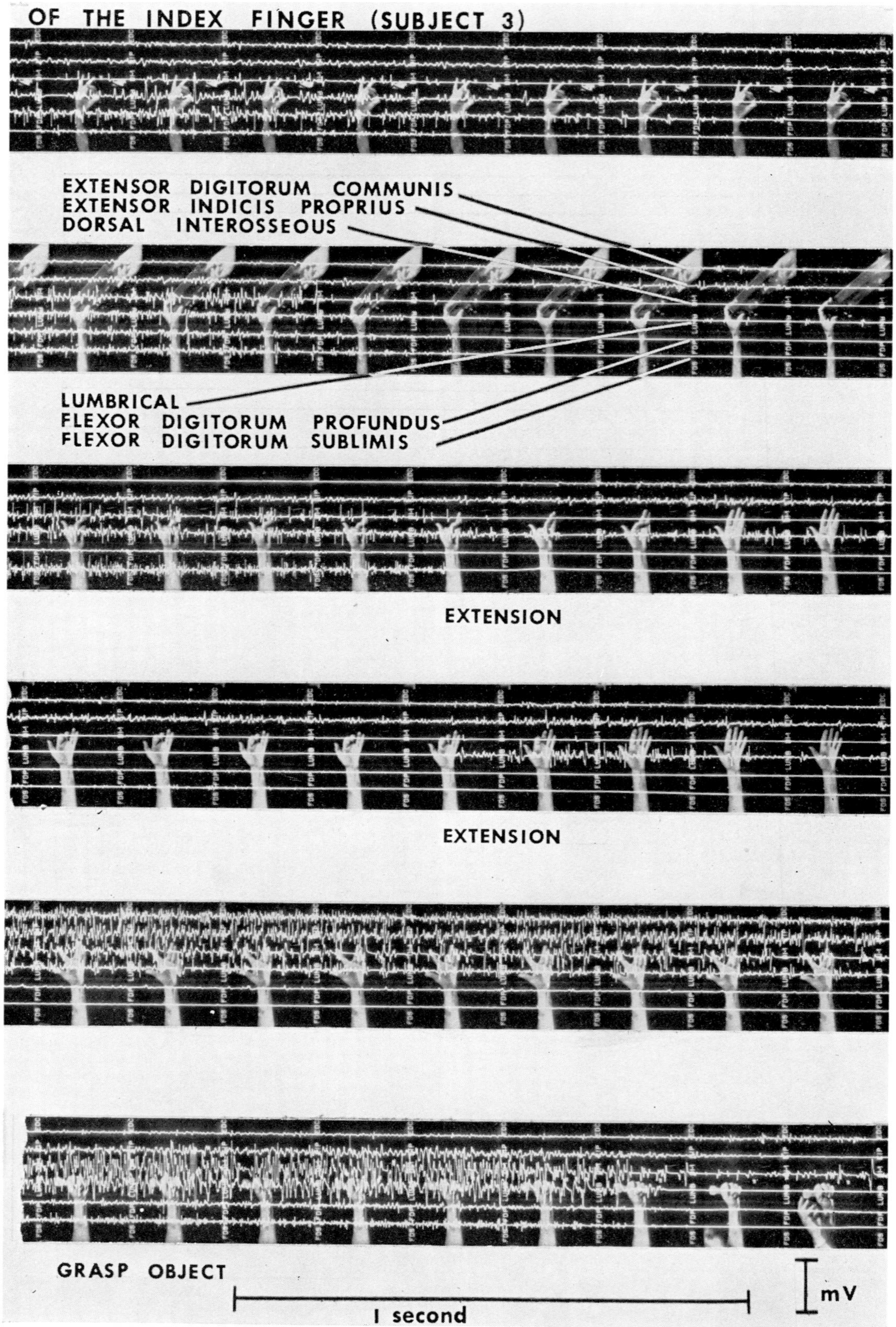
OF THE INDEX FINGER (SUBJECT 3)
EXTENSOR DIGITORUM COMMUNIS
EXTENSOR INDICIS PROPRIUS
DORSAL INTEROSSEOUS
LUMBRICAL
FLEXOR DIGITORUM PROFUNDUS
FLEXOR DIGITORUM SUBLIMIS
EXTENSION
EXTENSION
GRASP OBJECT
I second
mV

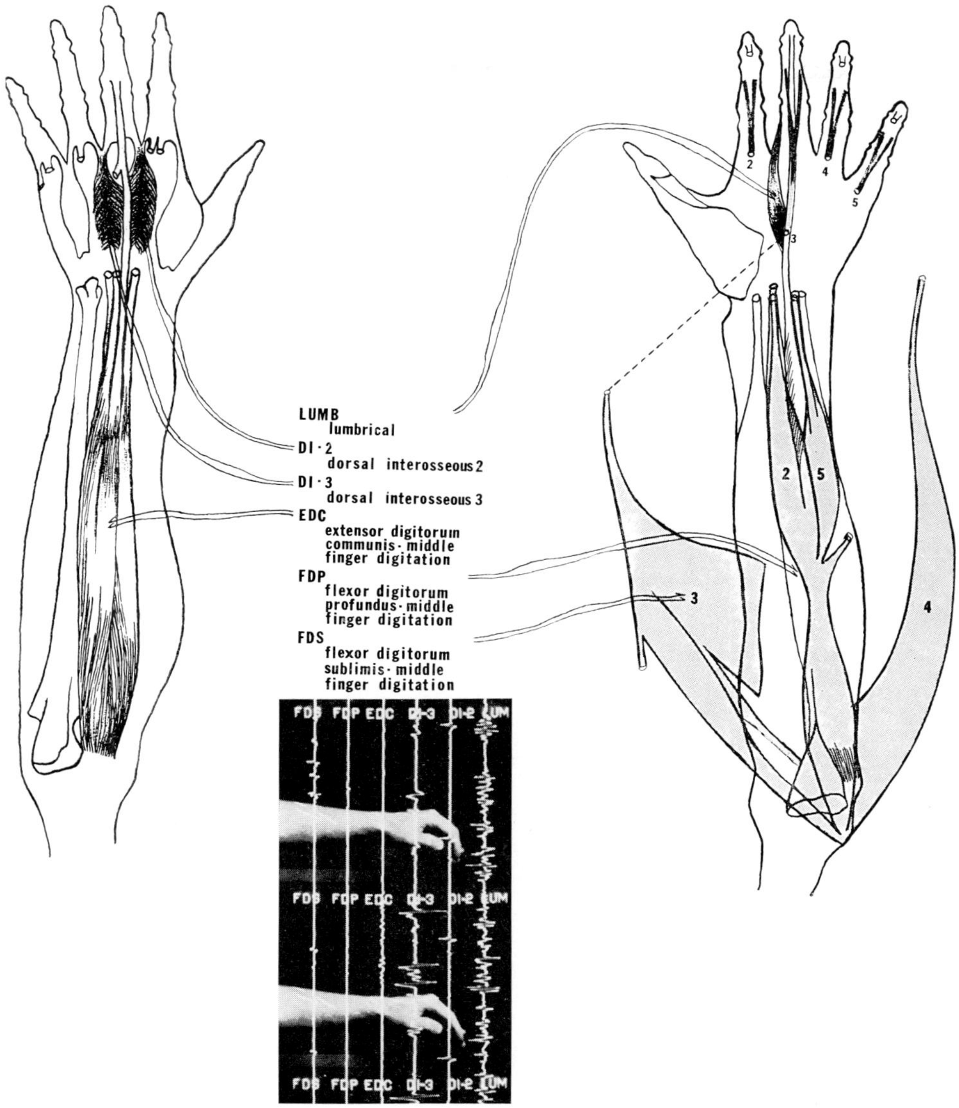

Figure 147. The muscles of the middle finger, showing electrode placement.

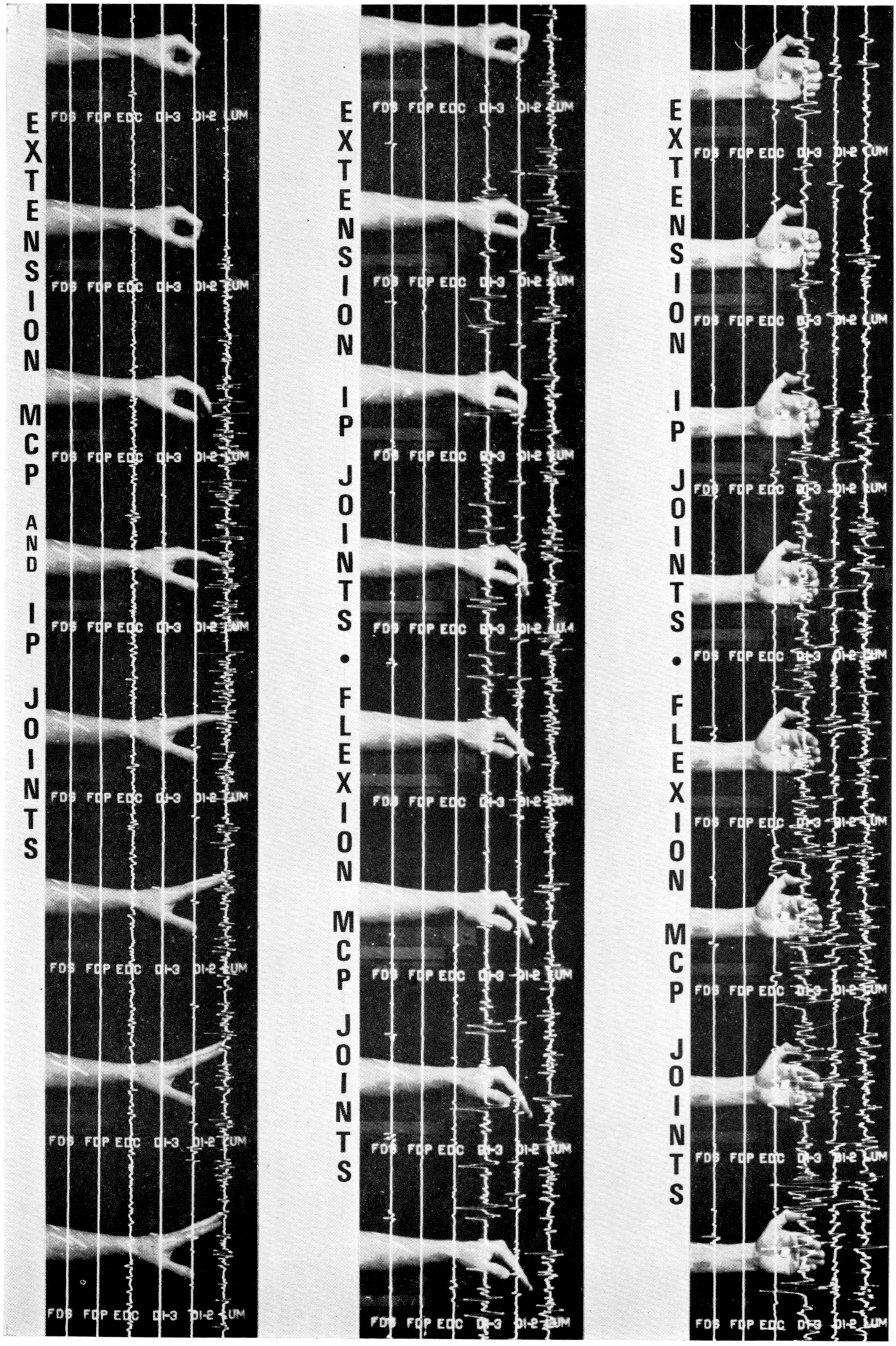

Figure 148. The actions of the muscles of the middle finger.

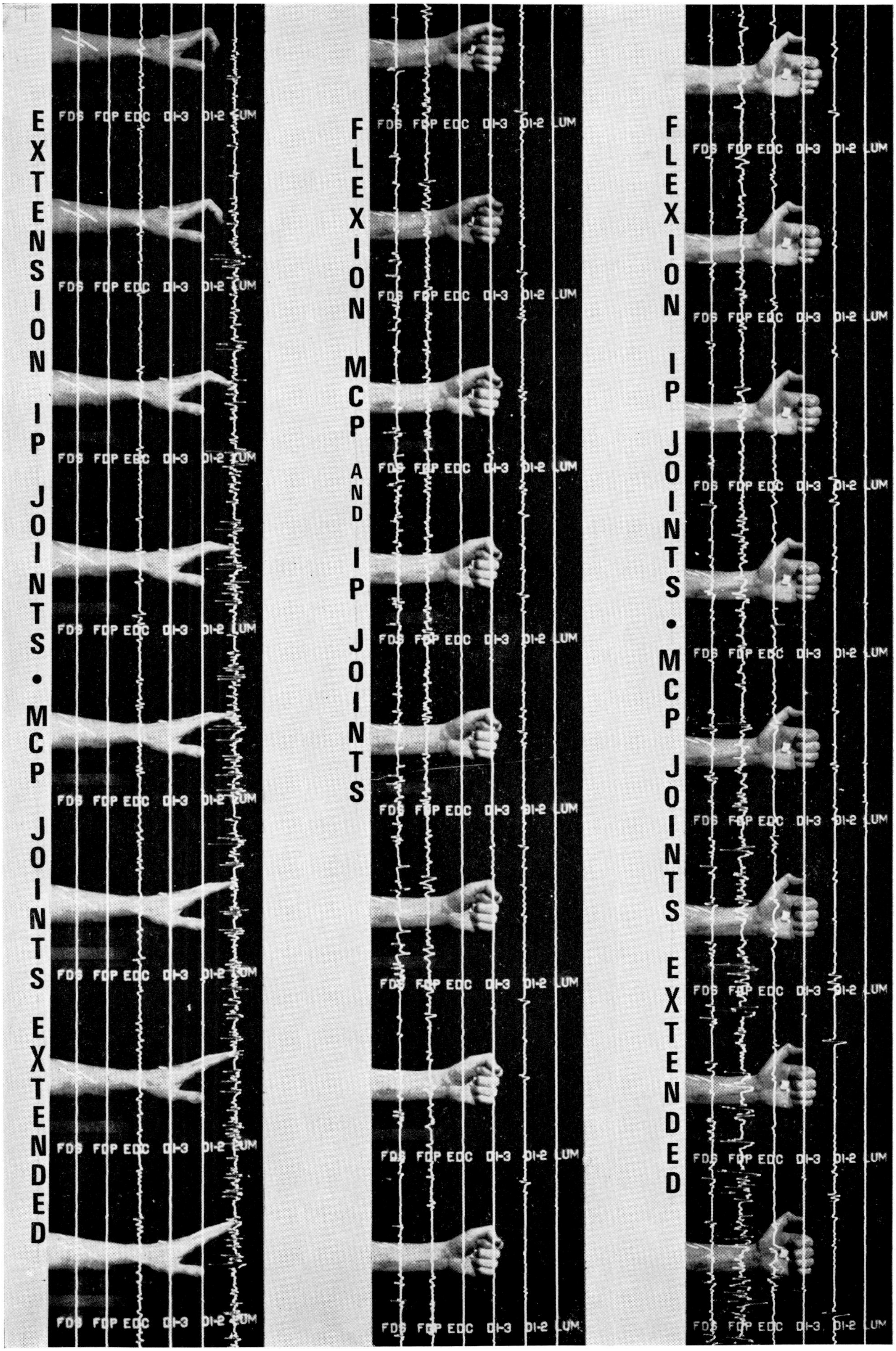

Figure 149. The actions of the muscles of the middle finger.

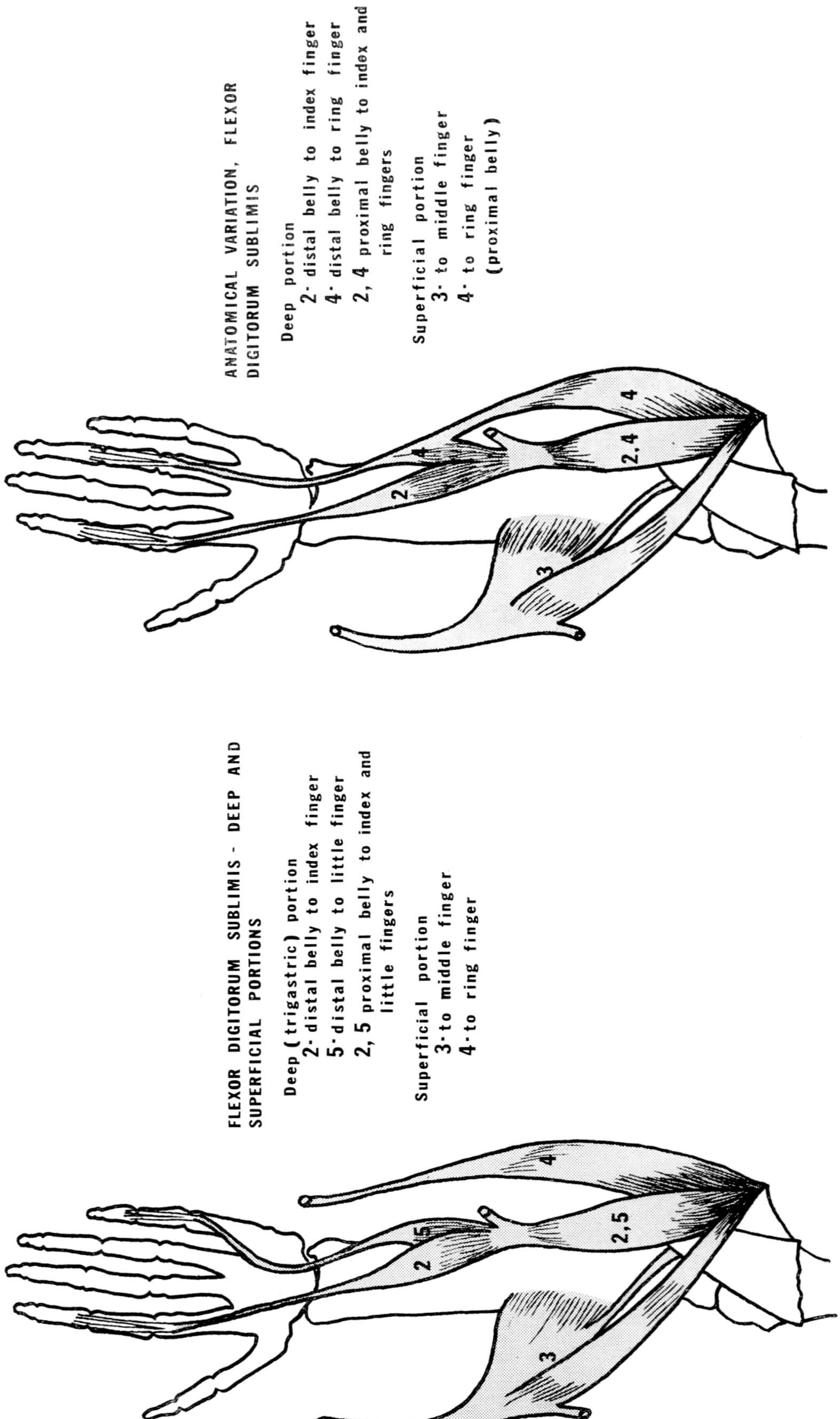

Figure 150. The components of m. flexor digitorum sublimis.

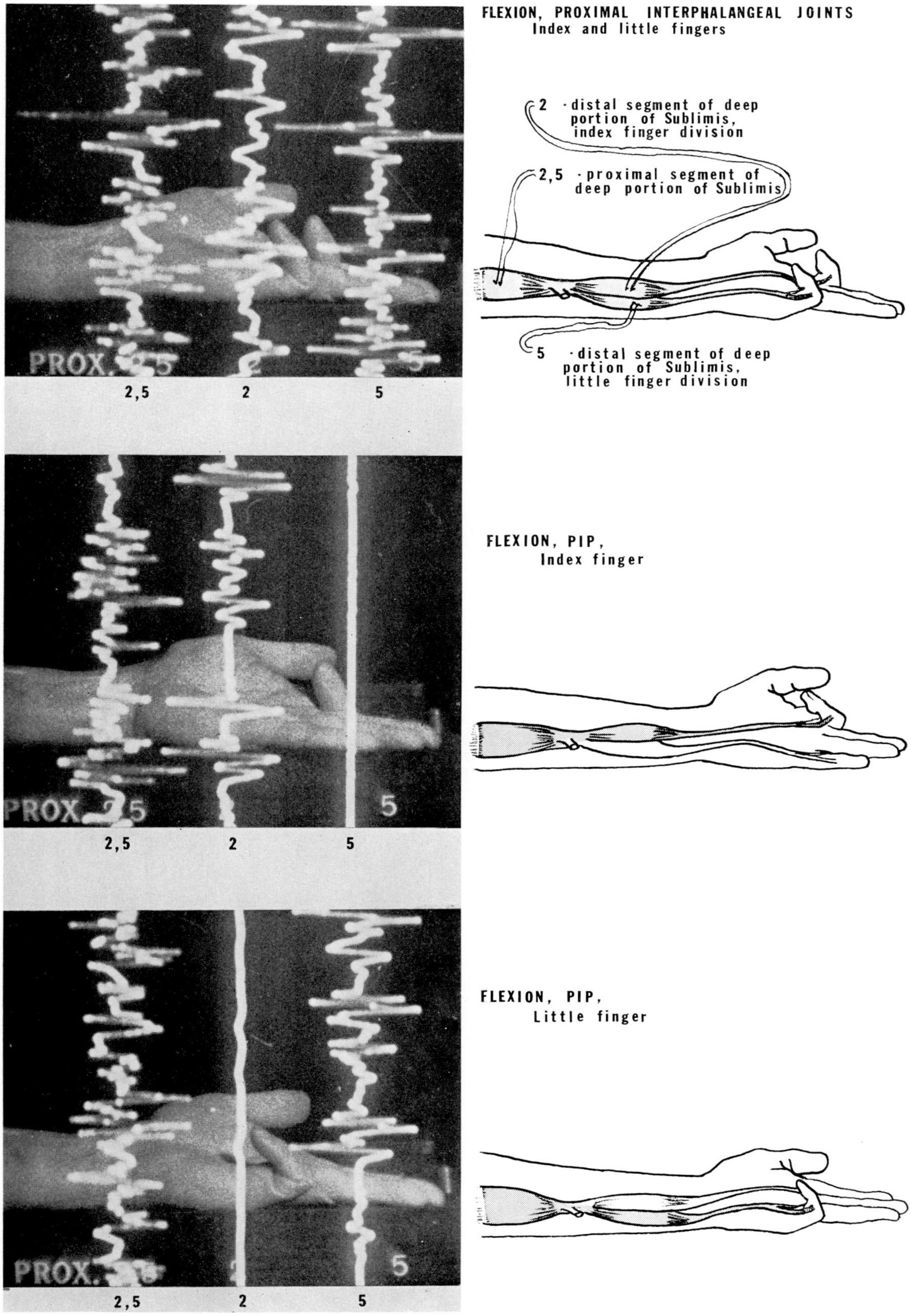

Figure 151. The three combinations of activity of the deep sublimis components.

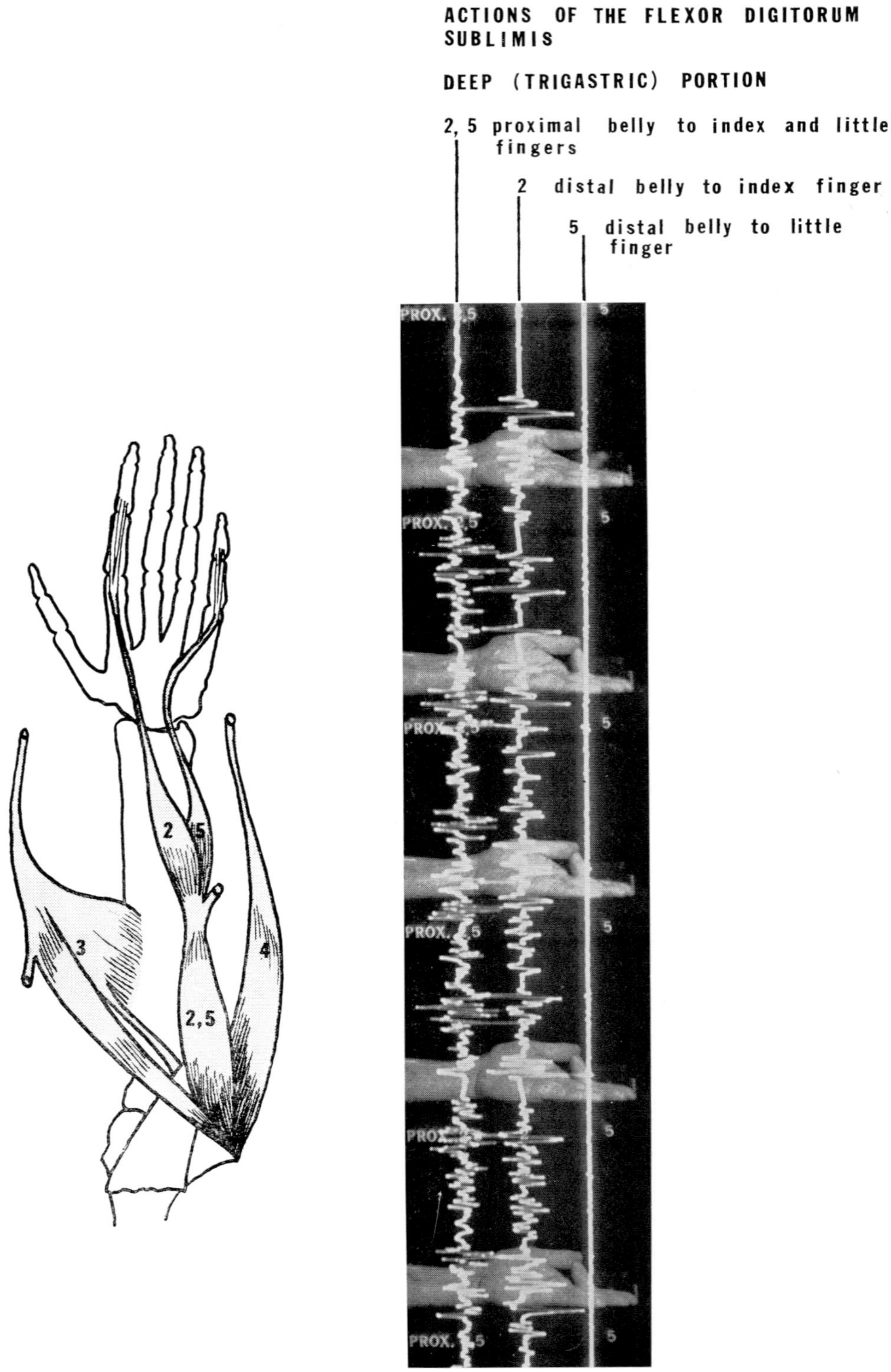

Figure 152. Deep sublimis component activity, continued.

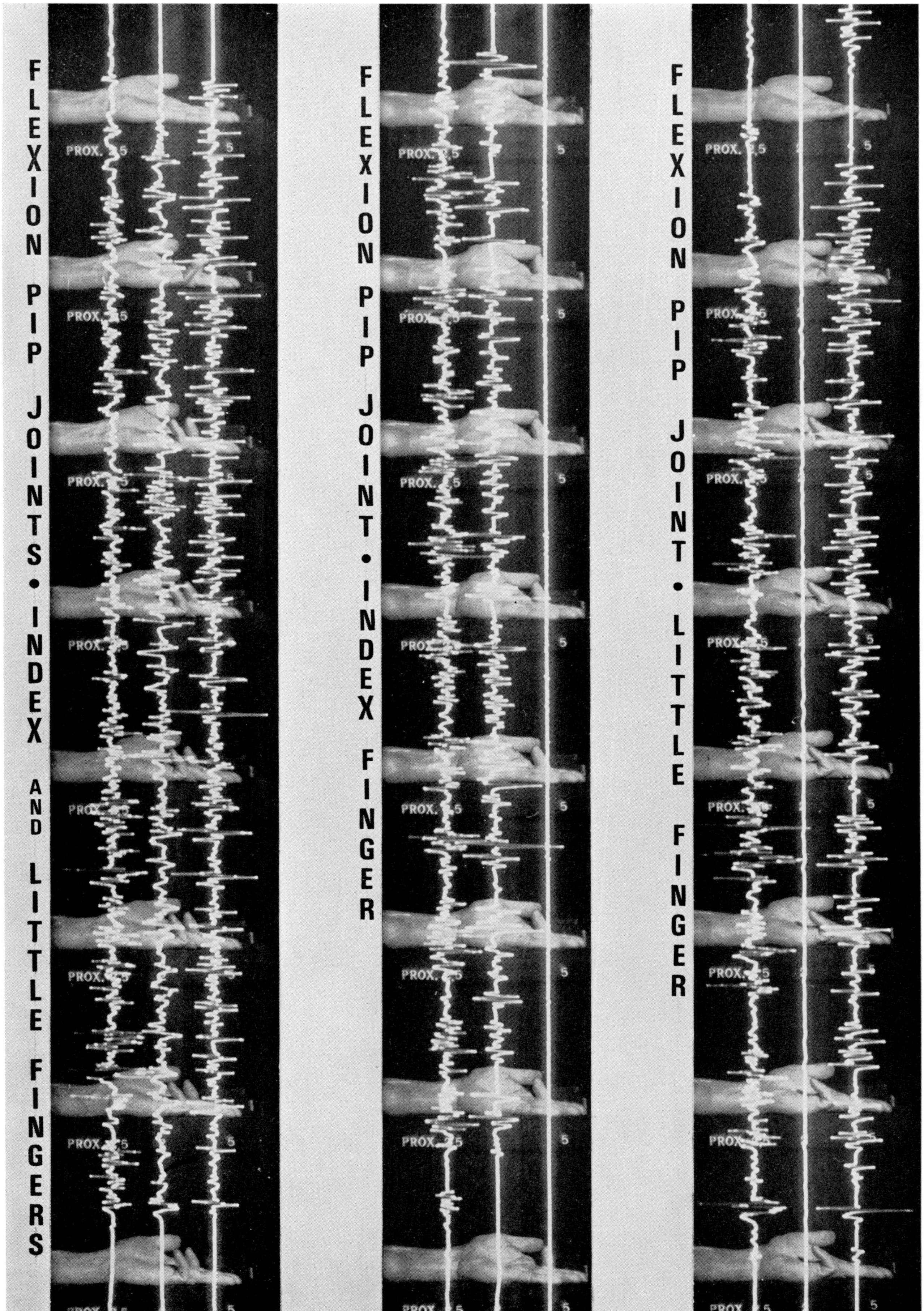

Figure 153. Deep sublimis component activity, continued.

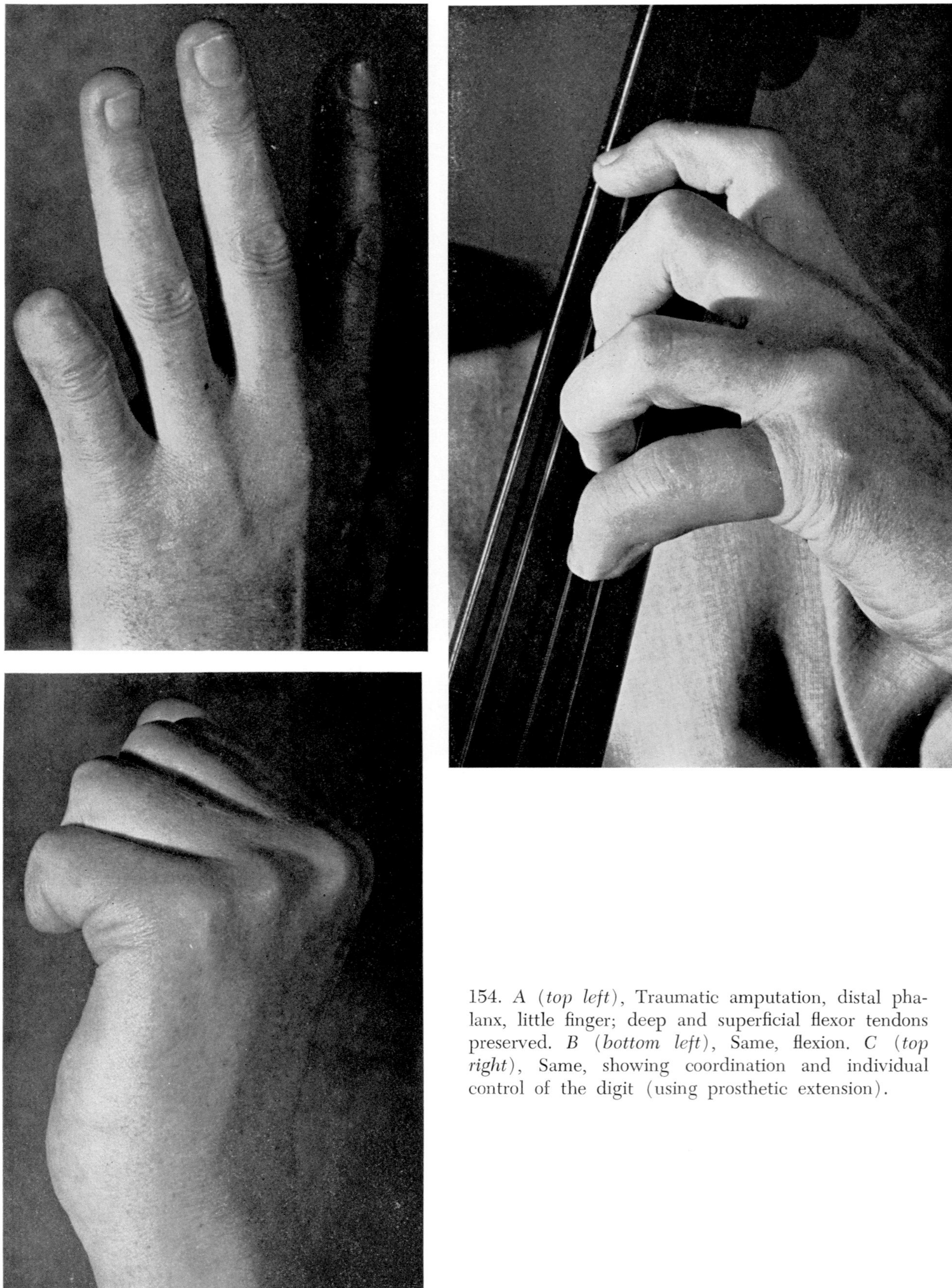

154. *A* (*top left*), Traumatic amputation, distal phalanx, little finger; deep and superficial flexor tendons preserved. *B* (*bottom left*), Same, flexion. *C* (*top right*), Same, showing coordination and individual control of the digit (using prosthetic extension).

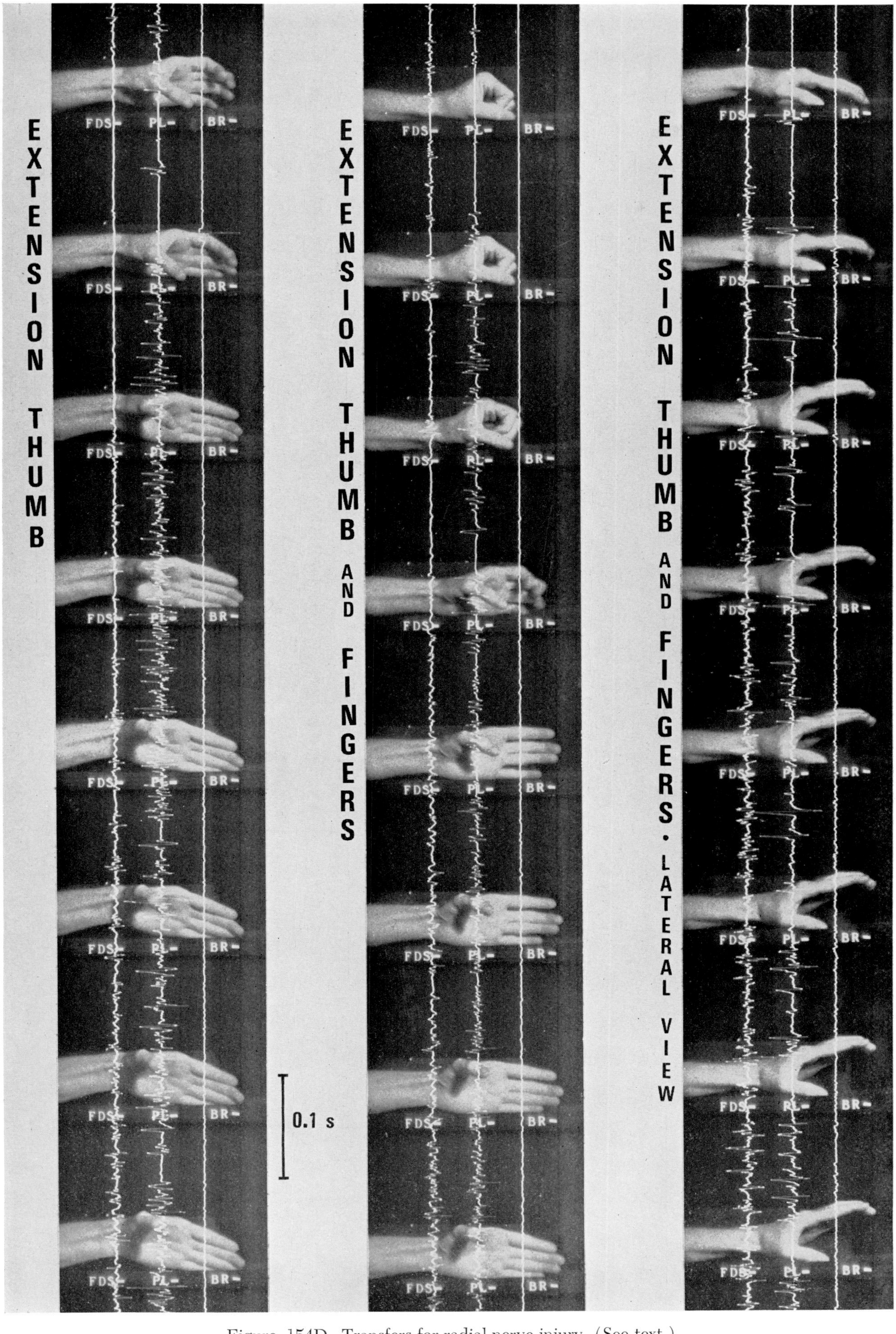

Figure 154D. Transfers for radial nerve injury. (See text.)

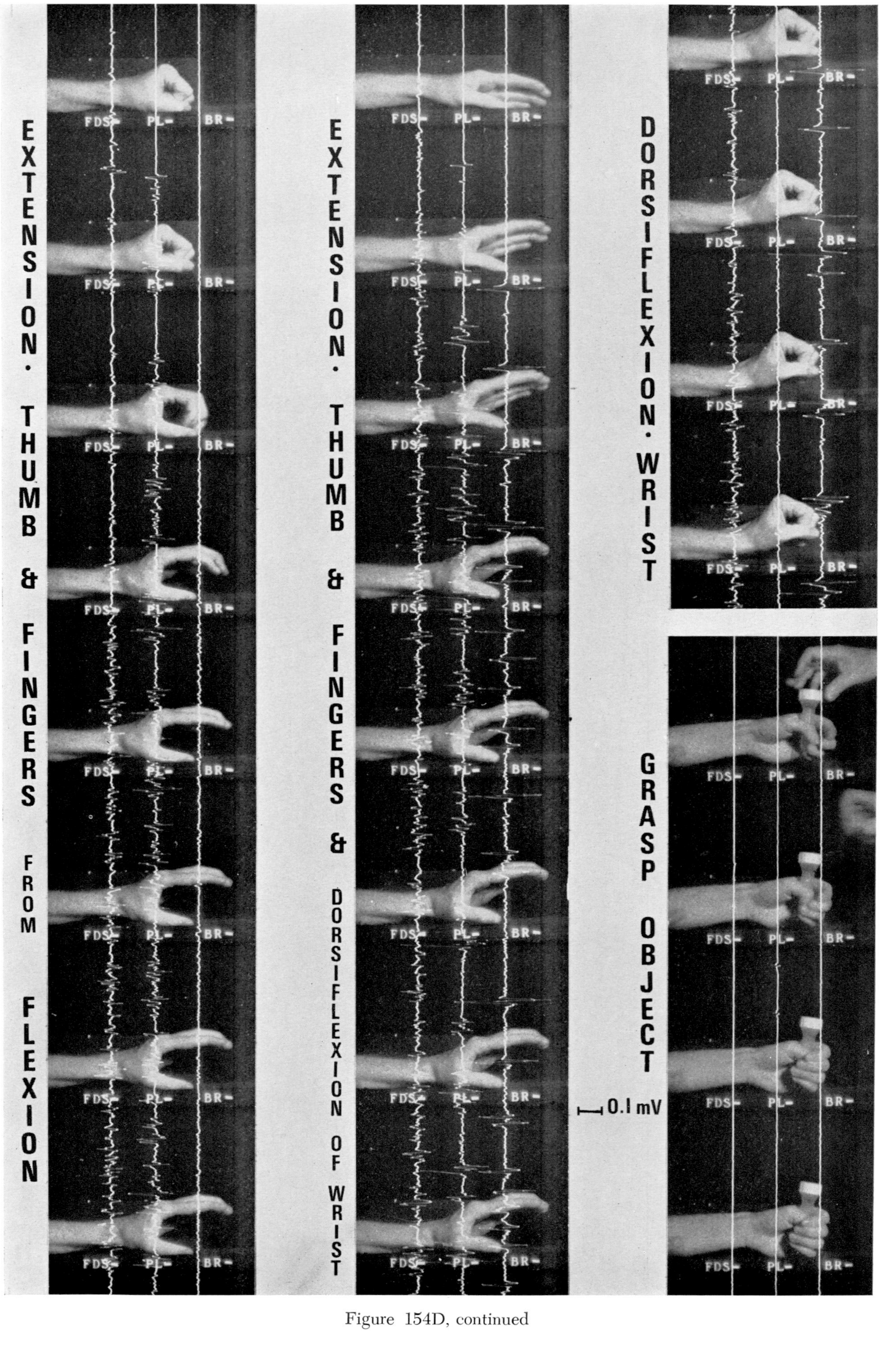

Figure 154D, continued

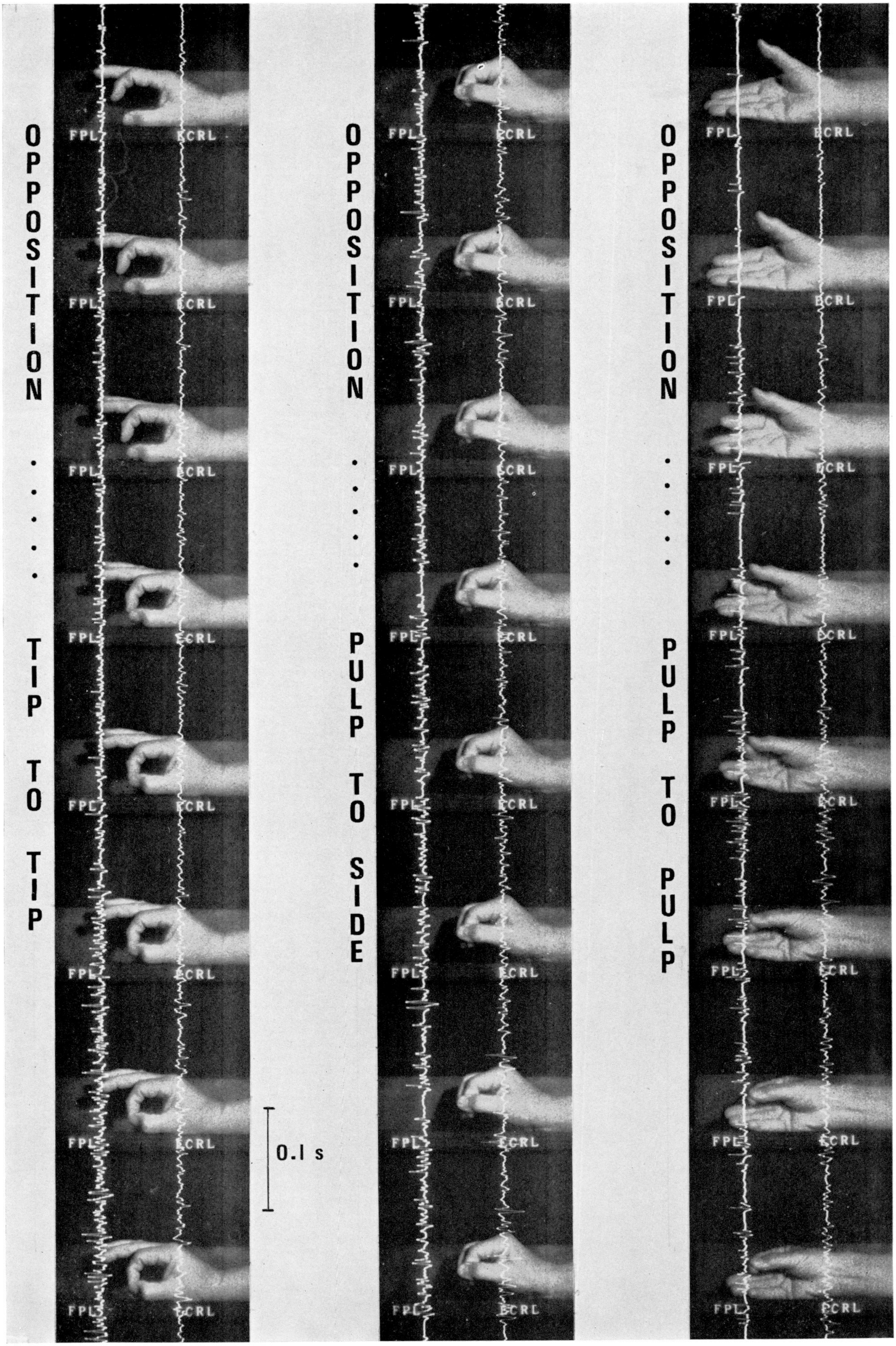

Figure 154E. A transfer to restore thumb flexion. (See text.)

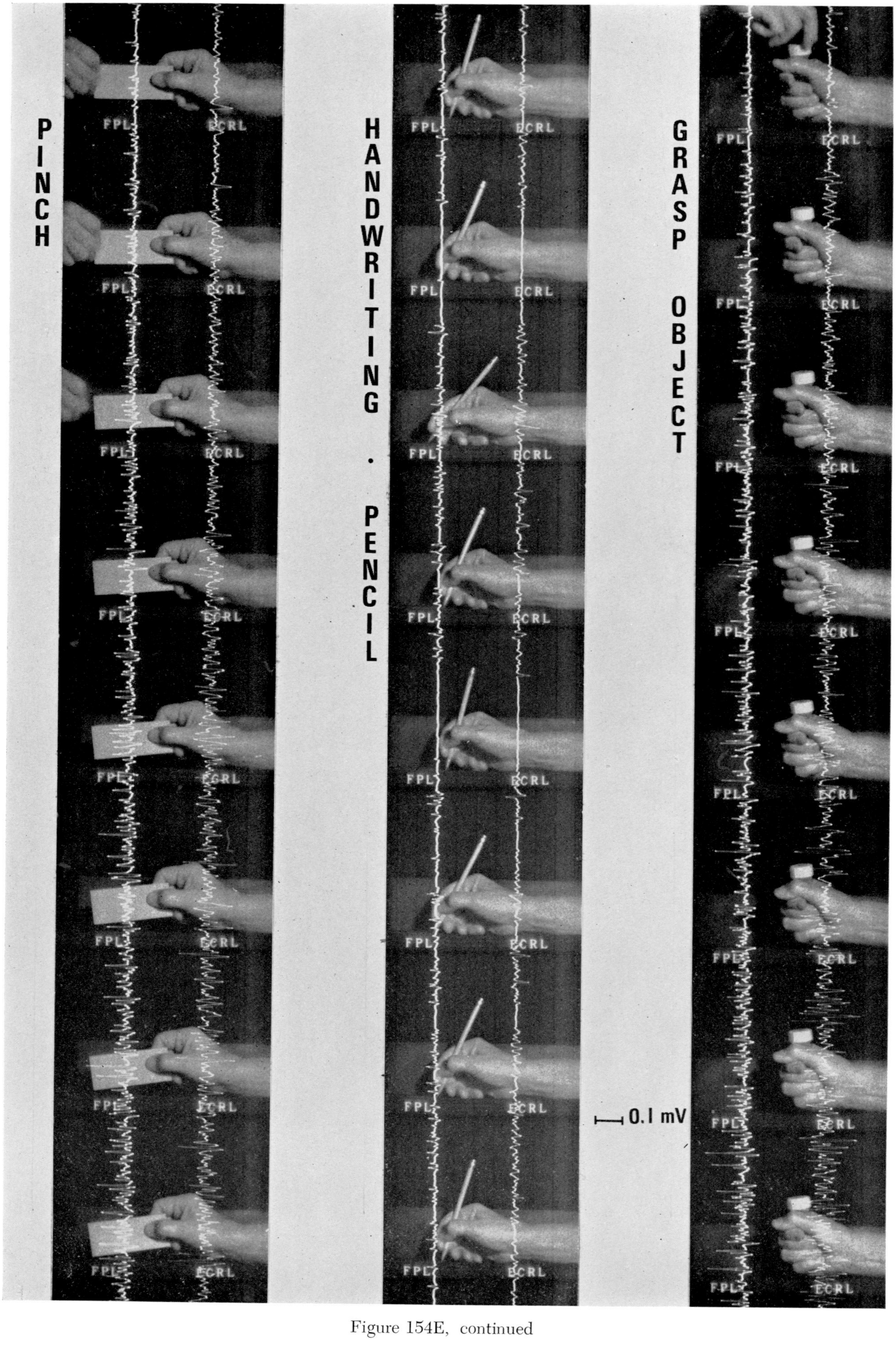

Figure 154E, continued

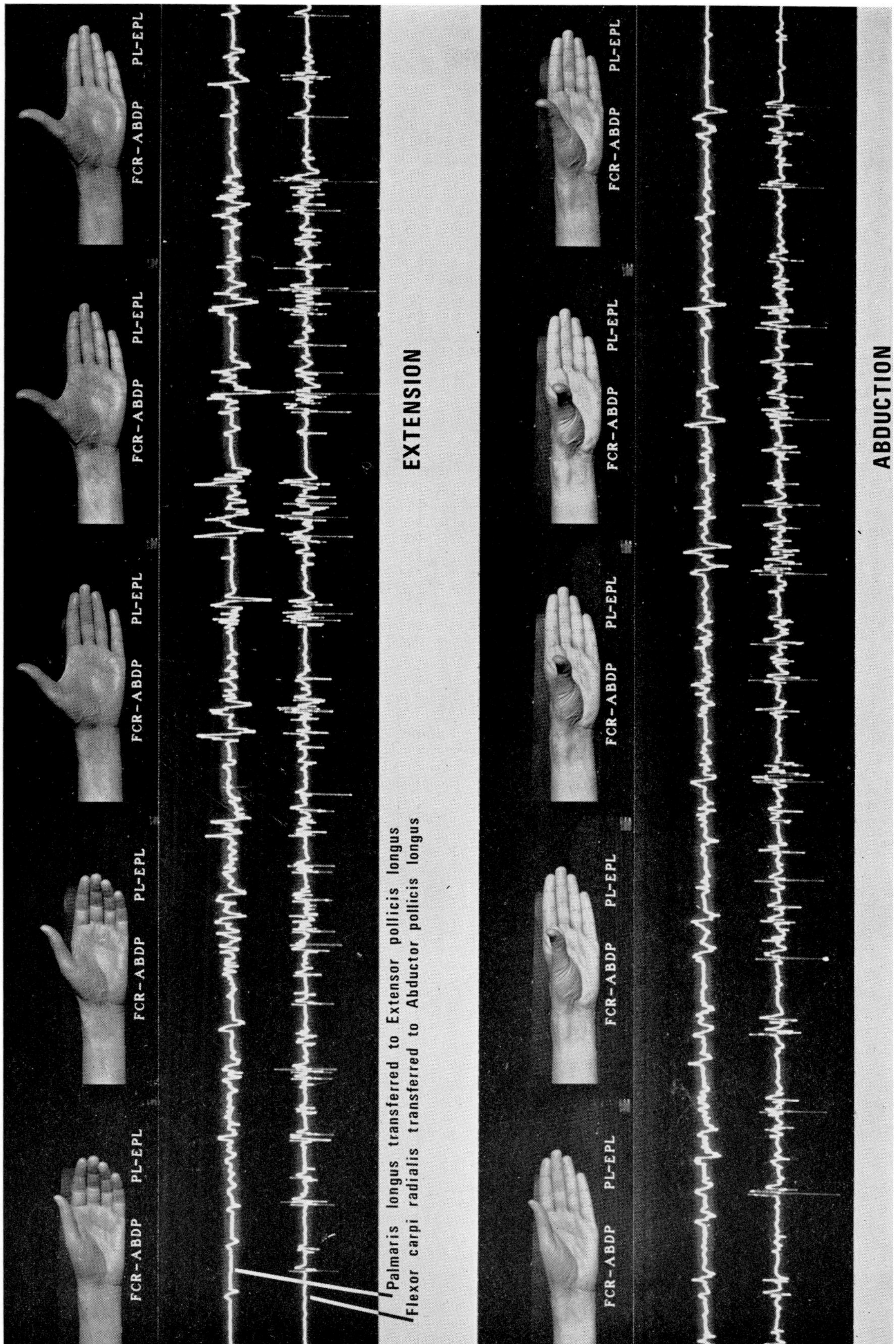

Figure 154F. Transfers to restore extension and abduction of the thumb.

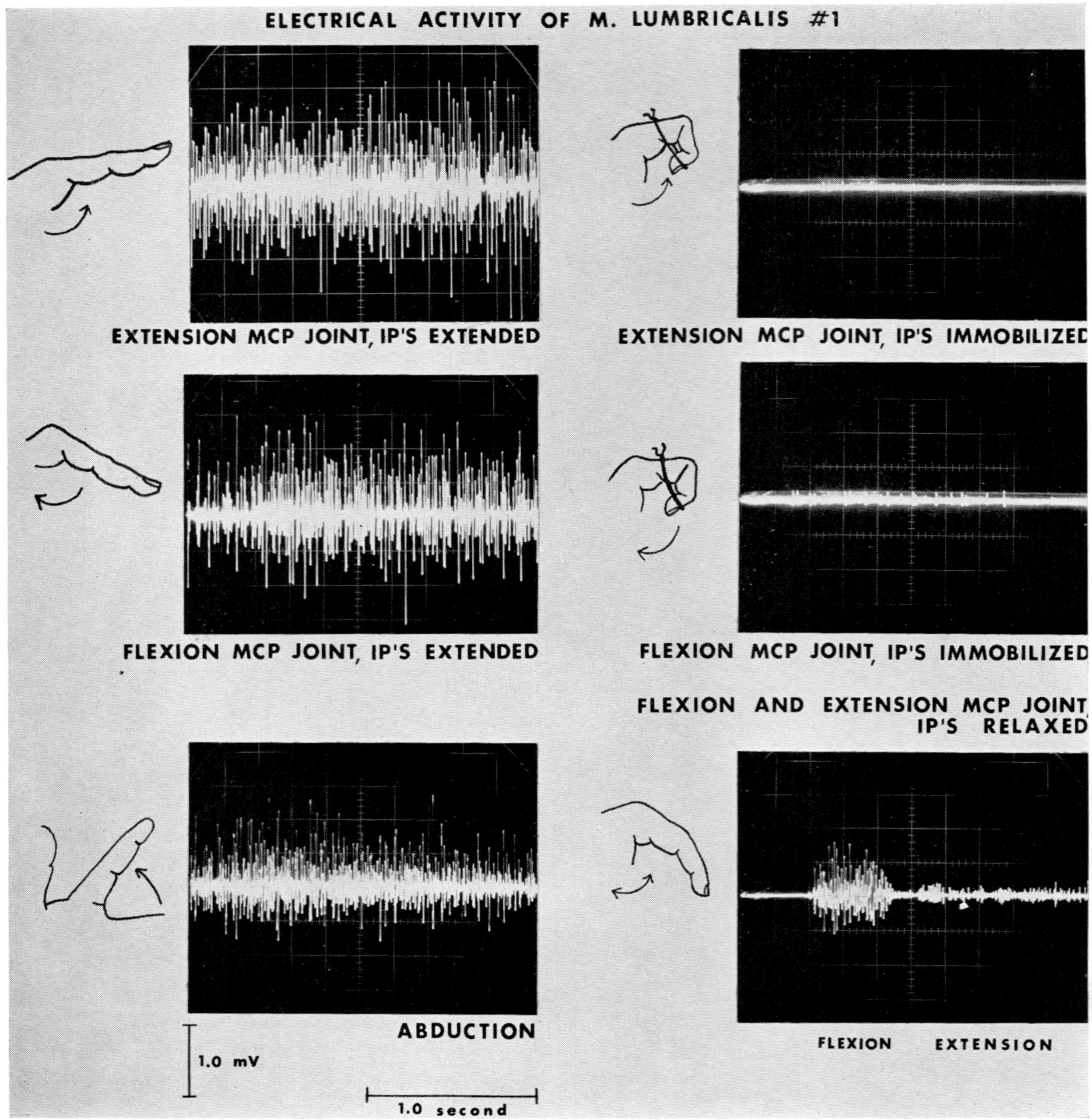

DISCUSSION

In recording the electrical activity of six muscles in synchrony with a prescribed list of motions, one encounters the possibility of a great many combinations of

1. Types of motions.
2. Wrist and palm positions.
3. Variation in intensity and frequency of action potentials.
4. The degree of muscular force used to perform the activity.
5. The restriction of the activity to a single digit (e.g. index finger) or observations of the activity of the muscles of that digit during simultaneous motions of the four fingers (Figs. 141, 142, 143).
6. Original positions of hand, e.g. when

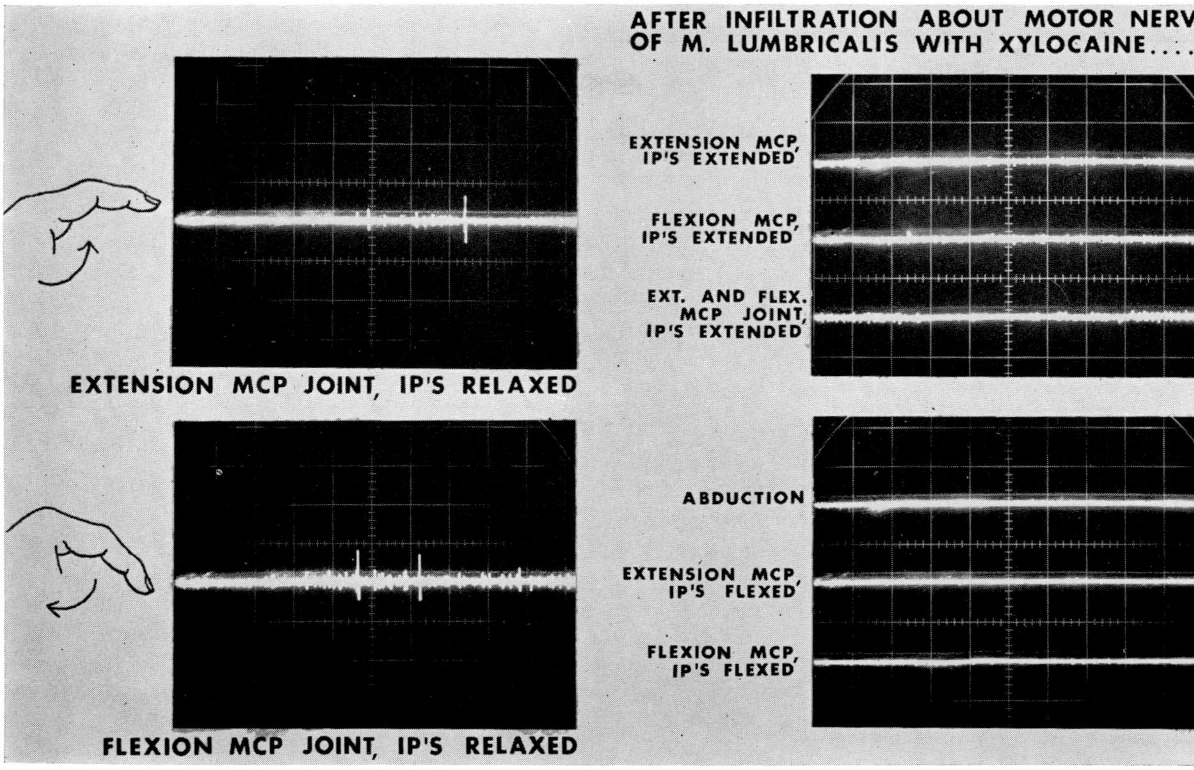

≪ Figure 155. Electrical activity of m. lumbricalis no. 1. (*From Journal of Bone and Joint Surgery*, December, 1969.)

index metacarpophalangeal flexion is studied, the other digits may rest in a position of flexion or extension (Figs. 141, 142, 143).

The endless number of possible combinations of conditions for study make some standardization and restriction of variables necessary. Thus, for example, most of the data reported were taken with the wrist in neutral position and were confined to single-digit activities. Variations are shown where the 16-mm moving picture film, taken to include most of the variations, has demonstrated their significance. A great deal of moving picture film was taken at each study. Only a small fraction of the material taken during a few of the studies can be demonstrated here. Tables IV, V and VI, however, reflect the results of review of all of this material.

Muscles of Individual Digits: Small Muscles, Digitations and Components of Larger Muscles

There are individual muscles anatomically and functionally assigned to certain digits, e.g. the flexor pollicis longus, completely separated from the rest of the deep flexor; the thenar muscles; the intrinsic muscles. But also to be considered are the digitations of the extensor digitorum communis and the flexor digitorum profundus, in which the digitation to the index finger results in more independent use of the muscle by the index finger than by the other fingers. The index tendon of the deep flexor is encumbered by only a single lumbrical muscle belly.[80]

The flexor digitorum sublimis, developed as a lamination of the flexor digitorum profundus, is anatomically and functionally more com-

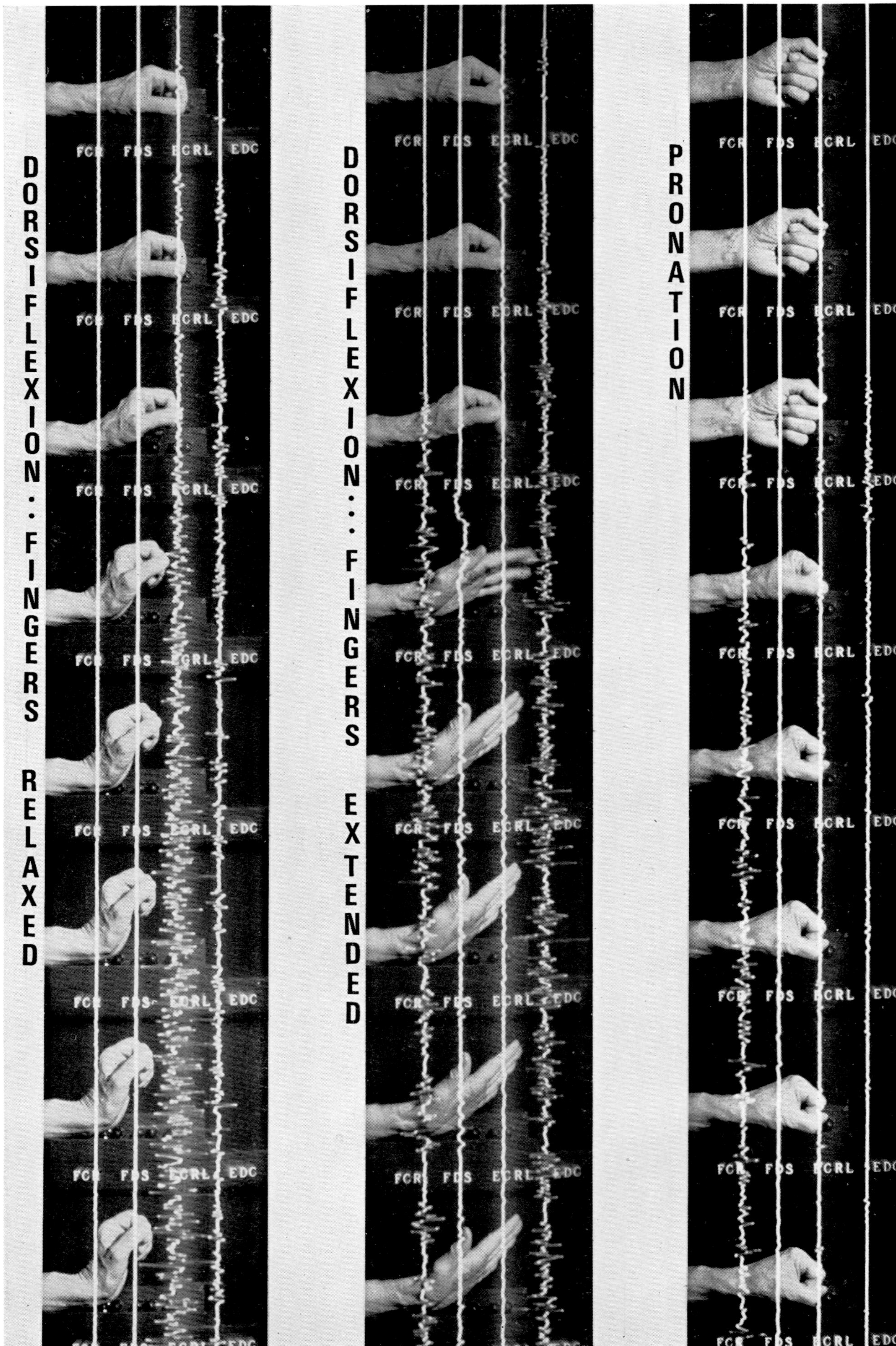

Figure 156. Electrical activity of four extrinsic muscles.

TABLE IV

SUMMARY OF THE ACTIVITY OF THE MUSCLES OF THE THUMB

MOTION:	APL	EPL	ADD	APB	OPP	FPL
EXTENSION — MCP and IP joints	M	F	O	M	F	O
FLEXION — MCP joint	M	M	M	O	O	O
MCP and IP joints	M	M	O	O	O	F
ABDUCTION	M	F	O	M	F	O
ADDUCTION	O	O	F	F	F	O
GRIP OBJECT	O	O	F	F	F	F

Full activity–F  Medium activity–M 

Zero activity –O ——————

pletely divided into discrete parts for each finger than the deep flexor. Many subjects readily demonstrate complete functional separation of the individual parts of the sublimis by independently flexing the proximal interphalangeal joint of each finger. This difference in control of the deep and superficial flexors, to which additional confirmation is given by these studies, may be of considerable clinical significance. Where the option exists to retain either the deep or superficial tendon of an injured digit, it is suggested that conservation of the proximal tendon of the muscle known to possess the greater degree of individual digit control may produce the better result.

The extensor indicis proprius, as its anatomical separation from the common extensor suggests, is capable of virtually isolated action, as is demonstrated by the tracings of Figure 141. As the index finger is extended, the extensor indicis proprius is active and the common extensor quiet. There is some lumbrical activity, beginning as extension is over one half completed. This late lumbrical activity is more apparent with the wrist in neutral position. The extensor indicis proprius appears consistently to initiate extension, with the lumbrical becoming active during extension.

The common extensor's digitation to the index finger is more extensive than its digitation to the other fingers. In recording from the muscles of the index finger, care was taken to place the electrodes in the index finger digitation of the common extensor.

These anatomical and functional considerations are essential to the planning of a muscle function study involving the muscles of individual digits. Where the groups of muscles are selected which are most closely related to the movements of each digit, these groups must include for each digit the individual muscles and parts of larger muscles most concerned with producing the controlled forces necessary for the complex and delicate movements of that digit. For example, among the obvious muscles that one might select as representative

TABLE V

SUMMARY OF THE ACTIVITY OF THE MUSCLES OF THE INDEX FINGER

MOTION:		FDS	FDP	LUMB	DI-1	EIP	EDC
EXTENSION							
MCP and IP joints	wrist neutral	O	O	F	O	M	tr
	wrist flexed	O	O	F	M	F	O
MCP joint		O	O	F	O	M	tr
FLEXION							
MCP and IP joints	wrist neutral	F	M	O	O	tr	tr
	wrist flexed	F	M	O	O	tr	tr
MCP joint		M	O	O M	F	M	M
ABDUCTION (wrist neutral)		O	O	F	F	F	F
PINCH		M	F	F	O	M	M
GRIP		F	F	O	tr	O	O
GRIP OBJECT		M	M	F	F	M	O

TABLE VI

SUMMARY OF THE ACTIVITY OF THE MUSCLES OF THE MIDDLE FINGER

MOTION:		LUMB	DI #2	DI #3	EDC	FDP	FDS
EXTENSION	MCP and IP joints	F	O	O	M	O	O
	IP joints	F	O	O	M	O	O
	IP joints with flexion of MCP joints	F	F	F	O	O	O
FLEXION	IP joints with MCP joints extended	O	O	O	tr	M	tr
GRIP		O	tr	O	O	M	M

movers of the index finger are the following:

1. Deep flexor (index digitation).
2. Superficial flexor (index component).
3. Common extensor (index digitation).
4. Extensor indicis proprius.
5. First lumbrical.
6. First dorsal interosseous.

Specificity is possible and is essential to a meaningful study:

1. *Deep flexor.* One can locate the index finger digitation, the best-differentiated part of this muscle, and record exclusively from this part of the muscle.
2. *Superficial flexor.* A component to each digit is well defined.
3. *Common extensor.* The index finger digitation is well enough defined to insert recording electrodes.
4. *Extensor indicis proprius.* This is entirely separate from the common extensor and is important to include.
5. *Lumbrical and dorsal interosseus.* The appropriate lumbrical and dorsal interosseous muscles may be located and proper insertions made.

Electrical stimulation clearly establishes the position of the electrodes in small, individual muscles as well as the components or digitations of larger muscles to the digit to be studied.

Anatomically, the index finger portion of the flexor digitorum profundus is more completely digitated than the portions to the other fingers. The index lumbrical is unipinnate, originating from a single deep flexor tendon. These are factors in promoting more independence of action of this finger than is found in the others.[80]

Tension

For the most part, the recordings were made during deliberate motions of the digits where force was not a consideration. Usually the muscular contractions were isotonic.

The direct relationship between the total electrical output of contractions of the isometric type and muscle tension is well known.[36,41] A similar relationship during isotonic contraction does not exist—at least not as a relationship of linear form. This gives us no license to abandon the possibility of some type of mathematical relationship between isotonic muscular tension and series of counts of action potentials.[36,41] Indeed, the level of electrical activity must be proportional to the transmembrane passage of ions involved in conduction of action potentials along the membrane of muscle fibres: Na^{+}, K^{+}, Cl^{-}, Ca^{++}. Such would be the case during contractions of any type. The sum total of electrical activity may thus be considered as proportional to the amount of ionic activity that is taking place. This reflects the degree of the muscle's participation in an event—though not always in terms of developed tension.

Although the relationship of tension to cross-sectional area is an approximation which holds only under specified conditions, it is a factor that could be included. All channels are normally equally calibrated. The calibration of each channel might be multiplied by a factor proportional to the muscle's cross-sectional area. Thus, if the cross-sectional area of m. extensor indicis proprius is considered to be one fifth that of m. flexor pollicis longus, the latter should be recorded at five times the sensitivity of the smaller muscle.

Such corrections of input channel calibration would result in a series of comparative recordings which would more closely reflect each muscle's effective participation. A method for arriving at a figure for cross-sectional area is described by Ralston *et al.*[144] Physiological cross section is arrived at by dividing muscle volume by length. There are some data available for making these calculations.[144] However, cross-sectional dimensions at certain levels of the extremities are also quite accurately reflected by the cross-sectional charts found in a number of atlases. Planimetric measurements of selected muscular cross sections found in such charts would serve as excellent approximations. Ikai and Fukanaga[99] made calculations relating tension to cross-sectional area of human muscle, using ultrasonic measurement.

M. Flexor digitorum sublimis

The use of electrical stimulation to establish electrode position led us to a more careful consideration of the anatomical divisions of the sublimis muscle. Unlike the deep flexor, in which the digitations to the four fingers are far less discrete, the superficial flexor is separated into components to a degree which makes possible individual voluntary control of the digitations of this muscle to each of the four fingers in many subjects. The m. flexor digitorum sublimis is composed of a deep and a superficial layer. As shown in Figure 150, the superficial layer consists of two sometimes completely dissociated muscles, components to the middle and ring fingers. The deep or trigastric[88] layer is unique. It consists of a proximal belly common to both the index and little fingers. At mid-forearm, the deep or trigastric layer of the sublimis becomes tendinous for a short distance. It then divides into individual muscle bellies for the index and the little fingers.

In individuals having no superficial flexor of the little finger, the distal belly intended for it is usually attached to the sublimis tendon of the ring finger (Fig. 150). Here, one would not select this component of the sublimis as an opponens transfer with the same anticipation of success as when it is a virtually isolated component.

If electrodes are placed into each of the three bellies of the trigastric portion of the sublimis (Figs. 150, 151, 152, 153), the anatomical arrangement of this muscle is demonstrated by the patterns of activity that are found on combinations of activity of the index and little fingers. As these digits are selectively flexed at their proximal interphalangeal joints, patterns of electrical activity result, according to which digit, or whether both digits, flex at the proximal interphalangeal joints (Figs. 151, 152, 153).

It would appear that the above knowledge of the arrangement of the sublimis and its variations may be essential not only to those who may wish to select the most favorable sublimis component for opponens transfer but also to those concerned with operative repair or reconstruction in the hand and forearm where this muscle is involved.

A difference in the time of action of the deep and superficial flexors of the finger can be clearly shown by multichannel equipment of the type shown in Figure 16. The differential technique with insertion of two wires into each muscle is essential to confine the recording to individual muscles. Electrical stimulation for verification of electrode position is essential.

A patient was studied several months following treatment for traumatic amputation of the distal phalanx of the left fifth finger. The patient, a concert cellist, had used the fingers of the left hand several hours daily for 25 years in forceful flexion to finger the strings of the cello.

Postoperatively, having only a superficial flexor to flex the fifth digit, she was unable to use that finger in playing the instrument. Moreover, as the patient forcefully used the index, middle and ring fingers, the little finger extended, failing completely to participate in the desired movement.

Further surgery was performed. The deep flexor tendon was recovered at the distal flexion crease of the palm and attached to the superficial flexor tendon. Following healing, there was prompt return of the patient's ability to control the fifth finger. She was again able to use it at the required time—when there was active flexion of the adjacent digits including their terminal phalanges (Fig. 154).

One must consider the conclusion, based on the finding in this single case, that only the deep flexor is primarily employed for some actions of the fingers. Prior to reattachment of the deep flexor tendon in the fifth digit, the patient was unable to control forceful flexion of that finger when the deep flexor muscle was forcefully flexing the distal interphalangeal joints of the adjacent three digits. Strangely, when asked to flex only the fifth finger, the patient was able to do so without difficulty. The central nervous system was apparently unable to employ effectively and simultaneously the deep flexor to the index, middle and ring fingers and the superficial flexor to the little finger.

To "finger the strings of the cello," the subject needs fingertip control—probably possible only with a complete set of profundi. One cannot expect the sublimis of the little finger to replace the profundus of the little finger and to coordinate with profundi of the index, middle and ring fingers in such a complex act. This would require that the proximal and little-finger portions of the trigastric (deep) layer of the sublimis (Fig. 150) coordinate with the remaining three profundi, since the little finger, though without a distal phalanx, must still act in "profundus" fashion to produce a fingertip action.

In surgical tendon repair, where one has the choice of retaining either the deep or superficial flexor tendon, the above case is strong recommendation for preservation of the deep flexor tendon. The superficial flexor tendon, though successfully attached, may fail completely to produce flexion of the finger, including the distal phalanx at the desired time, usually when the deep flexor muscle is producing flexion of the adjacent digits.

These muscles have been studied by electromyographic methods by Long and Brown,[120,121] who have made many recordings from the muscles involved in motion of the middle finger. They have consistently demonstrated a marked difference in the actions of the deep and superficial flexors. In addition, these workers have displayed the action of the extensor, lumbrical, and interosseous muscles. The technique of recording employed was based upon the fine, single-wire electrode (27μ) and the multichannel pen writer. In delicate, critical work of this type, the prevailing opinion is that electrical stimulation is essential for identification of electrode position, that the differential technique employing two electrodes in each muscle is essential in order to confine the recording to that muscle (Fig. 83) and that the same gain must be used for each channel if recordings are to be compared. It can be said, however, if the above investigators did not apply all of these standards, their work acquires a great deal of validity by virtue of the consistency found in their records.

Intricate movements requiring fingertip control of all of the fingers are apparently possible only when a complete set of profundi is present. The sublimis of the little finger cannot replace or substitute for the profundus of that digit and coordinate with the profundi of the ring, middle and index fingers, particularly in complex activities. The anatomical arrangement of the trigastric layer of the sublimis (Fig. 150) accounts for the inability of its components to substitute for individual components of the flexor digitorum profundus. This concept is reinforced by the records of the electrical activity of the three parts of the deep layer of the sublimis (Figs. 151, 152, 153).

The Lumbrical

In addition to observation of the electrical activity of the lumbrical, insight regarding its function results from study of

1. Its action when directly internally stimulated.
2. Its action when the ulnar nerve is stimulated internally at mid-forearm.
3. What deficiency (if any) results when its motor nerve is anesthetized.

Direct stimulation of the lumbrical using the inserted recording electrodes produced brisk extension of the interphalangeal joints and weak flexion of the metacarpophalangeal joint.

On one occasion, the ulnar nerve was stimulated internally at the mid-forearm level well distal to the innervation of the flexor digitorum profundus muscle. This resulted in contraction of the dorsal and volar interossei, the hypothenar muscles, the adductor pollicis and lumbricals no. 3 and no. 4 but not no. 1 and no. 2. The difference in the responses of the index and middle fingers, in which the lumbricals were not stimulated, and the ring and little fingers, in which lumbricals contracted, was consistent. Each of the four fingers extended at the interphalangeal joints, but the ring and little fingers flexed somewhat more decisively at the metacarpophalangeal joints.

Marked electrical activity in the index lumbrical is evident on metacarpophalangeal extension and flexion with the interphalangeal joints extended; to a lesser degree, on abduction of the index finger; and at lower levels, on flexion of the metacarpophalangeal joint with the interphalangeal joints relaxed. With electrodes in place in the lumbrical, the movements of the index finger and the associated electrical activity are observed (Figs. 141, 142, 143, 155). After xylocaine infiltration about the motor nerve to the lumbrical, virtually no electrical activity is observed as the same motions are repeated (Figs. 144, 145, 146, 155). A comparison of the ability of the subject to perform motions involving the index finger before and after rendering the muscle functionless is one method of assessing the normal lumbrical's contribution to the motion of that digit. Virtually no difference in performance could be observed. Upon flexion of the metacarpophalangeal joint as the interphalangeal joints were extended, metacarpophalangeal flexion and interphalangeal extension were very slightly weakened. This difference was not definite. Other motions of the digit in which the lumbrical may be expected to participate appeared unaffected. In general, it is suggested by this procedure that the lumbrical, though electrically active in a number of situations, makes no critical contribution to the motion of the digit. It is remembered that small dual copper wire electrodes, recording from a small volume of contracting muscle, may conduct large signals. In the case of the lumbrical, a signal of given amplitude is of far less significance than a similar signal from the extensor indicis proprius, a muscle of far greater cross section. The electrical signals of the lumbrical and of the extensor indicis proprius, in a state of maximal contraction, would be of comparable amplitude and configuration when recorded by this method, yet the influence on the movement of the digit by the extensor indicis proprius would be far greater.

It has been suggested[114] that the action of the lumbrical may be explained as follows. On contraction, it stretches the relaxed proximal profundus, releasing the passive tension in the distal profundus tendon, permitting the common extensor passively to extend the interphalangeal joints. It has been further postulated that the deep flexor is never active when the lumbrical acts[120] and that this is a basic pattern consistently found in many studies. The data of this study, based on observed activity of the index finger lumbrical and the index finger digitation of the deep flexor (Fig. 146A), do not support this theory. Activity, in several instances, is recorded from both the lumbrical and deep flexor at the same time. The studies demonstrated here are based upon tracings from the lumbrical, the extensor indicis proprius, and *index finger digitations* of the common extensor and deep flexor as demonstrated by electrical stimulation.

With the interphalangeal joints extended, it is noted that on *extension* of the metacarpophalangeal joint, the lumbrical is active and the first dorsal interosseous silent (Figs. 138, 141). During *flexion* of the metacarpophalangeal joint, the first dorsal interosseous is active, the lumbrical silent (Fig. 138). The remaining fingers are in a position of flexion. If the same motions are repeated with the remaining fingers extended (Fig. 138, center tracing), this pattern is altered, with lumbrical activity continuing throughout metacarpophalangeal flexion as well as extension.

MUSCLES OF THE FOREARM

A number of studies were made of the participation of the wrist extensors and flexors. These five muscles are listed above. Many of the recordings which included one or more of these muscles were made as part of a study which included selected extrinsic and intrinsic muscles of the digits (Fig. 156). In this figure, activity of the radial wrist flexor is demonstrated on dorsiflexion of the wrist as the fingers are extended. The muscle is inactive when the wrist is dorsiflexed without extending the fingers. The flexor carpi radialis, as demonstrated in Figure 156, participates in pronation.

In a number of subjects, the radial wrist extensors were observed to be silent on dorsiflexion of the wrist if the motion included extension of the fingers. Abundant activity of the common extensor was observed. The radial wrist extensors were active on dorsiflexion when the digits were not extended.

The use of the radial wrist extensors and the radial wrist flexor in radial deviation was repeatedly demonstrated. The activity of the ulnar extensor and flexor during ulnar deviation was also apparent.

MOTOR FUNCTION OF THE UPPER EXTREMITY COMPARED TO THE LOWER EXTREMITY

The actions of many of the muscles of the upper extremity have been studied and some of the data presented herein. This monograph deals primarily with the functional anatomy of the *lower* extremity where muscles behave during many of their functions according to a pattern which is more involuntary in nature compared to the muscles of the *upper* extremity. The comparison is most interesting and of great importance in the understanding of the reasons for specific differences in the behavior of the muscles of the *lower* and the *upper* extremity.

As stated in Chapter V, the muscles of the *lower* extremity behave according to a pattern of *phasic activity* which, for the most part, is almost involuntary during normal walking. This pattern of phasic activity is acquired by the infant as a normal development at a certain stage, as though it were an inherited complex body function. The muscles of the lower extremity, in general, have no individual common usage.

The *upper* extremity, required to provide actions of a finer nature, does not have clearly defined phases of motions. Motions of individual digits are possible. Many individuals have control of the metacarpophalangeal and proximal interphalangeal joints of each digit. Where complex actions are performed in series, the individual participating muscles are not likely to make the same contributions repeatedly, as do the muscles of the lower extremity during normal walking. The voluntary nature of usual upper extremity action explains this variation of participation.

We have seen (Chap. V) that the tendons of certain muscles may be transferred to a new insertion where the anticipated function may either closely resemble or differ from their original function. It was found that, in general, transfers within phasic groups were more likely to perform the desired function. Violation of the patterns of phasic activity in tendon transfer, in a number of instances, resulted in failure of the procedure (Figs. 97, 98).

In the upper extremity, one encounters different requirements and differences in neuromuscular behavior compared to the lower extremity. Wrist flexors are found to be almost uniformly capable of acting as dorsiflexors after transfer. Wrist and finger extensors have proved useful where thumb or finger flexion is required. It is noted that in patients where flexors have been transferred to the dorsum following loss of the radial nerve, wrist dorsiflexion and finger extension may be very satisfactory. Finely coordinated movements, however, were not observed. The patient extended all digits and dorsiflexed the wrist successfully as a combined motion. Control of individual digits in general was not observed.

In one case previously described (Figs. 154A, B, C), individual control of the little finger was of great importance to the patient. It is noted that the flexor digitorum sublimis (Fig. 150), with its unique trigastric configuration of the deep layer, was insufficient for individual control of the digit. Although this case is not an example of tendon transfer, the difference in the actions of the sublimis and profundus in the little finger is clearly demonstrated. It represents a situation in the upper extremity where a choice must be made between two muscles having a difference in function. In this sense, it is comparable to the choices of tendon transfer in the lower extremity where differences in muscle function are characterized by phasic activity (Chap. V).

Transfers for Radial Nerve Injury

The tracings of Figure 154D represent the electrical activity of three transfers used to restore extension of the digits and dorsiflexion of the wrist of a patient, age 21, with radial nerve palsy. The lesion resulted from a supracondylar fracture at age twelve. Either the innervation of the brachioradialis muscle was partially spared or the electrical activity demonstrated represents regeneration of this branch of the radial nerve. The tendon of the brachioradialis was transferred to the tendon of the extensor carpi ulnaris muscle. The palmaris longus tendon was transferred to the tendon of the extensor pollicis longus. The tendons of the flexor digitorum sublimis were transferred to the extensor digitorum communis tendons. The patient is able to dorsiflex the wrist, and finger extension is good. Extension of all of the joints of the thumb except the metacarpophalangeal joint is excellent. Hand and wrist function in general is very satisfactorily restored.

The records demonstrate lack of individual control of the transfers. When the patient extends the thumb, there is activity of the flexor digitorum sublimis as well as the palmaris longus (which has been transferred to the thumb). The brachioradialis is quiet. The brachioradialis, active on wrist dorsiflexion, demonstrates an abnormal pattern of electrical activity. Action potentials are larger than normal, but rate is reduced. All of the tracings were recorded at an oscilloscope sensitivity setting of 0.1 mV per division. This electrical pattern is consistent with regeneration and axon branching (Chap. IV).

During grasp, the transfer to the common extensor is quiet (Fig. 154D, tracing 7), not duplicating the action of the normal common extensor, which is uniformly active in the normal subject during grasp. The flexor digitorum sublimis muscle, its tendons transferred into the tendons of the extensor digitorum communis muscle, is no longer active during flexion.

A Transfer to Restore Thumb Flexion

The extensor carpi radialis longus was used to reinforce the flexor pollicis longus of a patient who six years previously injured his cervical spine when he fell from a ladder. Diagnosis of herniated cervical intervertebral disc, C5-C6, was made and cervical fusion performed. One year before the tracings of Figure 154E were made, the patient observed loss of use of the right flexor pollicis longus muscle. There was total loss of flexion of the interphalangeal joint. This was attributed to the cervical spine lesion or to local compression of the anterior interosseous nerve. A surgical procedure was done to free the interosseous nerve, which was exposed from its origin from the median nerve posterior to the pronator teres, to its penetration into the flexor pollicis muscle. The tendon of the extensor carpi radialis longus muscle was delivered to the anterior aspect of the wrist and transferred into the flexor pollicis longus tendon.

The records of Figure 154E demonstrate excellent use of the thumb. The extensor carpi radialis longus and the flexor pollicis longus muscles are active on all motions involving the thumb where activity of the flexor pollicis longus would be expected in the normal subject. There is no differentiation in the activity of the two muscles. The transfer has greatly improved thumb flexion, grasp and opposition. The extensor carpi radialis longus muscle is still active during dorsiflexion of the wrist, though transferred to serve as a flexor of the thumb.

The oscillograms show approximately equal activity of the two muscles in actions where both participate. High amplification (0.1 mV per division) was necessary in each case. Apparently the two muscles combined to restore function of the thumb. Clinical examination of the hand indicates function closely approaching that of the normal extremity.

Chapter VII

SINGLE ALPHA MOTOR AXON STIMULATION: THE ACTION POTENTIAL AND TENSION OF THE SINGLE MOTOR UNIT

INTRODUCTION

Accurate analysis of the characteristics of known single motor unit action potentials is essential to the understanding and interpretation of the commonly recorded electromyogram. Electrical action potentials recorded from muscle by conventional means are usually a simultaneous presentation of several to many motor units, and the tiniest probes usually fail to demonstrate with any certainty this single quantum of motor activity. There is actually no positive means for assigning a given action potential to any particular structure recorded by a blind technique.

In making careful, precise recordings from active skeletal muscle, one may employ and compare various methods. Although there is a known direct current potential difference between the outside and inside of the resting muscle or nerve fibre, a potential easily demonstrated by established microelectrode techniques, it is the rapidly fluctuating potential along the muscle membrane that is coincident with activity. The outer sides of the membranes of nerve and muscle fibres are positively charged in their resting state. Action potentials are the passage of negative charges down the membrane, a phenomenon found only in nerve and muscle. The membrane quickly recovers its positive charge, the fluctuation occurring with the passage of each impulse. The magnitude of this fluctuation, though much smaller than the transmembrane resting potential, is of primary importance, since it represents with great accuracy of coincidence the presence—the beginning and termination—of a given neuromuscular event. It is best demonstrated by alternating current input coupled instruments. In clinical studies, the comparatively large probes used normally lead to the recording of an indeterminable number of motor units, often representing many thousands of muscle fibres. The motor unit, thought to be the source of the solitary waveform of the electromyogram, must be defined and precisely delineated histologically in order for one to understand the quantum of motor activity[149]—that resulting from the discharge of the single alpha motor axon and the many muscle fibres it reaches. The spatial distribution of this unit of neuromuscular activity is yet to be determined.[1]

Avalanches of action potentials of single and very likely groups of synchronized motor units make up the "normal interference pattern" commonly seen on the usual tracing. Electrodiagnosis in muscular disorders often depends to a great extent upon the minute characteristics of the action potentials recorded from the muscle. Clinically, it would be desirable to see and to evaluate the action potential of the single motor unit of the muscle under study.[36] This action potential and its characteristics could then be compared to the single motor unit action potentials previously taken from known normals. The clinician would then not be forced to base his conclusions on what the preponderance of action potential characteristics might be, since the single motor unit action potentials often mask themselves or otherwise disguise themselves as described above.

In the study to be described, the action potential known to represent the firing of the single motor unit is observed. Under properly controlled conditions, identical single motor unit action potentials are observed as the single alpha motor neuron is stimulated at chosen intervals (Figs. 162, 163).

Since the action potentials of the conventional electromyogram increase in rate as the contraction increases in strength, their rate becomes so great that action potentials are crowded together and escape individual recording (normal interference pattern) (Fig. 5). The identity of discrete motor units, even the action potentials of a few combined motor units, is then lost. Quantum methods of measurement of total count then fail unless one can devise a method of identification (in the recording) of the single motor unit. Unquestioned, discrete single motor unit action potential records may be produced by single motor axon stimulation in isolated preparations. This technique may lead to a valuable study of the histopathology of the single motor unit—"the live biopsy." In order to evaluate the composite series of oscillations from whole muscle, we must first establish the magnitude and nature of the electrical action potential of the single motor unit. If an axon innervating a superficial motor unit is selected, a beautiful display of this structure is seen with the same stereomicroscope used in microdissection of the axon (Figs. 157, 158). The site chosen, most convenient for identification and stimulation of the axon, is the region just proximal to the beginning of axon branching.

If one could isolate individual alpha motor axons of a muscle-nerve preparation with its motor units intact, microstimulation would make possible the recording of the pure action potential of the single motor unit involved. Such a procedure would make it possible to visualize the single fluctuations of which the series of action potentials recorded from contracting muscle is composed. Here, we have employed the small surface electrode, 9 mm in diameter, in order to avoid penetration of and injury to muscle fibres and also to include all of the fairly widely distributed components of the single motor unit.

Krnjevic and Miledi,[110] in 1958, described action potentials recorded from the rat diaphragm, an *in vitro* study. Single axons were stimulated and muscle action potentials recorded by the differential method from small platinum wires (diameter 175μ) placed very close together (0.19 mm). These small applicators, fused in glass, were used to make a survey of the surface of the muscle.

The isolated muscle-nerve preparation in which alpha motor neurons are stimulated leaves no possibility for gamma loop (spindle) participation. We began our studies using the intact animal. The gamma system is expected to influence the recording after a few alpha stimuli have been made[77,79] and to produce changes in the tension curve, as will be discussed later in this chapter.

The investigation into the characteristics of the single motor unit yields further valuable incidental information. The waveform itself may be subjected to computer analyses and valuable information regarding frequency recording ranges obtained. It leads also to other data concerning the specifications and methods appropriate for diagnostic and experimental electrodiagnosis, electromyography, functional anatomical studies and studies of neuromuscular behavior.

The nature of the single motor unit action potential has evaded accurate description because of the virtual impossibility of *in vivo* anatomical isolation. This becomes obvious when one imagines the many structures involved; the many end plates, axon branches and muscle fibres of different lengths. The possibility of isolating a single *in vivo* motor unit by severing all other alpha motor neurons in the region fails because of axon branching from the one remaining axon. This takes place to such an extent, even during the brief interval while one is waiting for all of the denervated muscle fibres to succumb, that the usual "giant motor unit" appears.[36] Under conventional methods of recording, one may occasionally see single spikes possibly attributable in many cases to the action potentials of single motor units. Scientific conclusions must be based upon better evidence than this. We must remember the tendency of adjacent groups of motor units to behave in a synchronous fashion.[41] The tiniest probe placed among the digitations of a motor unit is inevitably influenced by other motor units in the area and their interdigitations, some of which are mixed with those of the motor unit concerned.[1]

Isolation of the single alpha motor neuron makes several observations possible. The single structure, upon microstimulation, finally reveals the action potential of the single motor unit thus innervated. Only by this method have we seen the uncomplicated, pure, single motor unit action potential lacking the usual interference of discharging neighboring motor units or interdigitating subunits. Such a single motor unit action potential study offers a variety of experimental opportunities. The surrounding biochemical environment of the extracellular fluid may be varied and the effects of certain common variations studied. After establishing the consistent form of the single motor unit of one muscle (anterior tibial), it is of great interest to proceed to other nearby muscles of different composition or ratio of muscle fibres to single motor nerve fibres. Each muscle is expected to demonstrate a different single motor unit action potential, depending upon the number of fibres in its motor units.

METHOD

The present procedure depends upon meticulous dissection of the individual components of the branches of the common peroneal nerve of the rabbit (Flemish Giant). The branches used are (a) the branch to the anterior tibial muscle and (b) the branch to the anterior tibial and extensor digitorum longus muscles, selecting the portion to the anterior tibial muscle.[47]

By the method reported, it is possible to approach many anatomical sites. Dissection more cephalad (e.g. in the thigh) suggests many advantageous sites where one can locate small nerve bundles about to penetrate well outlined and surgically undisturbed muscles. Such additional possibilities for study are the m. tenuissimus, m. semimembranosis, the m. gracilis, and other common muscles about the thigh.[47]

The live anesthetized rabbit (Flemish Giant) was first used in this laboratory for single alpha motor axon stimulation. The intact neuromuscular system was studied in order to preserve the Golgi tendon organ[142,162] and spindle (gamma loop) innervation,[77,78] as well as the circulation; and to collect data for comparison with that of isolated muscle-nerve preparations to follow.

There are two nerves that penetrate the anterior tibial muscle of the rabbit.[47] The more proximal one, a branch of the common peroneal nerve, is easily identified and isolated. An alpha motor axon penetrating the superficial portion of the upper section of the muscle was selected in order that the stimulated motor unit might be directly visualized as well as recorded from. The next branch of the common peroneal nerve in the rabbit innervates the lower section of the anterior tibial muscle and part of the extensor digitorum longus muscle.[47]

The upper nerve to the anterior tibial muscle can easily be followed into the muscle where its axons begin to give off branches. The most superficial alpha motor axon was selected and the remainder gently teased away from it with a minimum of trauma. This was done in order to eliminate undesirable ephaptic stimulation which would have defeated our attempt to excite a single motor unit. After the procedure was completed, the solitary alpha motor fibre used was carefully inspected at higher power (×80), in order to be certain that it consisted of only one filament. In some instances, stained preparations of the fibre used were mounted for microscopic verification (Fig. 158).

In order to stimulate the single axon and thus make possible recording from the single motor unit under direct vision, the experimental apparatus shown in Figure 157 was em-

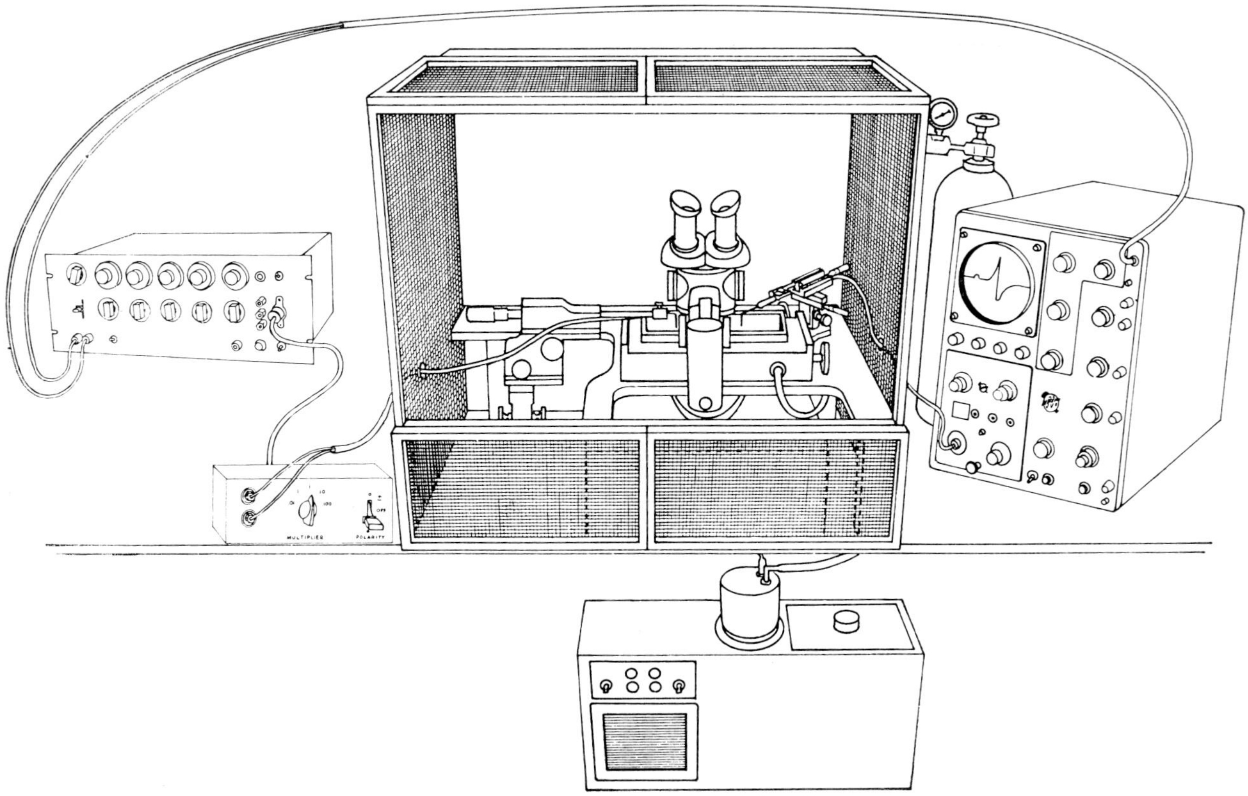

Figure 157, *Top,* Apparatus for microstimulation of and recording from neuromuscular preparation. *Bottom,* The experimental method for microdissection, selection of alpha motor neuron, and recording from the single motor unit of live animal.

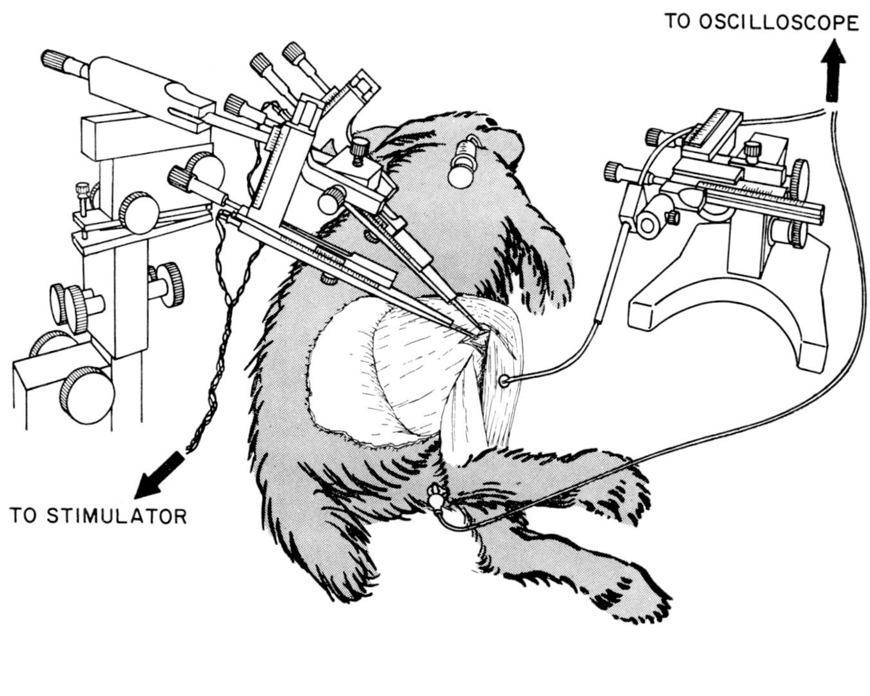

ployed. This apparatus was designed for work with neuromuscular preparations isolated from the animal. Tests were first made, however, on the intact animal after limited dissection of muscle and motor nerve.

The isolated neuromuscular preparation was placed in mammalian Ringer's solution. The end of the muscle and the tendon were sutured to posts placed for attachment. Oxygen and carbon dioxide (95%-5% mixture) was bubbled through the Ringer's solution from a perforated polyethylene tube lying on the surface of the chamber. The oxygen-carbon dioxide mixture was turned off during the recording. The muscle-nerve chamber was surrounded by an enclosed space through which water of controlled temperature was circulated. Desired temperature could thus be maintained, using the constant-temperature circulating bath which did not come into contact with the preparation.[48]

An alpha motor axon which penetrated a group of fibres near the surface was then placed on the stimulating electrodes. The electrodes were held in place by a Brinkman micromanipulator. A model 104A American Electronic Laboratories microstimulator was used, as well as the model 112 stimulus isolation unit.

Single stimuli of 1 to 10 v with time duration of 0.2 to 0.025 msec were used as single shocks. A second, smaller Brinkman micromanipulator was used to place a surface electrode over the region of the muscle innervated by the axon selected. A second electrode (wire tip) was placed in the inactive tendinous region. The two electrodes were connected to the input of a type-532 Tektronix differential oscilloscope preamplifier. This recording technique had the advantages of the differential system of amplification as well as the more realistic but difficult single-ended system. The value of the muscle action potential is compared to the nonfluctuating value found in the electrically inactive tendon. The muscle action potential was visualized on and recorded from the screen of the No. 532 oscilloscope. The small stimulus artifact preceded the muscle action potential. A frequency bandpass of 0.2 to 60,000 Hz was used. The oscilloscope sweep was triggered by the microstimulator just before the stimulus output (Fig. 159).

The stereo dissecting microscope previously used to select and manipulate a large alpha motor axon into position over the stimulating electrodes was then used to view the contraction. Provisions were made to expand its range of view by mounting it on a level platform under the stage supporting the preparation and bath (Fig. 157). The Zeiss stereo dissecting microscope was adaptable to stereo microphotography. The microscope and micromanipulators were grounded to the surrounding, shielding cage (Fig. 157). The enclosure itself along with the oscilloscope and stimulator were connected to the common ground.

Prior to single motor (alpha) nerve stimulation, the entire nerve trunk was stimulated. A bundle of fibres was then freed after carefully sectioning the outer fibrous sheath of the peripheral nerve. Records were made from the contracting portion of the anterior tibial muscle as the nerve bundle was stimulated (Fig. 159). It was then necessary to cut the tough fibrous sheath of the nerve bundle (fasciculus) in order to gain access to the single nerve filaments within it.

Records then made from the muscle in the region of visible activity, upon stimulation of the single alpha motor fibre, were very simple in character, had good amplitude and could be reproduced upon successive stimuli. The amplitude and other characteristics of the single motor unit action potential were unchanged as the magnitude of the stimulus was increased. The contraction of the motor unit could be easily visualized under low power ($\times 16$). The transverse dimension of the contracting region of muscle fibres was surprisingly large—over $\frac{5}{16}$ in. Thus, the motor unit in this species is undoubtedly divided into subunits, as described by Buchtal.[1] These subunits interdigitate with those of other motor units. This explains the difficulty in obtaining records representing a single motor unit in clinical studies using even a very tiny probe. The record is influenced by many nearby contracting fibres, some from other motor units, and is therefore a complex tracing in contrast to the simple record of the actual motor unit (Figs. 159, 160).

After recovery of the single alpha motor neuron, it must by some means be placed across two stimulating, bipolar, ground-isolated electrodes. This would be a most awkward task except for the double-electrode technique described below.

The procedure has been greatly facilitated by adaptation of a Brinkman C II micromanipulator with its precision micrometer which covers a range of ½ in, moving a carefully fitted rod 6 in. in length; its diameter is 3⁄16 in. On this adjustable rod are mounted two additional adjustable devices—two RP III x-y units—in such a way as to permit identical adjustment, and each x-y unit bears one pair of electrodes, as demonstrated in Figures 157 and 175.

A flat silver electrode is positioned directly on the right anterior tibial muscle surface after visualizing the region of contraction as the chosen fasciculus is stimulated. As explained below, this is clearly the *recording* electrode.

Its position may have to be adjusted after the precise location of the active region following stimulation of the single alpha motor neuron is observed. Since a differential amplifier is used, a second electrode is placed in an area of known negligible activity or complete electrical inactivity. Such a site is the Achilles tendon. Another useful site is the calcaneus, into which a small conducting screw may be inserted. This is the *indifferent or reference* electrode. Since the display on the oscilloscope screen is equal to the difference between the potential at the active portion of the muscle and that at the indifferent electrode, the displayed result is most readily understood when the indifferent electrode is placed at a site of zero or nonfluctuating potential. In this manner, one may take advantage of the artifact-cancellation effect of the differential amplifier, yet demonstrate absolute potential changes at the point of interest (Fig. 166).

This type of electrode placement where one side of the differential amplifier is connected to the area of interest and the other side to an isoelectric region is useful in studies such as this where a single motor axon is stimulated. This technique is clearly useless in muscle function studies; specificity is lost, since signals at the region of the indifferent electrode become unavoidable. This should be considered the only acceptable method in electrodiagnosis. Here, fibrillary potentials may be recorded from the site of the probing electrode. Small potentials at this electrode are not expected to influence the indifferent electrode, though the tervening tissue acts as a volume conductor. The high impedance of the tissues comprising the volume conductor suppresses high-frequency signals, such as the fibrillary potential, at the indifferent electrode.

Our clinical studies have demonstrated the single-ended technique under many conditions, with results identical to the above-described application of the differential mode where B is the indifferent electrode.

The differential mode, with the electrodes about one centimeter apart, remains the ideal technique for muscle function analysis. The recording is confined to the muscle under study. Amplitude becomes less important and represents the difference between the signals from two active electrodes.

The fasciculus which is chosen and placed first upon the RP III x-y micromanipulator bearing the first set of electrodes is stimulated at the usual voltage and time duration, approximately 8 v and .05 msec. Care is taken to maintain the environment about this neuromuscular preparation by constant application of small amounts of Ringer's solution. Using the second set of electrodes, which can be easily manipulated with the second small RP III x-y micromanipulator mounted also upon the C II micromanipulator, one of the axons freed from the fasciculus on the first set of electrodes is then easily manipulated into place on the second pair of electrodes. Next, an electrical switch directs the stimulus from the first set of electrodes to the second set, and the single alpha motor neuron may be stimulated. Redundant Ringer's solution is carefully removed by suction just prior to stimulation through the microelectrodes (Fig. 157). An alternative to Ringer's solution, which may reduce the possibility of spread of the stimulus due to the effect of the action of tissue as a volume conductor, is mineral oil. This was used in some experiments in a series of 24 cats, a study involving recording of the single motor unit and its tension, reported later in this chapter.

The motor unit and nerve selected must be superficial, minimizing surgical trauma and exposure. The anterior tibial muscle may be exposed, the recording electrode put in place, and motor nerves to superficial motor units isolated without dissecting the structures at all. The common peroneal nerve and a branch to the region of the upper one third of the anterior tibial muscle are superficial and are close to the desired area for stimulation. It is essential to stimulate close to the penetration of the axon into the muscle. This prevents the complication of ephaptic stimulation.

The identity of the single axon as such was established by direct microscopic vision at 80 to 200 power. In some instances, the structure was recovered after the experiment, fixed, and stained for compound microscope visualization (Fig. 158). Gradual increase of stimulus voltage was also used to ascertain that the structure stimulated was a *single* axon. Increase of stimulus voltage from the least voltage required to stimulate the motor unit to much larger stimuli does not change the characteristics of the action potential if the structure stimulated is a solitary axon.

At certain sites, the ephapse, in this mode of recording, cannot complicate the result. It is after selection of the appropriate branch of the common peroneal nerve, the branch more cephalad of the anterior tibial muscle innervation, that one must take the greatest care in the dissection to follow. Ideally, a single alpha motor neuron is followed into the upper portion of the anterior tibial muscle and eventually placed upon the second pair of stimulating electrodes with minimal dissection. At this point, ephapse would not be recorded even if produced later. It would not influence the recording, since the single motor unit, with its innervation very close to the electrodes, would be displayed long before any possible ephaptic stimuli could pass over their long routes proximally and back to the muscle. Occasionally, we have observed small action potentials 6 or 7 msec after the single motor unit display has been completed.

Through the entire procedure, meticulous care must be taken in handling each of the involved structures, particularly the myelinated axons. The myelin of the alpha motor neuron turns out not only to be an excellent insulator but physically protects the axis cylinder inside from the tools of the dissector. At this point, the special fine instruments—the smallest obtainable—adjustable hand rests, full arm rests and favorable body position are essential to the procedure.

Since it then becomes necessary to work at 40 power, the small micromanipulators bearing each set of electrodes make the rest of the work possible. These fine instruments, so uniquely adapted, make it possible actually to do the work of this type under higher power, a most desirable development. It is not unusual to transfer the single axon from the fasciculus to the second set of electrodes at 80 power.

It may be necessary to repeat the process of isolation and stimulation of single axons. It is repeated until an axon is found which upon stimulation activates superficial muscle fibres of the anterior tibial muscle. It is not possible by this method to ascertain the depth of the motor unit involved; it is seen only in relation to the surface of the muscle. During some of the experiments it was not unusual to isolate 15 to 20 single axons before alpha motor neurons with suitable muscle fibre distribution were found.

Some of the most typical of these structures were photographed, using the Contarex Special Camera adapted to the Zeiss dissecting microscope. The 6.5 × 9 cm Zeiss Attachment Camera with Polaroid film pack was also used because of larger picture size. The Attachment Camera eliminated the longer exposures required by the slower 35 mm film. Because of vibration at the end of a long tube, shutter

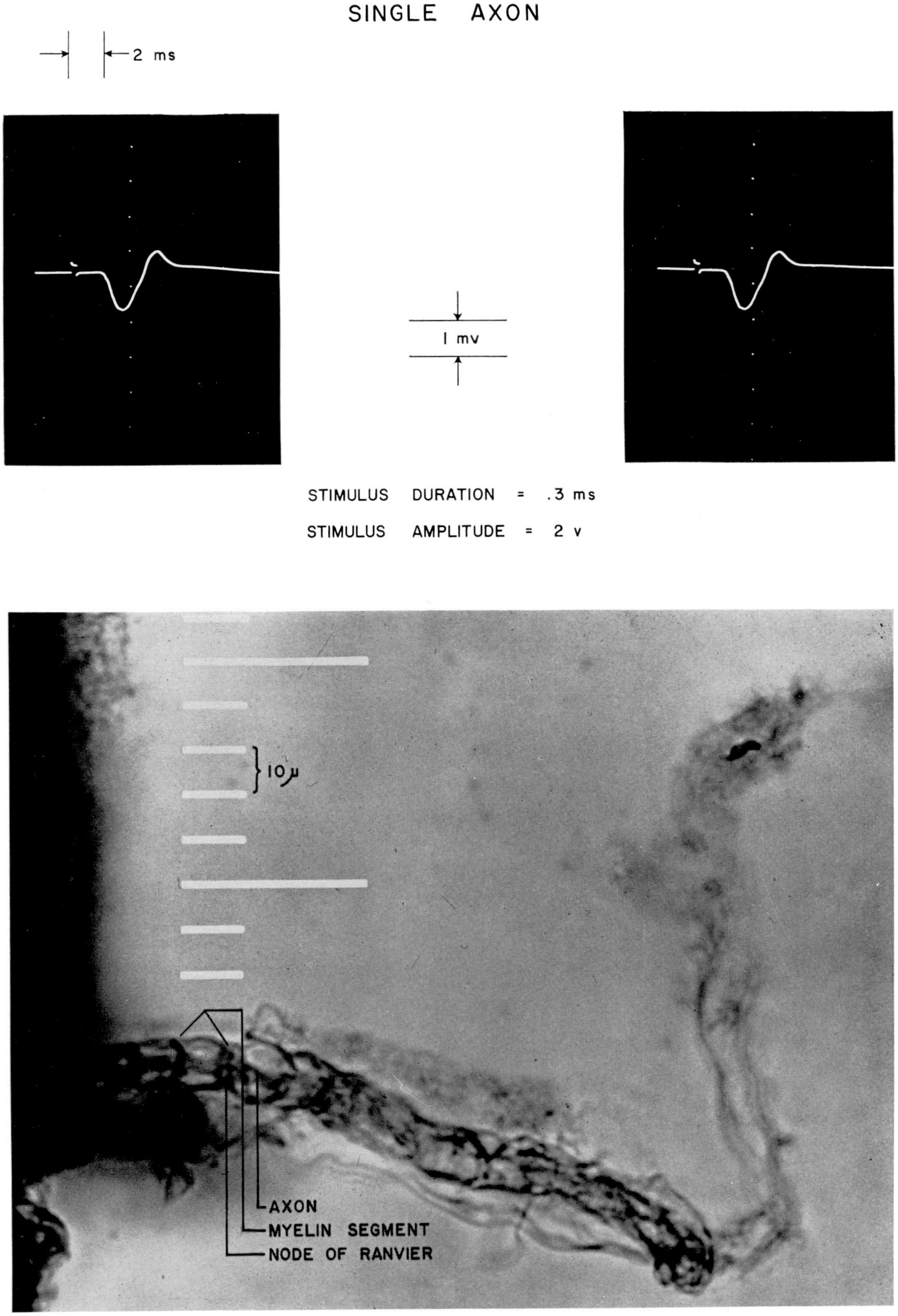

Figure 158. Single motor unit action potentials recorded following stimulation of isolated alpha motor neuron. The single axon isolated and stimulated is demonstrated.

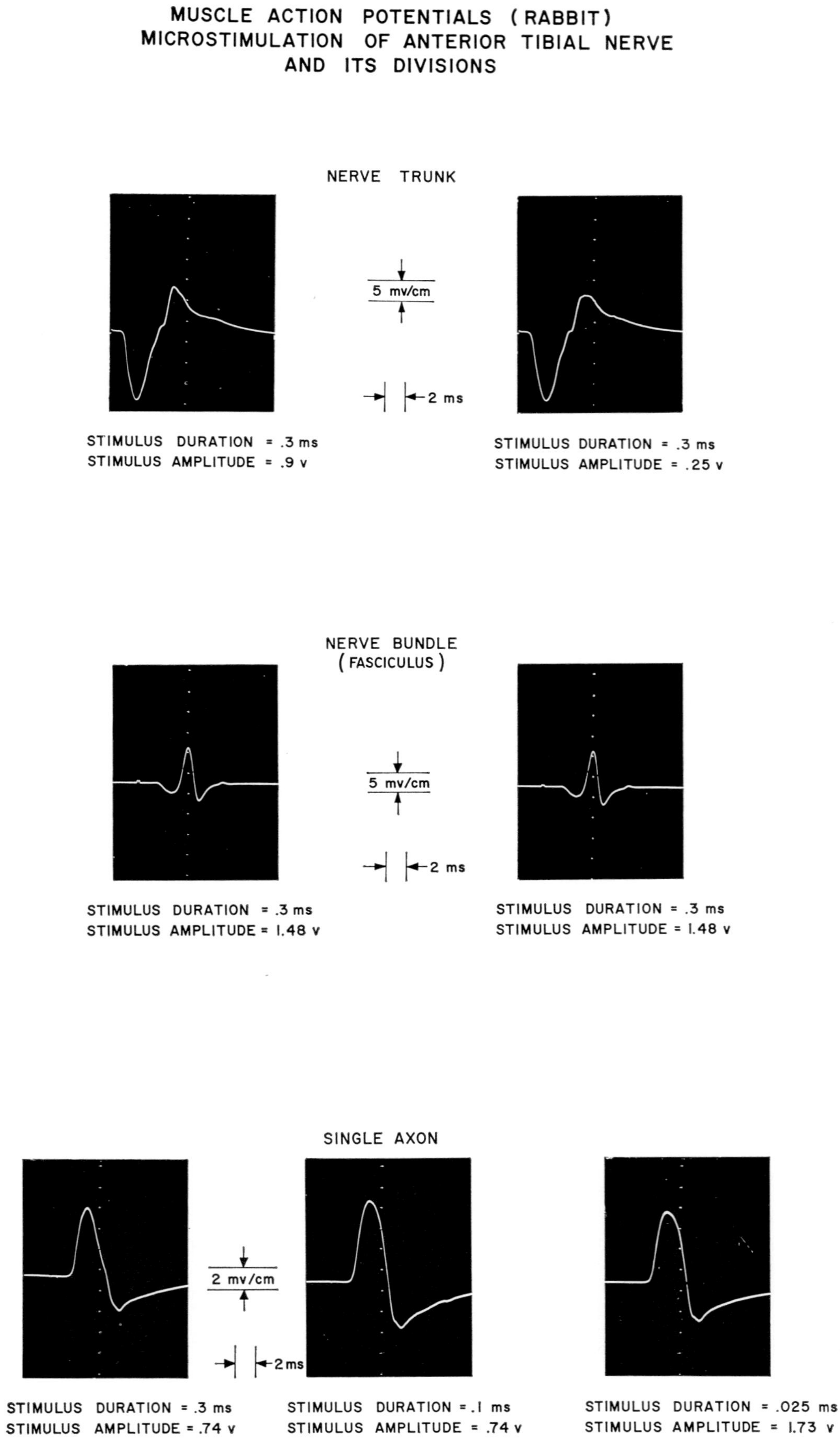

Figure 159. Single muscle action potentials obtained by microstimulation of whole nerve, nerve bundle, single alpha fibre (anterior tibial muscle, rabbit.)

speeds of 1/10 or 1/15 second possible only with the Attachment Camera were most helpful. Such photographs were taken particularly when a series of matching action potentials had been photographed from the tube of the cathode ray oscilloscope. The recording instrument used is the Tektronix Oscilloscope 530 Series with differential *alternating current coupled* preamplifier with input 10 megohms paralleled by 50$\mu\mu$f. The oscilloscope screen is photographed using the Tektronix Oscilloscope Polaroid camera.

Electrical stimulation by square wave impulses of the desired magnitude and time duration is provided by the American Electronics Laboratory Stimulator (104-A), employing an interposed Stimulus Isolation Unit Model 112. The stimulus isolation unit is connected to the stimulating electrodes. The larger pair on the first portion of the micromanipulator are 175μ in diameter and are placed 3 mm apart.

The second pair of microelectrodes, stainless steel or copper wire of 75μ diameter and glass enclosed, are placed on the second RP III x-y Brinkman micromanipulator, itself mounted on the micrometer arm of the larger CII Brinkman micromanipulator. The tips of the second pair are curved slightly upwards. They are 1.5 mm apart. These known diameters in the field help to select and identify structures by acting as calibrators, in addition to the ocular calibration reticule.

When using the 6.5 × 9 cm Attachment Camera, calibration of image size is directly visible in the horizontal focusing eyepiece.

Action potentials are displayed on and recorded from the 5-in. screen of the oscilloscope with a P 11 phosphor. The sweep is externally triggered by the stimulator 3 msec before the stimulus occurs. The necessary adjustments are made for stimulator voltage output, stimulus time duration and the oscilloscope adjustments of frequency recording band and sensitivity. The oscilloscope sweep speed is set at 2 msec/cm for most of the recordings.

The Zeiss stereo dissecting microscope employed has a range of ×2.5 to ×200. Two of the most useful dissecting instruments are needles with rectangular tapered stainless steel handles with small pin chucks into which are inserted half-curved cutting needles (e.g. Anchor brand, #1821–20). These needles are easily replaced. They can be used at magnifications of ×32 to ×80, using hand rests. The micromanipulating arrangement described above or previously greatly enhances the power to work at even greater magnification. Objective to object distance equals *8.0 cm*. Objective to object distance with ×2 objective lens added equals *2.8 cm*.

The animal, microscope, handrests, micromanipulators, electrodes and microscope light are screen-shielded. The microscope light is turned off during recording (it is preferable to disconnect it). Compartments of the shield will open. They may be closed for recording. Both the stimulating and recording electrodes are shielded by the screen. The screen is grounded to a previously tested water pipe or conduit. The oscilloscope and microstimulator chassis are grounded at the same point. Thus, all of the instruments within and without the screened enclosure are grounded to the screen, which in its entirety has a common ground.

The circular silver metallic recording electrode, with a diameter of 9 mm, was selected in order to assure the inclusion of all of the activated surface fibres. Accurate dimensions of the motor unit are unknown. It was desirable that the single motor unit action potential recorded represent all of the subunits of which the single motor unit is composed.

One must consider the possibility that the use of an electrode of this shape might result in the recording of a distorted action potential, one of incorrectly long duration compared to that of a narrow, transverse electrode (Fig. 157). A few preliminary tests using a narrow transverse recording electrode demonstrated very little difference in the waveform.

The recording from the circular electrode, in direct contact with surface fibres, may be influenced only by the fibres of a single motor unit. The recording appears to reflect more than the potential change at one point of contact. The duration of the pulse compared to

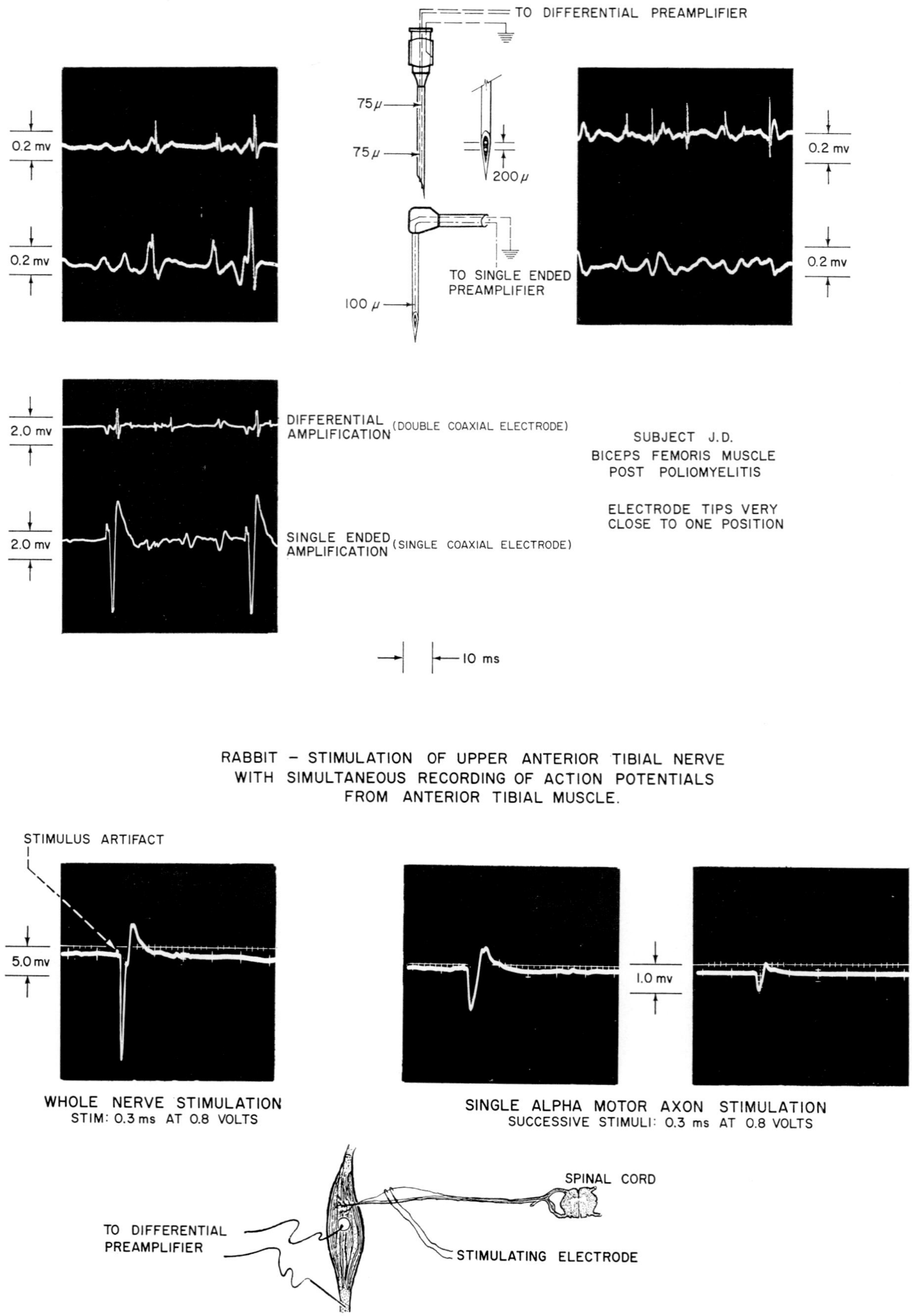

Figure 160. Recordings using double coaxial and single coaxial electrodes (*upper three records*). The third tracing is from polio muscle. Records from rabbit anterior tibial muscle following microstimulation (*lower three records*).

the time easily calculated that is required for the impulse to pass an electrode of this size is greater. Therefore, the impulse begins to influence the recording at some time between the activation of the motor unit and the time at which it reaches the contact disc electrode. This is expected, since the activated portion of muscle behaves as a volume conductor.[122]

As the action potential passes the recording electrode, its influence begins to diminish. It is not known whether the beginning and end of each pulse are determined by the times of occurrence of these events or by the extent to which the tissue acts as a volume conductor.

Of significance is the rather abrupt onset of each recorded waveform, indicating that the volume conductor effect is not a consideration here, since a single motor unit is activated. This is unaltered as amplification is markedly increased. At faster sweeps, the onset of the action potential curve is demonstrated to be less abrupt (see Fig. 168), which demonstrates recordings made at a sweep speed of 1.0 msec per division rather than the usual sweep speed of 2.0 msec per division (Fig. 162).

RESULTS

The many single motor unit action potentials recorded by the previously described method are of remarkable consistency (Figs. 162, 163, 167, 168). They are reproducible, provided the environment or surrounding fluids remain approximately the same. Many preparations, using the same single axon, in this manner produced similar action potentials up to 20 in number over a period of 1 hour. Consistently, the amplitude observed was 4 to 8 mV. The time duration of the entire complex varied according to the input type used—alternating or direct current coupled—but the time duration of the primary initial single spike, in either case, was approximately 2.0 msec. This was followed by a negative deflection (hyperrepolarization) of low amplitude and very short time duration (Fig. 167). The initial primary deflection demonstrated more variation in amplitude than time duration. Duration was almost precisely constant. Since motor units are known to vary in size, some variation in amplitude is expected. The configuration and dimensions of the recording circular metallic electrode could have influenced the characteristics of the action potential.

It was considered necessary to demonstrate that no portion of the motor unit action potential recorded was conducted by the reference electrode. Reference electrode artifact (here) may be attributed to the ability of the tissues of the extremity and other body parts to function as volume conductors. In this experiment, known isoelectric regions were chosen for placement of the "indifferent" electrode. Studies were made clearly demonstrating the absence of reference electrode influence on the action potential. For this demonstration, a tracing was taken by the usual method, using the circular metallic disc recording electrode in contact with the anterior tibial muscle at the point of maximum activity. The right Achilles tendon was used as a reference electrode. Without changing any of the other conditions, the point of contact of the reference electrode was changed to the left tendo Achillis. As is demonstrated in Figure 166, there is no discernible difference between the traces of the two action potentials thus obtained. Comparisons were also made by stimulating the same neuron of the anterior tibial muscle but with recording electrodes in the right tendo Achillis and left tendo Achillis; each electrode was lead to the differential amplifier. Here, no difference in electrical potential was found; a straight line resulted on the oscilloscope screen (Fig. 166).

Further comparisons of a similar type were made where the reference electrode B was inserted firmly into the *right* calcaneus. The resulting record closely approximated that made when B was inserted into the left calcaneus. Records were similarly made comparing the right tendo Achillis to the right calcaneus and to the left calcaneus (Fig. 168).

The above are realistic applications and tests of the differential technique. One may take advantage of its artifact cancellation feature and record absolute potential changes if one electrode is in an isoelectric site. This may represent the most dependable and meaningful clinical method for electrodiagnosis (see Fig. 4A).

Earlier studies demonstrate action potential complexes using the 530 series Tektronix oscilloscope with low-level *alternating current coupled* preamplifier. The low-frequency setting previously tested was known to be 33 Hz. This setting was selected to eliminate base-line drift. However, the series capacitor necessary in such an input caused artifact. This artifact consisted of delay in return to the isoelectric point as the motor unit action potential completed its terminal deflection below the base line (Figs. 162, 163).

Records made on the 502 Tektronix oscilloscope, using the *direct current coupled* differential input, were quite similar in configuration, except for what can be considered immediate return to the base line, as the terminal negative deflection promptly returned to the base line (Fig. 167). The artifact attributed to the series capacitor is due to the initial positive spike, which charges it. As the initial positive spike terminates, the capacitor discharges, altering the portion of the waveform that represents hyperrepolarization in the case of *alternating current coupled* recorded potentials. The *alternating current coupled* and *direct current coupled* curves of Figure 167 demonstrate a difference —a delay in final return to the isoelectric line— attributable to *alternating current input coupling* condensers. This sometimes-undesirable feature of the *alternating current coupled* differential preamplifier is often compensated by the ease with which control or stability of its trace is attained.

However, the *direct current coupled* differential preamplifier at high sensitivity is difficult to control; its trace often disappears prior to stimulation. It appears to have great advantages in accuracy of reproduction for single traces. This would not be true where it is required to produce a long series of action potentials, as required for diagnostic examination and muscle function analysis.

The failure of the *alternating current coupled* recorded trace to return as rapidly to the base line as the *direct current coupled* trace (Fig. 167), is due to the condenser in series required for the lower frequency setting employed.

Records made to show the isoelectric state of the tendo Achillis or calcaneus are shown in Figures 166 and 168. It is possible to show very easily that the single motor unit action potential is virtually the same when recorded with the A electrode (disc) placed over the muscle and the B electrode placed in the tendo Achillis, or with the B (indifferent) electrode placed in the calcaneus. Furthermore, direct comparisons of the potential at the tendo Achillis to the potential at the calcaneus by using one as the A input and the other as the B input demonstrated complete absence of electrical activity at these sites. The method thus makes use of the desirable differential artifact-cancellation effect, yet gives an absolute value of the true potential at the A electrode, since, as we have shown, the B electrode may be placed in an isoelectric region. Since the action potential is unaltered by changing the site of the indifferent electrode from one calcaneus to the other, the volume conductor (the intervening tissue) does not contribute to the form of the action potential.

If such a technique—the use of recording and indifferent electrodes (using the differential mode)—were used to record action potentials produced by voluntary contraction (clinically), the above would not hold. The volume conductor would include many active muscles between the electrodes. This would produce a hopelessly confused record. The resulting signal, A–B, would represent the difference between two unknown values. However, this technique is ideal for diagnostic purposes in recording fibrillary potentials from *resting* muscle. It is essential that all intervening musculature be inactive (Fig. 4A).

Repetitive demonstrations of single motor unit action potentials indicate several characteristics of this waveform. The conditions essential for standardization of the stimulation and recording techniques are described below.

Eccles and Sherrington[58] found that a significant number of motor axons of several limb muscles of the cat dichotomize in the nerve trunk several centimeters proximal to the myoneural junction. The nerve to the medial head of the gastrocnemius was found to contain 393 motor fibres 60 mm from the muscle; there were 482 motor fibres 6 mm from the muscle. Thus 82 percent of the alpha motor neurons represent an anterior horn cell and are part of a single motor unit in this muscle if selected 6 mm from the muscle, assuming that dichotomy is not a significant factor proximal to 60 mm. The remaining 18 percent do not represent complete motor units. For true motor unit value, stimulation should be done 5 or 6 cm proximal to the myoneural junction. Rarely, there is even division proximal to this point.

The results of this study provide further information regarding the frequency spectrum of the action potential of the motor unit. Such information is essential in the design of amplifiers for recording of the action potential. The characteristics of the waveforms of the single motor units obtained dictate the frequency range of the preamplifiers necessary to record the event with highest fidelity (Fig. 161).

The records from the single motor unit (Fig. 160) indicate that the single sharp spike, with a duration of 1.8 msec, is the significant portion of the action potential of this motor unit. The primary spike represents depolarization and repolarization. The accessory accompanying oscillations obtained clinically along with a sharp spike are attributable to subunits of other interdigitating motor units and subunits in the volume of muscle near the electrode tip. Accessory fluctuations are thus not part of the action potential of the single motor unit. The sharp spike of the single motor unit is characteristically followed by a *single* oscillation of opposite sign representing hyperrepolarization (Figs. 159, 168).

The essential information gained from this study is the fundamental character of the action potential of the single motor unit. The critical dimensions of the sharp spike of the action potential—its amplitude and time duration—varied somewhat. This is to be expected, since the technique is subject to some variation and the motor units themselves, as well as their axons, may vary in size.[58] Every effort was made to stimulate the main axon proximal to the beginning of branching. However, the axon may sometimes branch proximal to the site of stimulation, even in the spinal cord—the recurrent collaterals of the axon.[149] We therefore do not expect the characteristics of the action potential to be the same in each experiment. We have, at any rate, collected data on which we base the calculations that follow. The variations in time duration and amplitude were not large. All of the data presented here were taken from the live, anesthetized animal.

In Figure 160, action potentials are shown following successive stimuli of the single alpha motor axon. The diminished amplitude in the second recording suggests that only part of the motor unit of the axon stimulated responded. This leads us to question the application of the all-or-none law to the motor unit. As the action potential of the axon becomes smaller, it seems reasonable to conclude that it may excite only a portion of the motor end plates of the stimulated axon's branches.

The validity of the all-or-none law regarding its application to the motor unit has been questioned by Ralston. He and his co-workers have observed variations of action potential size at different muscle lengths. They have suggested that as the muscle is subjected to increasing passive stretch, some of the fibres of the motor units fail to contract. This results in a smaller motor unit action potential. We have demonstrated that the onset of tension is delayed as passive tension is increased (Fig. 176).

It is known that cholinergic nerve fibres, even during periods of inactivity, discharge "packets" of acetylcholine molecules at a very low rate, perhaps one per second. When the fibre is activated, the rate of discharge of packets of acetylcholine is greatly increased in the order of many thousands per millisecond.[107]

If the rate of discharge of acetylcholine is subject to significant variation, this could sometimes result in failure to reach all of the end plates, defeating the all-or-none concept.

DETERMINATION OF OPTIMUM RECORDING FREQUENCY LIMITS BY THE APPLICATION OF THE FOURIER TRANSFORM TO THE ISOLATED TRIANGULAR SPIKE

The upper frequency limit of electromyographic equipment is an important consideration in the accuracy of the reproduction of action potentials. If the recording sensitivity is low, at high frequencies, sharp spikes are dulled and drawn out. The objective of this study is to establish an optimum recording frequency range. The pulse is first analyzed to determine how much of the high-frequency component can be discarded before the loss becomes significant. The pulses analyzed are action potentials of the single motor unit.

Several action potentials are shown in Figures 159 and 160. They were obtained from rabbit anterior tibial muscle. A surface electrode ⅜ in. in diameter was placed in direct contact with the exposed muscle over the portion active on stimulation. An indifferent electrode was inserted into the tendon (differential mode). The records were obtained by photographing the sweep of the oscilloscope. The oscilloscope itself had maximum frequency response of .2 Hz to 60 kHz. The artifact of the stimulating pulse is visible slightly in advance of the action potential.

A representative impulse is shown in Figure 161. It is characterized by an initial sharp spike of 1.8 msec duration followed by a much

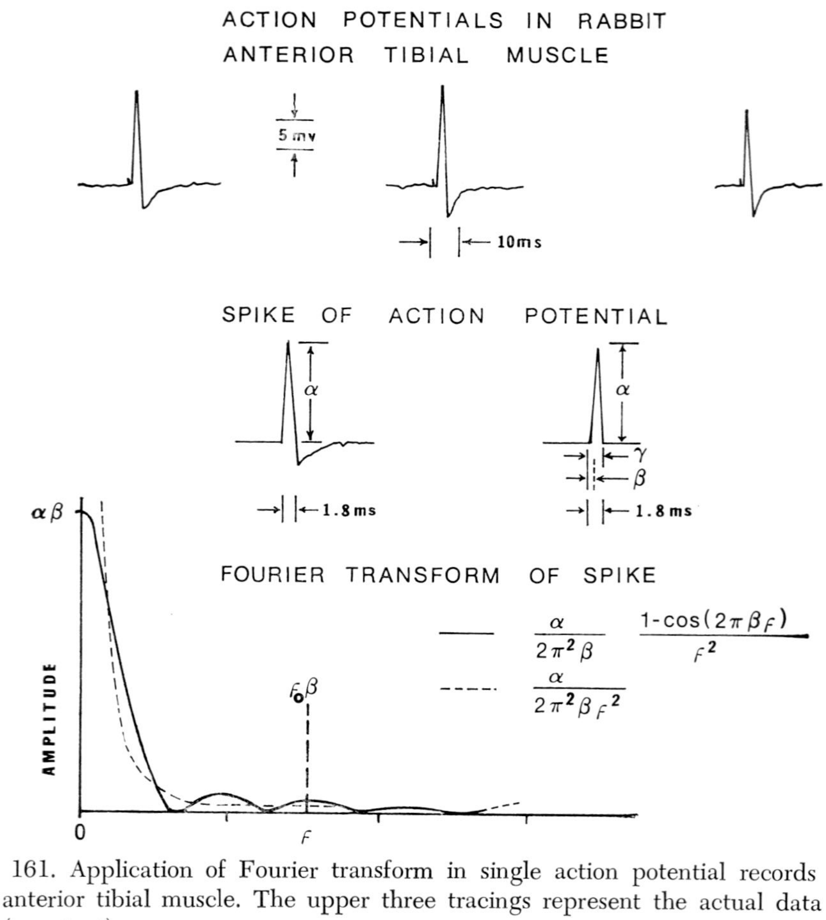

Figure 161. Application of Fourier transform in single action potential records from rabbit anterior tibial muscle. The upper three tracings represent the actual data analyzed (see text).

slower recovery effect lasting about 5 msec. If this impulse were analyzed for its frequency components, it would be expected that the sharp spike would have more significant high frequency components than the recovery effect. Hence the slower effect can be neglected and the analysis can be carried out on the isolated sharp spike (Fig. 161).

The Fourier transform A(f) of a waveform $\psi(t)$ is

$$A(f) = \int_{-\infty}^{\infty} \psi(t)e^{-2\pi i f t}\,dt$$

where f is the frequency, e is the basis of Naperian logarithms, 2.718 . . ., i is $\sqrt{-1}$, and t is the time variable. In this case:

$$\psi(t) = \begin{cases} & t < -\beta \\ \alpha^0 + \dfrac{\alpha}{\beta}t & -\beta \leq t \leq 0 \\ \alpha_0 - \dfrac{\alpha}{\beta}t & 0 \leq t \leq \beta \\ & t < \beta \end{cases}$$

where α is the amplitude of the spike and β is the half-width of the spike, as in Figure 161.

The Fourier transform of this function is

$$A(f) = \frac{\alpha}{2\pi^2\beta}\,\frac{1 - \cos \beta f}{f^2}$$

and its graph is shown in Figure 161. When the Fourier transform is applied to a waveform, it yields a function which relates the amplitude of a frequency component to that frequency. Referring to the graph, we find that the greater percentage of the components is found in the higher frequency lobes.

If the equipment has an *upper frequency* limit of f_0, the frequencies represented by the curve to the right of f_0 (Fig. 161) will be lost. To determine the amount lost, the integral of the transform from zero to infinity can be compared to the integral from f_0 to infinity:

$$\int_0^{\infty} \frac{\alpha}{2\pi^2\beta}\,\frac{1 - \cos(2\pi\beta f)}{f^2}\,df = \frac{\alpha}{2}$$

When βf is larger than 10 or 20, the average value of $(1 - \cos 2\pi\beta f)$ may be taken as 1. Hence,

$$\int_{f_0}^{\infty} \frac{\alpha}{2\pi^2\beta}\,\frac{1 - \cos(2\pi\beta f)}{f^2}\,df \approx$$

$$\int_{f_0}^{\infty} \frac{\alpha}{2\pi^2\beta}\,\frac{df}{f^2} = \frac{\alpha}{2\pi^2\beta}\,\frac{1}{f_0}$$

If the loss must be less than y, then:

$$\frac{\alpha}{2}y = \frac{\alpha}{2\pi^2\beta f_0}$$

It follows that:

$$f_0 = \frac{1}{\pi^2 y\beta}$$

or, since $\gamma = 2\beta$,

$$f_0 = \frac{2}{\pi^2 y\gamma}$$

The minimum *upper frequency* limit for a loss of less than y part of the waveform due to high-frequency attenuation is f_0. It must be emphasized that this loss is not evenly distributed over the entire waveform. This would preserve the shape. Rather, this loss results in a distortion of the waveform more than a diminution. The corners are rounded off and the pulse assumes a bell shape.

For the representative impulse of Figure 161, a loss of less than 5 percent will imply:

$$f_0 = \frac{2}{\pi^2 \cdot 0.05 \cdot 1.8 \cdot 10^{-3}} = 2.3 \text{ kHz}$$

and for a 10 percent loss,

$$f_0 = \frac{2}{\pi^2 \cdot 0.1 \cdot 1.8 \cdot 10^{-3}} = 1.1 \text{ kHz}$$

It is apparent now that the upper-frequency limit must be at least four to five times the "fundamental" frequency of $\frac{1}{\gamma}$, (0.6kHz).

The permissible *lower-frequency* limit (f_1) may be calculated similarly. Where f is small, A(f) is approximately $\alpha\beta$. Then the content lost is $\alpha\beta f_1$. Stipulating that this be less than y part of the integral of A(f) from 0 to ∞,

$$\alpha\beta f_1 = \frac{\alpha}{2}y$$

Then

$$f_1 = \frac{y}{2\beta} = \frac{y}{\gamma}$$

For y = 5 percent and $\gamma = 1.8$ ms,

$$f_1 = \frac{.05}{1.8 \cdot 10^{-3}} = 28 \text{ Hz}$$

A permissible frequency bandpass, based upon the data of this experiment, is approximately 30 Hz to 5 kHz.

The need for standardization of clinical recording equipment is obvious. Units incapable of reproducing the above range produce distorted, misleading records. There is now no question that we must uniformly employ equipment capable of recording well over 10,000 Hz if we are to produce realistic records of motor unit action potentials.

The above data and the permissible frequency recording band apply to the motor unit of the *rabbit anterior tibial muscle.* The permissible frequency range for recording the fibrillary potential of pathological human muscle is not included in this discussion, since the frequency spectrum described above is based on data taken from rabbit muscle. It is anticipated that similar data taken from human muscle will indicate a frequency spectrum extending into a considerably higher-frequency range.

The stimulus voltage appropriate for the procedure was found to be about 8 or 9 v. This is after the correction necessary due to the stimulus isolation unit. The time duration of the stimulus should be less than 0.1 msec. Stimuli of short time duration are less injurious to small nerves than longer stimuli, though lower voltage is required. This is especially true when the single alpha motor neuron is stimulated. Here it is almost imperative that stimuli be less than 0.1 msec, since longer stimuli applied a few times produce nerve destruction, making successive stimuli more subject to variation. After a few shocks, the structure is often too attenuated to permit normal physiological identity. It is therefore a desirable procedure to use a series of short stimuli using the data from only the first two or three in the series. Stimuli in one experiment were performed at time intervals of equal duration and the resulting action potentials compared (Fig. 163). The stimuli, fired at 10-min intervals, were of the same voltage and duration. The single motor unit action potential is repetitive to a striking degree.

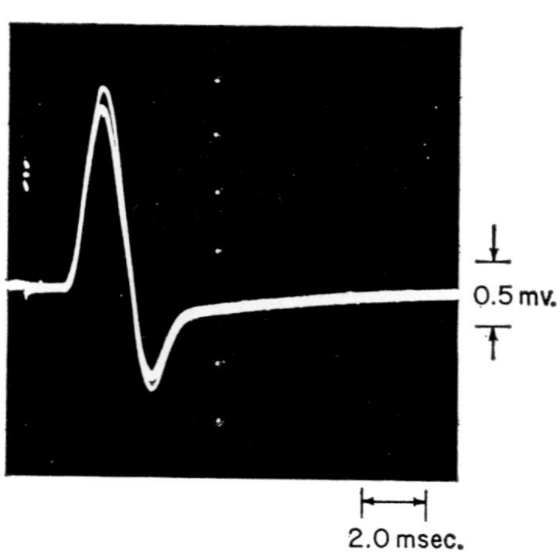

Figure 162. Single motor unit action potentials recorded from the same motor unit at 10-min intervals, recording conditions unchanged. (From Close, Maurer and Poor, *Clinical Orthopaedics and Related Research, 42:*171–190, 1965).

Deterioration of the action potential is soon observed when high frequency repetitive pulses are used to stimulate the *single* axon. The persistence of discreet action potentials, as the contraction becomes tetanic,[150] applies to composite nerves. The rate of the stimulus used was 67 per second.

The photographed action potential readily demonstrates any variation in magnitude. The magnitude of the action potential in millivolts is immediately obvious, due to calibration of the oscilloscope graticule in centimeters. Although the camera reduction is 1 to 0.9, the 1-cm intervals are clearly identified. On one graticule, a row of points are placed 1 cm apart (Fig. 164). On another illuminated graticule, horizontal and vertical lines are placed 1 cm apart (Figs. 167, 168).

The readily available stimulating electrodes are maneuvered with ease and convenience very close to the point of penetration of the alpha motor neuron into the muscle. As shown in Figure 164, one may work at a point very close to the myoneural junction. No known difference would exist with the stimulating elec-

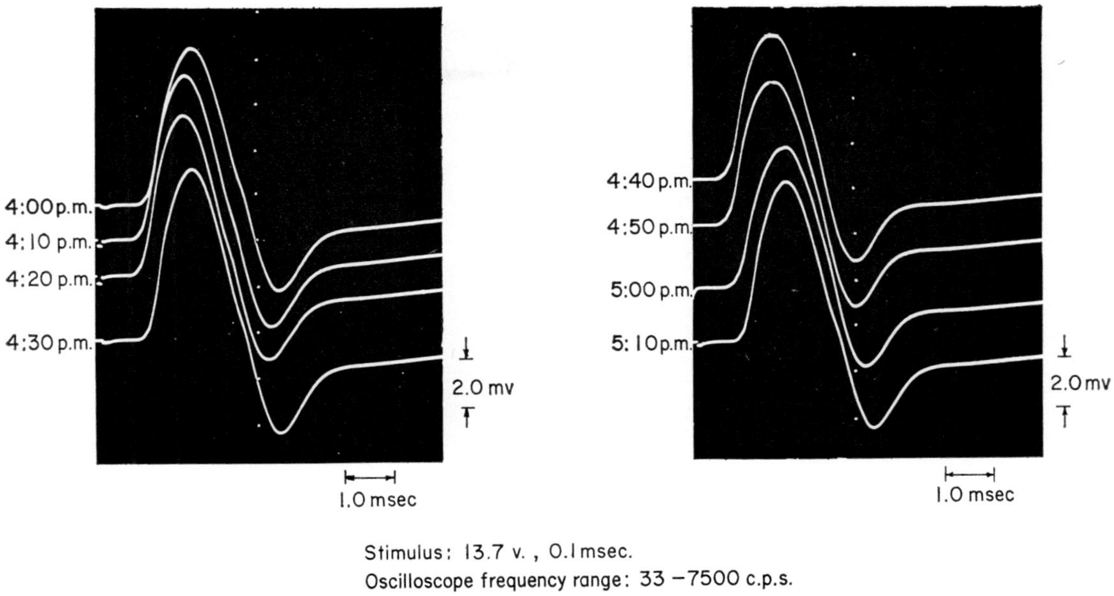

Figure 163. Muscle action potentials recorded at 10-min intervals, over a period of 70 minutes. (From *Clinical Orthopaedics and Related Research, 42,* 1965.)

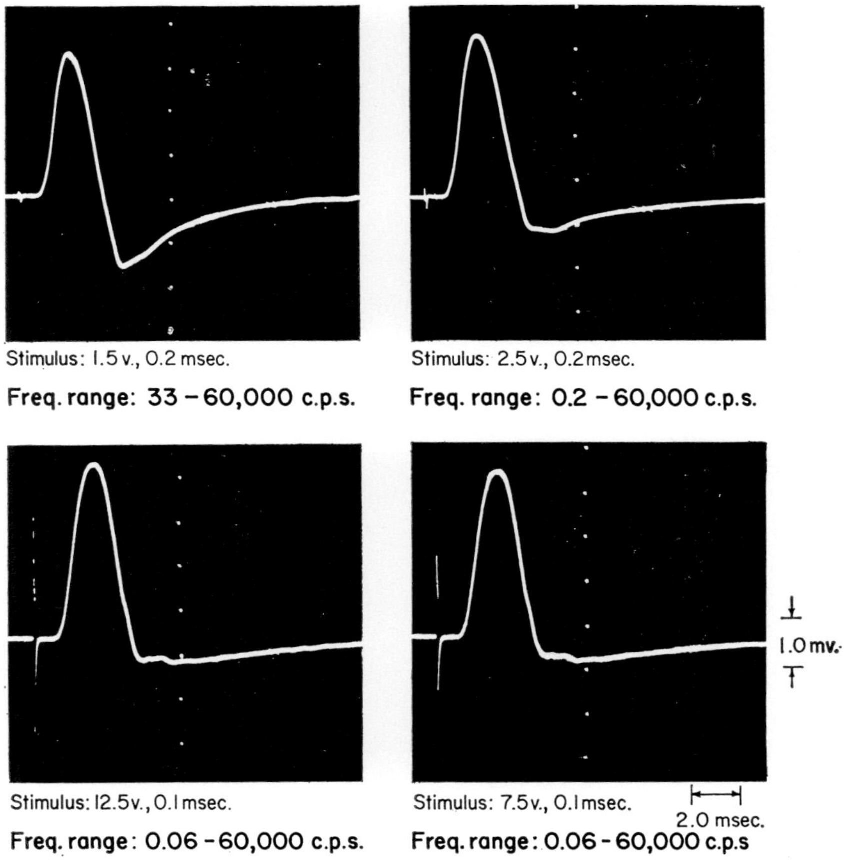

Figure 164. The effect of lowering the lower frequency bandpass setting. (From *Clinical Orthopaedics and Related Research, 42,* 1965.)

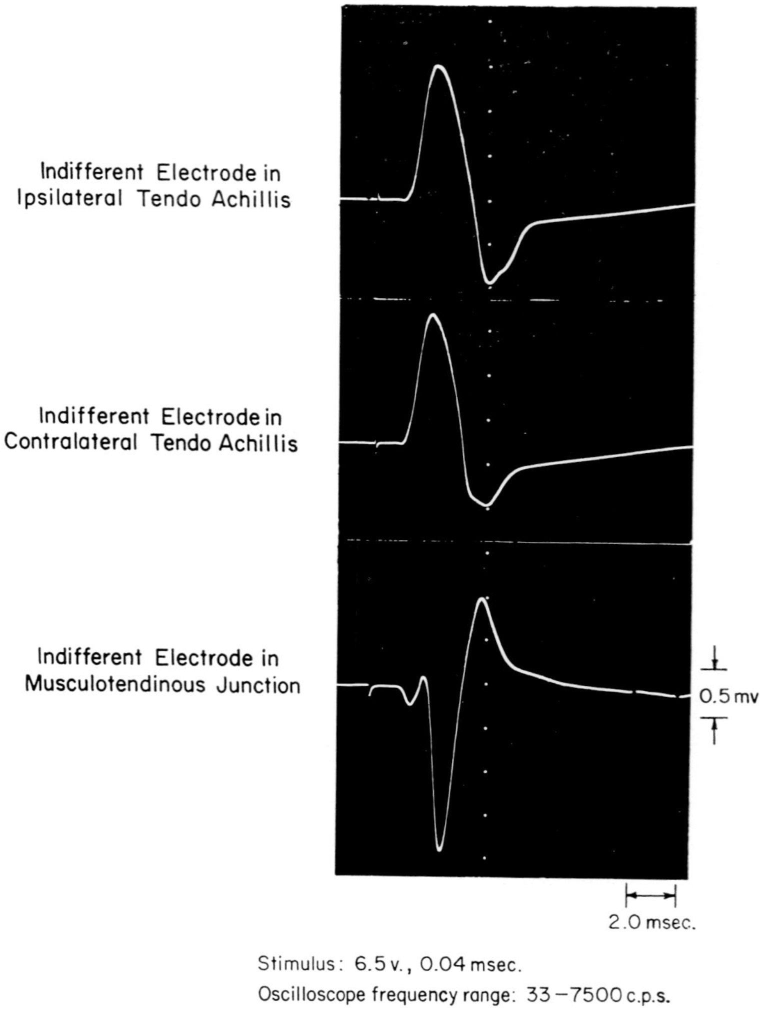

Figure 165. Single motor unit action potentials, indifferent electrode in (a) ipsilateral tendo Achillis, (b) contralateral tendo Achillis, (c) musculotendinous junction; position of the recording electrode is not changed. (From *Clinical Orthopaedics and Related Research, 42,* 1965.)

trodes at somewhat greater distances (Fig. 166, 167) if ephapse as well as confusion due to the effect of the volume conductor could be ruled out. Ephapse may be very difficult to rule out, and we have taken steps to assure stimulation of the single motor neuron where it is intact and not in a position to interfere or "produce a short" with adjacent motor structures. The possibility of retrograde transmission of the impulse by an ephaptic pathway to the spinal cord and back to the myoneural junction is, of course, ruled out by the short time interval demonstrated between stimulus artifact and the appearance of the motor unit action potential. Since the nervous system is intact, evidence of ephapse is a possibility as an event occurring after our photographed oscilloscope sweep. We would increase the likelihood of recording it by displaying a much longer record (slower sweep). In the records of action potentials presented, under the experimental conditions described, the trace is completed and the record terminated much too soon to include evidence of ephapse through afferents.

Another objection to longer stimuli is the possibility of spread, especially from sites not too close to the myoneural junction. Obviously, the longer the stimulus, the greater the oppor-

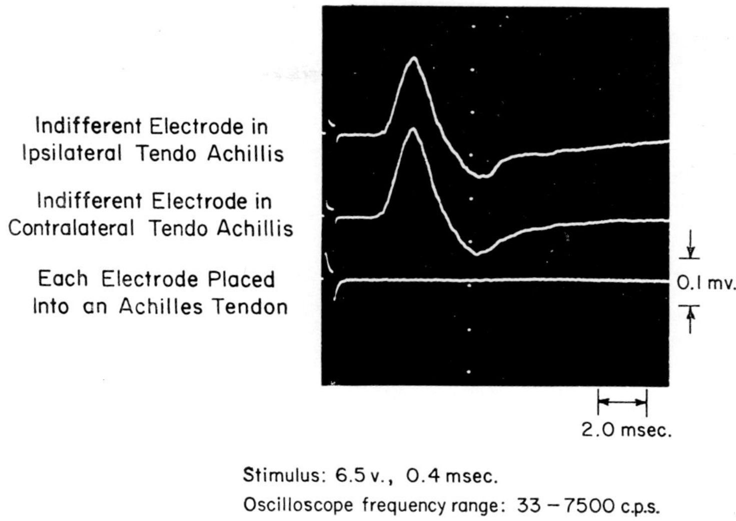

Figure 166. Single motor unit action potentials, recording electrode at anterior tibial muscle. (From *Clinical Orthopaedics and Related Research, 42,* 1965.)

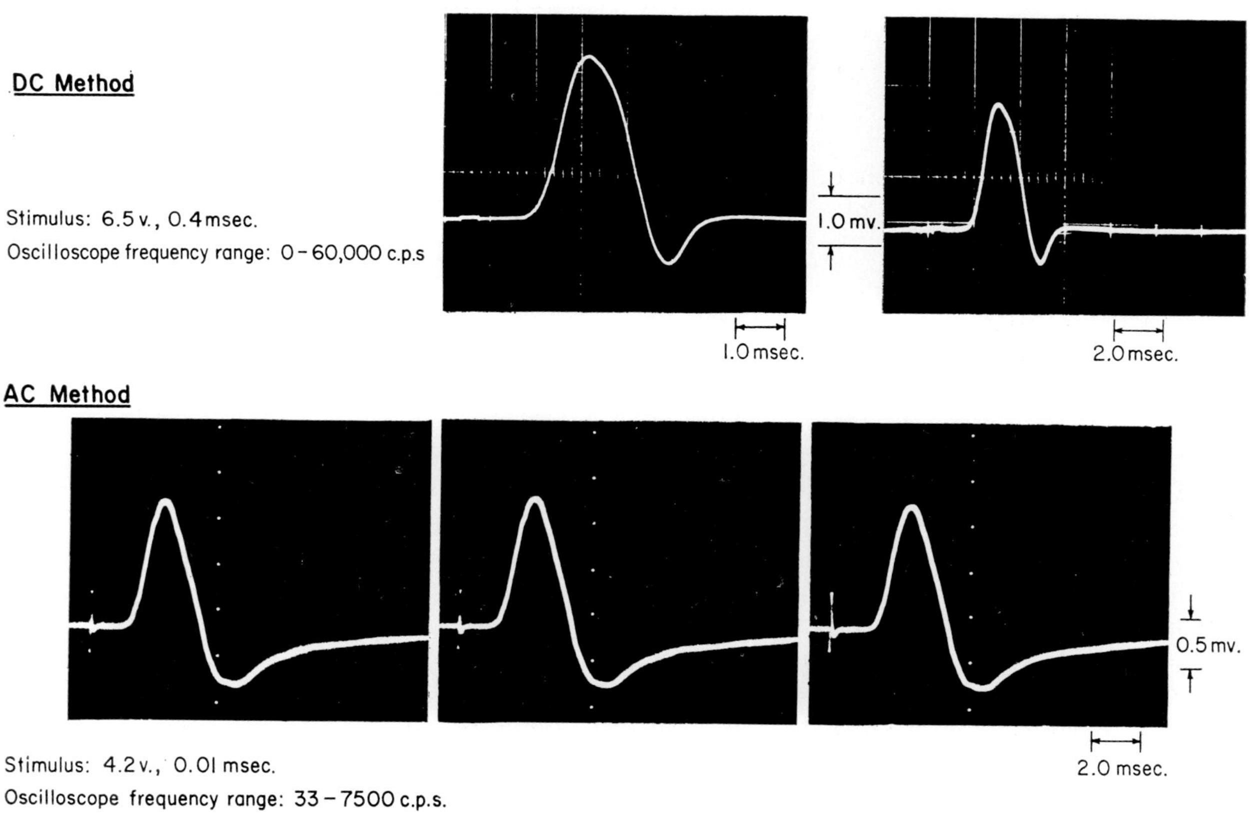

Figure 167. Traces recorded from single motor unit using *direct current coupled* amplifier (*upper*) and *alternating current coupled* amplifier (*lower*). (From *Clinical Orthopaedics and Related Research, 42,* 1965.)

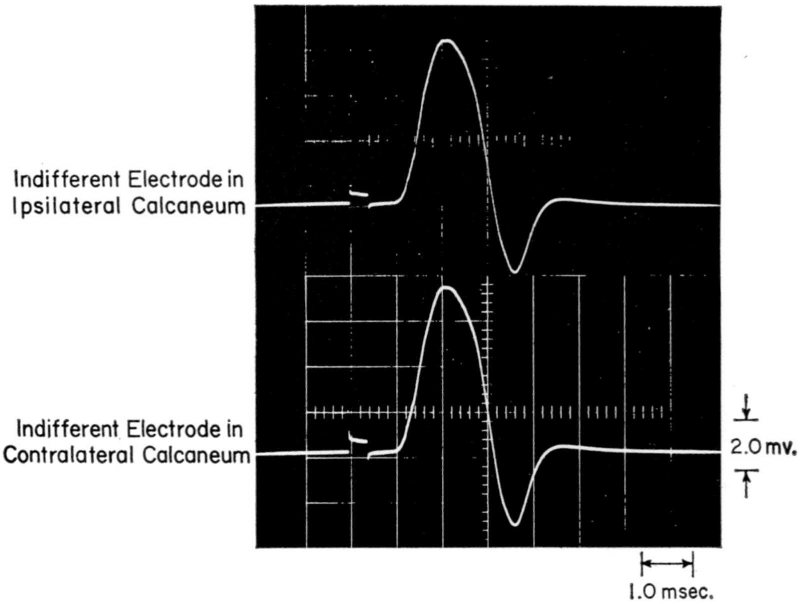

Figure 168. Single motor unit action potentials with indifferent electrode in (a) ipsilateral calcaneus, (b) contralateral calcaneus; recording electrode unchanged. Recordings made using *direct current coupled* amplifier. (From *Clinical Orthopaedics and Related Research, 42,* 1965.)

tunity for spreading electrical charges dispersed in the field to reach other structures; some charges would not be of sufficient amplitude to confuse the record. This is apparently less true of action potentials of the magnitude of those of nearby motor units. The greatest confusion would result from such inadvertent action potentials from unintentionally stimulated additional neighboring motor units.

An impulse of relatively high voltage but very short time duration (less than 0.1 msec) has a far better chance of resolving the action potential of the single structure on the recording system.

The Fourier transform described here, as well as the Fourier analysis presented below, are applied as methods for arriving at the frequency spectrum of the single waveform. The methods described are for direct analysis of the written form of solitary waveforms. Electronic averagers are useful in the analysis of electrophysiological data. Some, such as the Fabri-Tek, readily produce the histogram of a basic waveform from *series* of waveforms recorded by, for example, the method shown in Figure 4A. Thus, data are analyzed, the composition of which may represent series of many action potentials recorded randomly from a number of motor units. The Fabri-Tek signal-averaging system with the addition of the PDP-8 series general-purpose computer will produce Fourier transform data based on a single waveform. The single waveform is first collected with the signal averager. It was the purpose of the microdissection techniques described herein to stimulate the single alpha motor neuron and record the action potential of the single motor unit first to identify the single motor unit action potential. The Fourier transform and Fourier analysis could then be applied directly to the known basic waveform—that of the known single motor unit—in order to calculate the frequency spectrum of the single motor unit as well as to record its amplitude and duration.

DETERMINATION OF THE FREQUENCY SPECTRUM BY FOURIER ANALYSIS (COMPUTER)

What information may be gained from the action potential of the single motor unit now accessible? In addition to amplitude, duration and configuration, its frequency spectrum is of foremost concern to us. This curve designates the frequency band of the action potential and the relative energy distribution at the frequencies of the spectrum. A(f) indicates the amplitude of a frequency represented by a point on the frequency spectrum curve (amplitude $\sim$ voltage).

The frequency spectrum may be approximated by analysis of the single spike representing the Fourier transform. This approximation, used previously in our work,[36] is useful where repetitive series of wave forms are not found. It is represented by:

$$A(f) = \frac{1}{\sqrt{2\pi}} \int_{-\infty}^{\infty} \psi(t) e^{-2\pi f i t} dt$$

where f is frequency in CPS, t is time, i is $\sqrt{-1}$, $\psi(t)$ is the action potential as a function of time and A(f) is the frequency spectrum as a function of frequency.

The Fourier transform provides a very satisfactory approximation of such wave forms where high-speed computers are inaccessible.

The action potentials of motor units can never occur as regularly repeated wave forms, such as the pulses of true frequencies. It is possible that regularly spaced pulses of electronic stimulators may elicit equally time-spaced single motor unit action potentials. Under natural conditions, biological factors govern the spacing of muscle action potentials. Such intervals between biological events are necessarily unequal in duration.

The beginning of each wave form is delayed until repolarization is almost complete. Successive single motor unit action potentials are therefore usually expected to be complete, even at the maximal frequencies of which the central nervous system is capable. The new action potential may begin while tension is still present; tension is of much greater duration than the action potential. A repetitive cycle of excitation of actomyosin may begin while tension, due to the last response, is terminating.

As the rate of stimuli conducted by the axon to the motor unit steadily increases, fusion of the mechanical response to these stimuli results (Fig. 176). Tetanus may occur, though discreet muscle action potentials persist in regular (periodic) occurrence, as observed by Ruch and Fulton.[149] As the frequency of stimuli to a small fasciculus is increased, the muscle action potentials finally lose their repetitive near identity. We have observed, by stimulating a fasciculus of the nerve to the anterior tibial muscle of the Rhesus macaque, that muscle action potentials show irregularity of form and low amplitude at stimulus frequencies approaching 200.

These action potentials are in response to artificially induced nerve impulses. Intervening periods of equal length occurring regularly between series of identical single motor unit action potentials are not found under natural conditions, since the motor unit is receiving its signals in less regular fashion from the intact nervous system.

The "all-or-none" law of the motor unit, to the extent that it exists, applies to the action potential rather than the mechanical response. The membranes of the muscle fibres govern the timing of the action potential. Repolarization of the muscle membrane must be almost complete before another action potential may begin.[149] This applies to natural CNS-controlled stimuli. Action potentials resulting from electronic stimulators, as frequency is increased, show evidence of steady encroachment upon the membrane controlled refractory period.

Since we find neither strict periodicity nor identity of waveforms in series of single motor unit action potentials, this analysis is confined to the single waveform. This curve, Figure 162, may be virtually duplicated, provided the environment is unchanged and time for "stabilization" is allowed. Attempts to produce series of like, equally spaced action potentials by artificial means have failed (Fig. 160). The method, however, is by single alpha motor neuron stimulation. Other findings referred to above are undoubtedly based upon stimulation of whole nerve or direct stimulation of muscle. It is significant that in this work, successive stimuli of the single axon produced progressively smaller single motor unit action potentials (Fig. 160). A period for "stabilization" was required (Figs. 162, 163), after which the action potential regained its original form.

It is recognized that these pulses, though truly representing the single motor unit, are not likely to be found naturally. The stimulus of the motor unit single axon permits recording of the single motor unit action potential. The confusion of interdigitating subunits of adjacent motor units is eliminated.

Fourier analysis is normally employed for calculation of the frequency spectrum of like wave forms having periodicity. It assumes that a basic fundamental frequency (sine wave) is common to all of the forms of the sequence, with the remainder of their composition consisting of integer multiples of this basic frequency. The Fourier theorem states that the frequency spectra of such curves may be determined by adding the sine and cosine functions of the fundamental frequency to the sine and cosine functions of the integer multiples of that wave form.

The single motor unit action potential has neither (a) the precise repetitive characteristic of true frequencies or (b) a basic fundamental frequency that might apply to a series duplicating the single waveform.

Nevertheless, by adaptation of cosine and sine functions, after careful measurement of the amplitudes of several hundred points along the curve, each separated from the next by a like interval on the horizontal dimension, the curve may be *digitized*. The *digitized* curve, though having no periodicity or mathematical equation, may be "integrated" in very close approximation.

This very close approximation of the integral delineates the frequency spectrum of the action potential. The procedure is possible due to the high-speed computer.*

After the measurements of the many points along the curve ($\psi_{0,1,2}$) have been made with corresponding values for the horizontal distances between them ($h_{0,1,2}$) (Fig. 169), these values are processed in sequence. The corresponding values of ψ and h are recorded numerically by keypunch, with 8 pairs on each card. The IBM computer is programmed according to the following:

$$\frac{h}{3\sqrt{2\pi}}[\psi_0 \text{ COS } (2\pi f\cdot 0) + 4\psi_1 \text{ COS } (2\pi f\cdot h) + 2\psi_2 \text{ COS } (2\pi f\cdot 2h) + \ldots\ldots + \psi_{12} \text{ COS } (2\pi f\cdot 12h)]$$

and

$$\frac{h}{3\sqrt{2\pi}}[\psi_0 \text{ SIN } (2\pi f\cdot 0) + 4\psi_1 \text{ SIN } (2\pi f\cdot h) + 2\psi_2 \text{ SIN } (2\pi f\cdot 2h) + \ldots\ldots + \psi_{12} \text{ SIN } (2\pi f\cdot 12h)]$$

f equals one frequency selected for the analysis; h equals the horizontal dimensional unit and is used instead of the usual unit of the x-axis time, t.

These equations are derived from the Simpson approximation of the integrals representing the curve

$$\frac{1}{\sqrt{2\pi}}\int_{-\infty}^{\infty} \psi(t) \text{ COS } (2\pi ft)\, dt$$

and

$$\frac{1}{\sqrt{2\pi}}\int_{-\infty}^{\infty} \psi(t) \text{ SIN } (2\pi ft)\, dt$$

The complete procedure is quickly performed by the computer. It is performed at chosen increments of frequency, such as:

0, 10, 20, 30, ... 20,000 CPS

or

0, 100, 200, 300, ... 30,000 CPS

or

0, 250, 500, 750, ... 30,000 CPS

* Ernest O. Lawrence Radiation Laboratory, University of California, Berkeley.

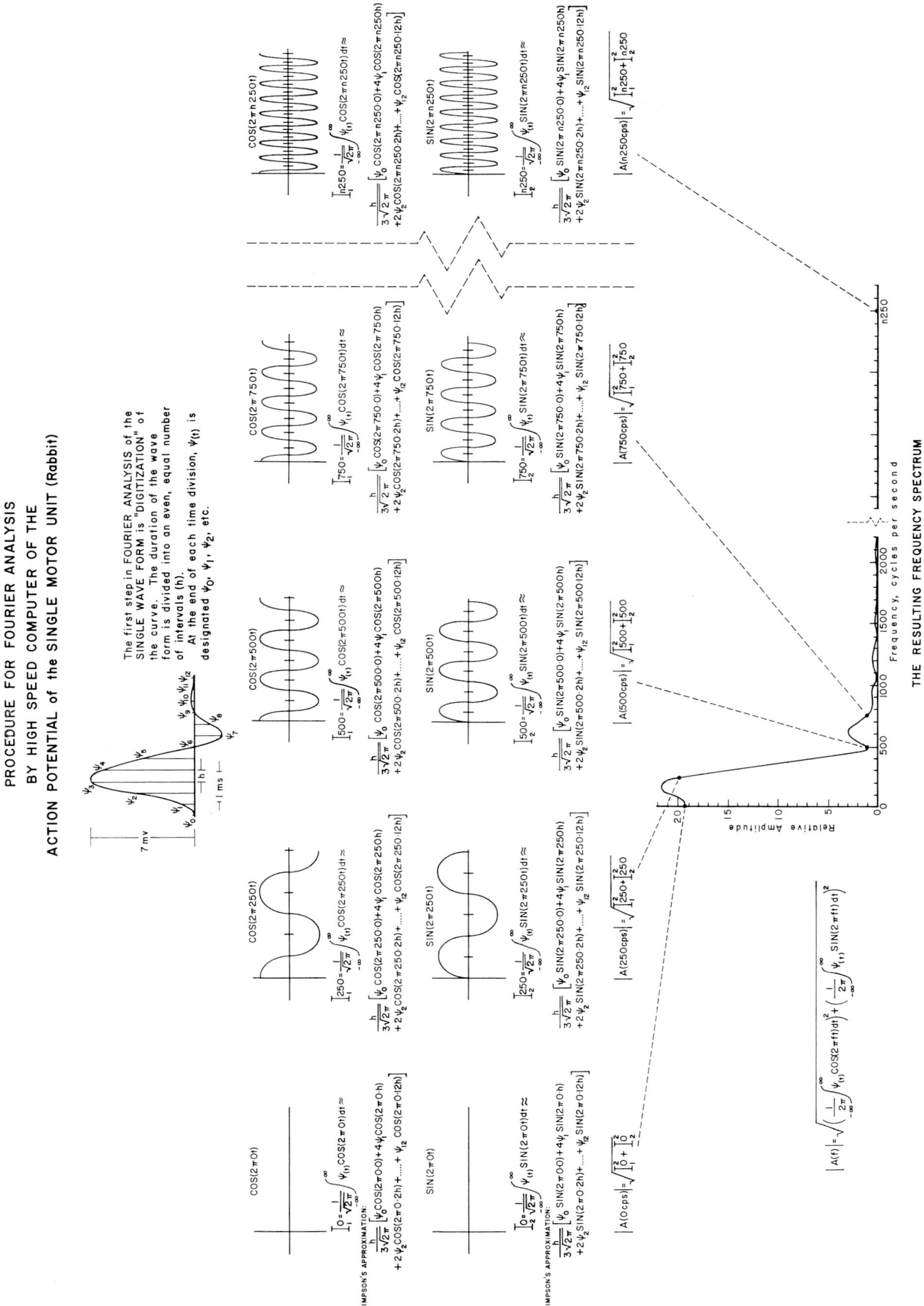

Figure 169. The method of calculation of the frequency spectrum of the action potential of the single motor unit by Fourier analysis using the high-speed computer. (From *Clinical Orthopaedics and Related Research 42,* 1965.)

For each frequency at which the entire series of cards (values of ψ and h) is run, e.g. O, 250, 500, 750, etc., a point, $|A(f)|$, representing the amplitude of the curve at that particular frequency is found and later plotted. Repetitive computer runs at each value of the frequencies used thus result in the frequency spectrum (Fig. 169).

In this way, the computer processes each corresponding ψ and h value in sequence. Programmed according to the above cosine and sine integrals, it calculates the amplitude of the points on the frequency spectrum. Successive frequencies analyzed represent integer multiples of the first basic frequency chosen for the analysis. Using successively each measured ψ and h found on the entire original curve, the computer, for each of the frequencies studied, solves the equation:

$$\left|A(f)\right| = \sqrt{\frac{1}{2\pi}\left[\left(\int_{-\infty}^{\infty}\psi(t)\cos\omega t\,dt\right)^2 + \left(\int_{-\infty}^{\infty}\psi(t)\sin\omega t\,dt\right)^2\right]}$$

$$\omega = 2\pi f$$

At each frequency, the amplitude thus determined is placed on the curve of the frequency spectrum (Fig. 169).

It will be recalled that ω is expressed in radians/sec., f is expressed in cps, $\omega = 2\pi f$; ψ represents amplitude or voltage in this consideration.

Previous determinations of the frequency spectrum of the single motor unit action potential employed Fourier analysis of the single triangular spike:

$$A(f) = \frac{1}{\sqrt{2\pi}}\int_{-\infty}^{\infty}\psi(t)e^{-2\pi fit}\,dt$$

where f is frequency in cps, t equals time, i equals $\sqrt{-1}$, $\psi(t)$ is the action potential as a function of time and A (f) is the frequency spectrum as a function of frequency.

The high order of values for the frequency spectrum previously reported,[36] 83 cps to 10.7 kc, results from the expression of frequency in radians. The values found throughout this report are based upon the frequency in cps, f, which is equal to $\omega/2\pi$. Results are comparable if this difference is considered.

$$A(f) = \frac{1}{\sqrt{2\pi}}\int_{-\infty}^{\infty}\psi(t)e^{-i\omega t}\,dt$$

$$\omega = 2\pi f$$

The Fourier transform (Fig. 170), with analysis of the frequency spectrum of the single triangular spike, produces values somewhat greater than direct analysis of the complete wave form. This is due in part to an artificially abrupt beginning of the spike and its sharp apex.

Since $\omega = 2\pi f$, the Fourier transform used in this analysis is

$$A(f) = \frac{1}{\sqrt{2\pi}}\int_{-\infty}^{\infty}\psi(t)e^{-i2\pi ft}\,dt\ldots\ldots e = 2.71828$$

$$i = \sqrt{-1}$$

since $e^{-i2\pi ft} = \text{COS}\ 2\pi ft + i\text{SIN}\ 2\pi ft$.

The transform may be written:

$$A(f) = \frac{1}{\sqrt{2\pi}}\int_{-\infty}^{\infty}\psi(t)\,[\text{COS}\ 2\pi ft + i\ \text{SIN}\ 2\pi ft]\,dt.$$

$$I_1 = \frac{1}{\sqrt{2\pi}}\int_{-\infty}^{\infty}\psi(t)\ \text{COS}\ (2\pi ft)\,dt +$$

$$I_2 = \frac{i}{\sqrt{2\pi}}\int_{-\infty}^{\infty}\psi(t)\ \text{SIN}\ (2\pi ft)\,dt$$

The absolute value of A(f)

$$|A(f)| = \sqrt{I_1^2 + I_2^2}$$

The expression for $|A(f)|$ may now be employed, since the high-speed computer, IBM, 7094 B,* makes possible the calculation of the absolute value of A(f):

---

* Ernest O. Lawrence Radiation Laboratory, University of California, Berkeley.

$$\left| A(f) \right| = \sqrt{\frac{1}{2\pi}\left[\left(\int_{-\infty}^{\infty} \psi(t) \cos \omega t \, dt\right)^2 + \left(\int_{-\infty}^{\infty} \psi(t) \sin \omega t \, dt\right)^2\right]}$$

$$\omega = 2\pi f$$

Each integral is approximated by Simpson's method. The process is duplicated at progressively increasing frequencies, each an integer multiple of the first frequency studied (Fig. 172). At each frequency, a value, $|A(f)|$, the relative amplitude at that frequency, is found and plotted as a representative point on the curve of the frequency spectrum.

The frequency spectra of several single motor unit action potentials are shown in Figures 170, 171 and 172.

In Figure 161, the Fourier transform of the single spike is shown. The spike is an approximation of the major deflection of the waveform, neglecting the terminal smaller deflection of opposite sign. The frequency spectrum by this method may be found by mathematical means. The results by mathematical means are compared to those produced using the high-speed IBM 7094-B computer (Fig. 170).

Two of the waveforms recorded from single motor units using the *alternating current*

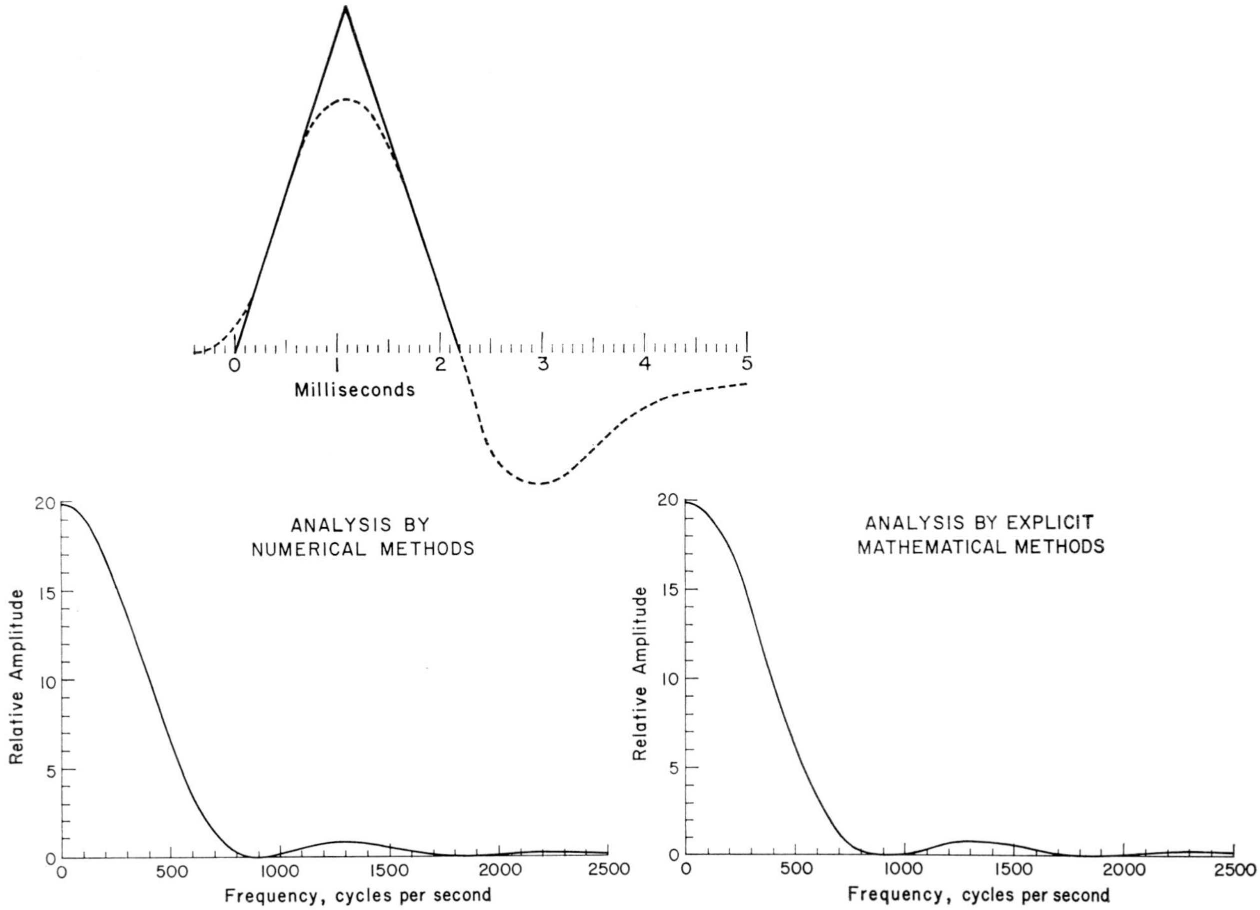

Figure 170. Determination of frequency spectrum of action potentials by Fourier analysis of triangular spike. *Top,* transformation of single triangular spike: Waveform No. 1, recorded by *alternating current coupled* method. *Bottom, left,* frequency spectrum by Fourier analysis (IBM 7094B). *Bottom, right,* frequency spectrum by Fourier transform, mathematical methods. (From *Clinical Orthopaedics and Related Research, 42,* 1965.)

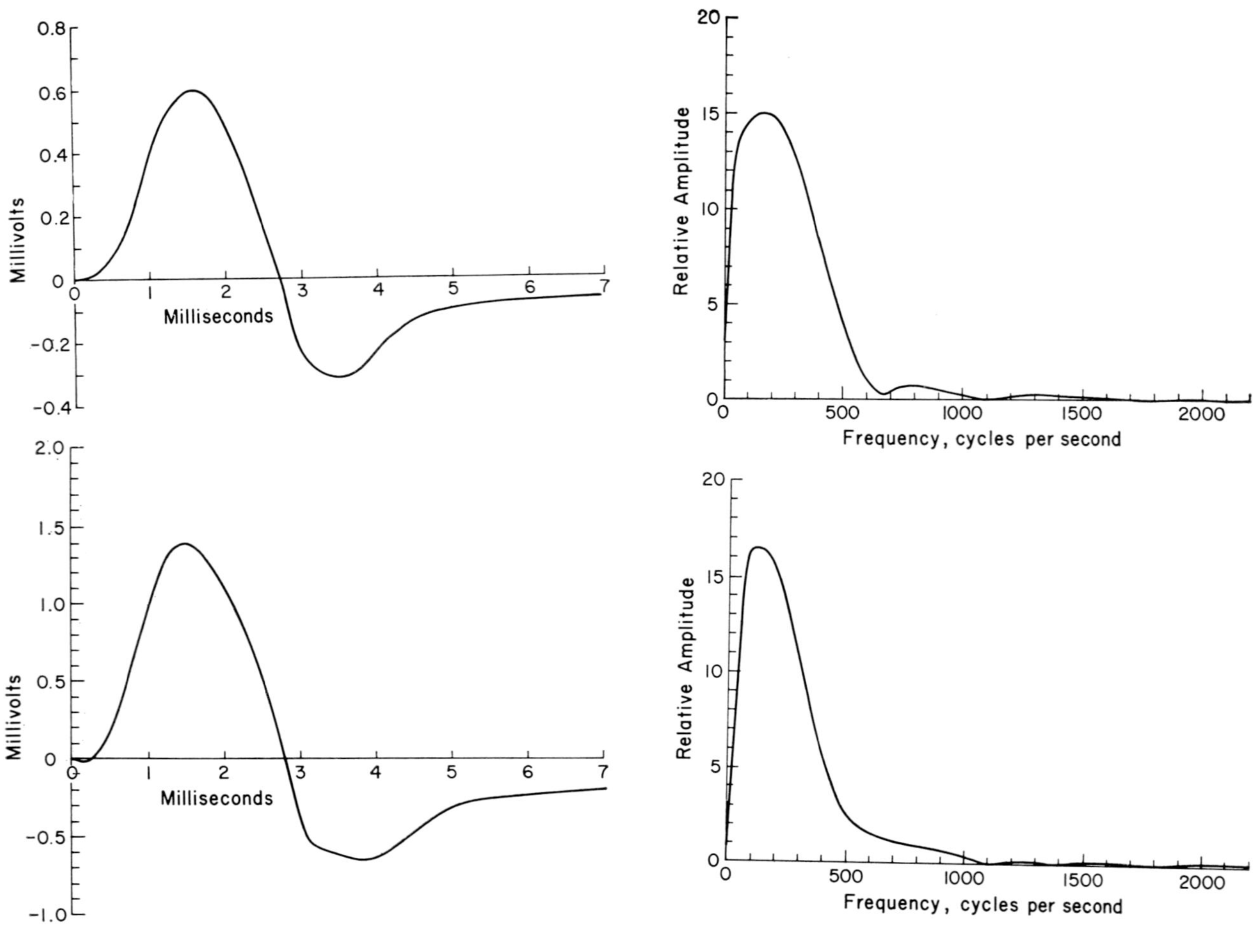

Figure 171. Determination of frequency spectrum by Fourier analysis (computer) of action potentials recorded by *alternating current coupled* method. *Top left,* waveform No. 1. *Top right,* the frequency spectrum of action potential of waveform No. 1, determined by Fourier analysis (IBM 7094B). *Bottom left,* waveform No. 2. *Bottom right,* the frequency spectrum of action potential of waveform No. 2, determined by Fourier analysis (IBM 7094B.) (From *Clinical Orthopaedics and Related Research, 42,* 1965.)

coupled input amplifier are shown in Figure 171. Their frequency spectra, obtained with the aid of the IBM 7094-B computer, are shown. These waveforms demonstrate an artifact delay in return to the base line of the terminal portion of the curve. This delay is due to the input coupling capacitor which is necessary in *alternating current coupled* amplifiers.

Two *direct current coupled* records are shown in Figure 172. Prompt return of their terminal portions to the base line is shown. The frequency spectra of the two *direct current coupled* waveforms, determined by Fourier analysis using the IBM computer 7094-B, are shown. These data show the significant amplitude at zero cycles per second not usually appreciated due to the use of the *alternating current coupled* instruments with higher settings of the low-frequency recording band and to our greater interest in phenomena found at the higher frequency levels. As is well known, instruments capable of reliable display of impulses such as the fibrillary potential should extend into at least the 10 kHz range.

Although the *direct current coupled* mode provides the above advantage and probably others, e.g. a more precise reproduction of the action potential, it is often very difficult to use.

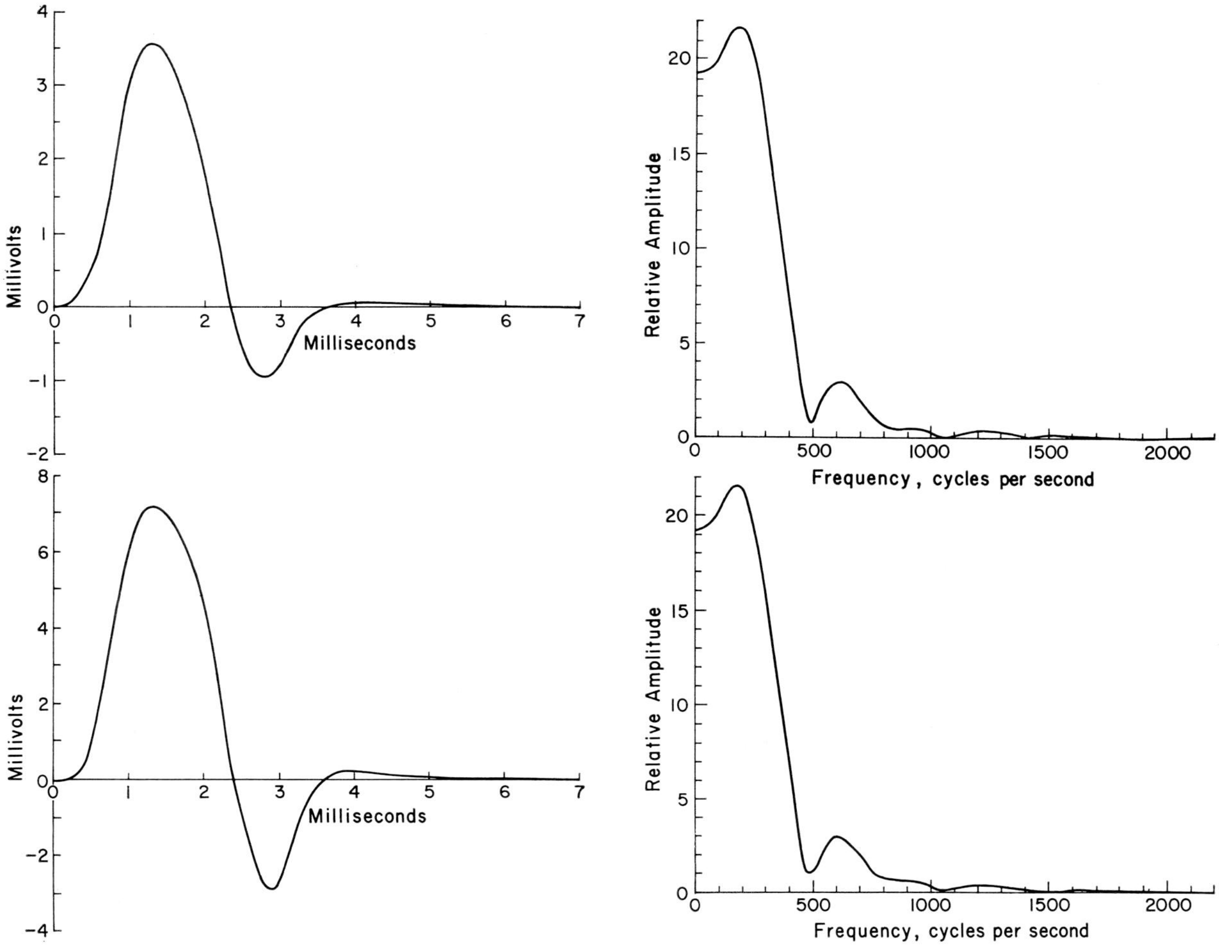

Figure 172. Determination of frequency spectrum by Fourier analysis (computer) of action potentials recorded using *direct current coupled* amplifier. *Top left,* waveform No. 1. *Top right,* frequency spectrum of waveform No. 1 by Fourier analysis (IBM 7094B). *Bottom left,* waveform No. 2. *Bottom right,* frequency spectrum of waveform No. 2 by Fourier analysis (IBM 7094B). (From *Clinical Orthopaedics and Related Research, 42,* 1965.)

The trace is unstable, requiring frequent adjustment of position. The *alternating current coupled* mode is by far the more practical one.

Observing comparative curves of the frequency spectrum (Fig. 173), we see considerable agreement of each with the inverse square of frequency, $1/\omega^2$. The spectrum of the single triangular spike resembes this curve to a greater extent than does the *direct current coupled* curve. The *direct current coupled* curve, in turn, resembles the curve of $1/\omega^2$ more closely than does the *alternating current coupled* curve.

This generalization regarding the relationship of amplitude to frequency makes possible valuable estimates of the percentage of loss due to frequency restrictions in the recording band.

The high-speed computer has made possible the Fourier analysis of the waveform of the single motor unit. The resulting frequency spectrum appears to be the most important source of information regarding the action potential of the single motor unit. The time and effort spent in this tedious dissection, previously yielding such data as amplitude, duration and general configuration of the elusive single potential, yields unanticipated reward

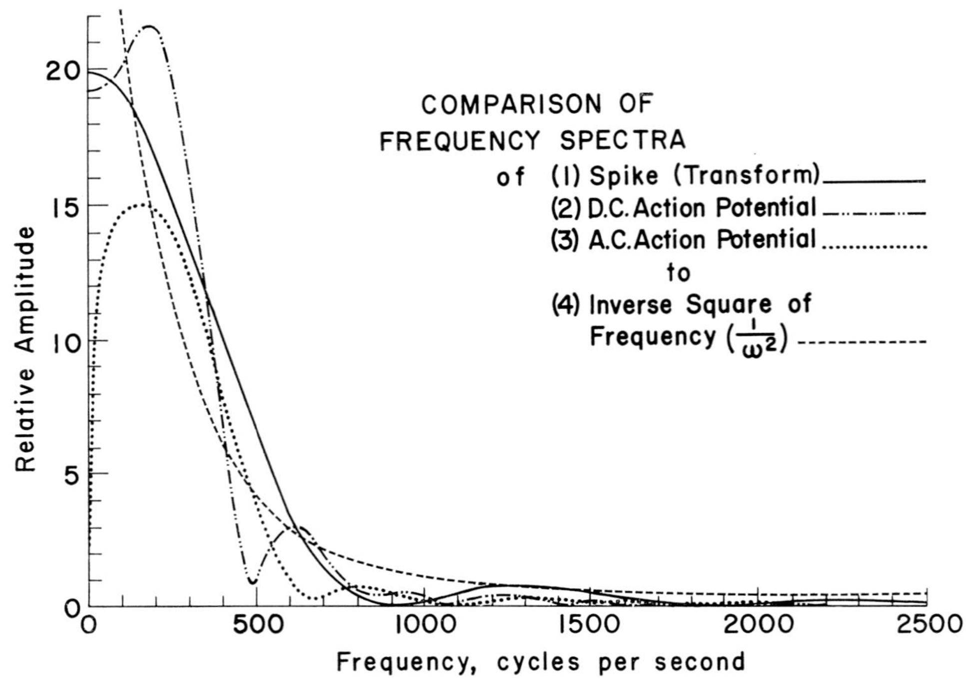

Figure 173. Comparison of frequency spectra of (a) triangular spike (transform), (b) waveform recorded by *direct current coupled* method, (c) waveform recorded by *alternating current coupled* method to (d) inverse square of frequency $1/\omega^2$. (From *Clinical Orthopaedics and Related Research, 42,* 1965.)

from the high-speed computer—the complete frequency spectrum.

It is hoped that the technique as described for production of the single motor unit action potential will be applied to other muscles and to the muscles of other animals. The procedure can now be done in a very short time. It is a feasible, safe procedure that may be adapted to the human operating room. Similar data from human musculature will be of great value.

These recordings were made using a small circular disc as the recording electrode and an indifferent electrode placed in an isoelectric position. Changing the site of the indifferent electrode about the body did not significantly affect the action potential, provided the site chosen is itself isoelectric. The question has arisen whether the circular disc electrode causes distortion or difference in the action potential compared to the action potential that might be obtained from a narrow, band-like electrode oriented transverse to the fibres. Since the velocity of propagation of the action potential of mammalian muscle is accepted as 5 meters/sec, an electrode 9 mm in diameter would be in direct contact with a portion of the muscle that would require 2 msec for the action potential to traverse. It will be necessary to find out if the condenser effect of the muscle membrane changes the duration of the recorded action potential. Any loss of higher frequencies due to the circular metallic electrode (approximately 9 mm in diameter) being used as the recording electrode (p. 213) must be determined.

The frequency spectrum which neuromuscular recording systems are capable of recording is a major factor in determining the fidelity of the record produced. Motor unit action potentials are complex electrical waves. The frequency bandpass is of the same general order as that of the human voice. The high-frequency level is of a lower order in the muscles of the experimental animals used (rabbit, cat). Such complex wave forms are composed of various frequencies over a fairly wide range. Equipment capable of accurate reproduction of such waveforms must be capable of uniform ampli-

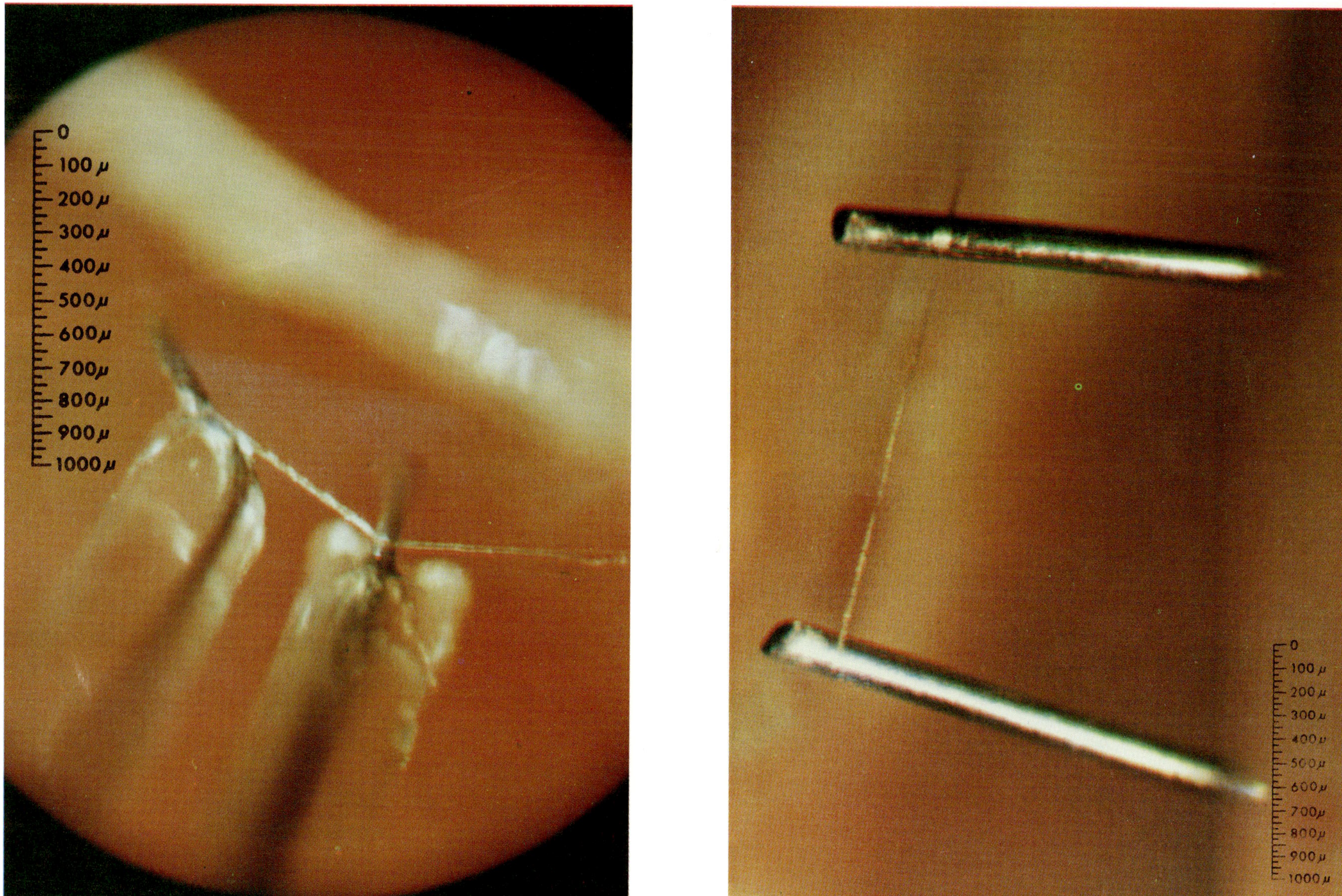

Figure 174. A single alpha motor neuron of the anterior tibial nerve of the rabbit *left* and of the cat *right* is placed on the stimulating electrodes. Circular interruptions (nodes of Ranvier) are seen at regular intervals. The fasciculus from which the single axon was removed is seen in the background.

fication at the frequencies contained in the waveforms. It is essential to determine the range of frequencies of which human muscle action potentials are composed. The results are considered very important fundamental as well as clinical information.

It is well known that the complex sustained continuous action potentials recorded from contracting muscle are combinations of the individual repetitive waveforms of the motor units of the volume of muscle contributing to the recording. The composition of the volume of muscle from which the record is made is difficult or impossible to identify in terms of numbers of motor units.

Since the frequency spectrum or range of composite frequencies of a single impulse is the same as those comprising the frequency spectrum of a dense random cascade of such impulses, we may conveniently arrive at the frequency band of dense series of motor unit action potentials by careful analysis of the frequency spectrum of the action potential of the *single motor unit.* There is constant addition, mixture and overlap in the usual recording with lack of positive identity of single deflections when they are found. This simultaneous display of pulses during muscular contraction is called the *interference pattern.* Individual analyses here become difficult and unreliable. Single pulses, if observed, can hardly be reliably identified. These potentials present confusion due to interdigitation among motor units.

These factors leave us no choice if we wish to examine the characteristics of the *known* motor unit. The single motor unit alone must be activated at a point of stimulation common to *all* of its fibres to the exclusion of the muscle fibres of any other motor unit.

This presentation is concerned primarily with isolation and analysis of the action potential of the single motor unit. Because of the virtual impossibility of histological isolation of this structure, insufficient knowledge of its morphology and lack of accurate documented data regarding the precise nature and identity of its action potential where recordings are, of necessity, by blind techniques, we have considered it timely and important to ascertain the nature of this waveform in several muscles about the body. A modest beginning was made, confining our work to a single muscle, the anterior tibial muscle of the rabbit, during most of our studies. The method has been developed to the extent that it may be used on many other muscles and possibly on humans in the operating room when certain advantageous procedures and exposures take place. Simultaneous recording of the tension of the single motor unit, described later in this chapter, was done on 24 cats. Practice with the technique has rendered the procedure harmless. It depends for success upon development of technical skills a few steps beyond our present-day techniques of peripheral nerve repair under microscopic control. In peripheral nerve repair, for many years, we have attempted to match the "cables" or fasciculi of the trimmed ends of peripheral nerve. This technique has recently been improved by Edshage.[59]

By careful transverse division of the severed ends of peripheral nerve, Edshage[59] has demonstrated microscopically that fasciculi may be placed in a more favorable relationship by careful transection and reapproximation, using a very sharp cutting edge and special holder. He has demonstrated the failure of fasciculi to find their mates after conventional types of repair. He has also been able to show with photomicrograms the ability to produce a greater percentage of continuity of fasciculi by using the procedure that he has outlined.

Magun and Esslen,[124] in an electromyographic study of reinnervated muscle and of hemifacial spasm, have shown the probable reason for neuromuscular confusion resulting after some nerve repairs or nerve recovery. Regenerating axons and fasciculi may not necessarily reach the same muscles that they originally innervated. Consequently, an inappropriate facial expression may result instead of the intended one. One patient involuntarily winked when another facial expression was intended. The necessity of matching fasciculi is thus clear. Microscopic surgery is necessary for this. The next step is an attempt to direct individual axons. This ambitious goal seems more formidable than did direction of fasciculi a few years ago. Axon control or axon matching, challenging as it seems at the present time, represents the ultimate to be achieved in nerve repair. It is not an insurmountable goal and is worthy of most-painstaking labors.

Even a modest percentage of success in general reapposition of motor axons due to axon branching would bring great improvement in our results following peripheral nerve suture. Conventional methods of nerve repair are often of value only because a few of the motor nerve filaments in the proximal portion happen to find their way across the junction, some of these by chance finding a suitable distal structure to follow. The fortuitous process of *axon branching* then occurs, with viable motor axons not only rescuing one motor unit but often sending "axon sprouts" to many other nearby muscle fibres. Thus, 15 to 20 percent return of alpha motor fibres may result in apparent complete motor recovery. This is an indication of the value of tedious attempts at optimum replacement of as many motor axons as possible.

The method for study of the single motor unit action potential as described is ideal for assessing the value of various pharmaceutical agents. It presents an opportunity to apply agents directly at the neurophysiological point of interest and to observe their effect on the lowest common denominator of muscular activity and structure—the motor unit. The effect of changes in the electrolytic environment, as well as changes in surrounding temperature, are easily studied.

Using the indwelling intravenous catheter, the effect of many solutions may be studied by direct addition to the circulating blood. In this way, the ionic distribution about the muscle membrane and its ionic permeability may be studied. The profound effect of additions of large amounts of potassium salts compared to the far lesser effects of equivalent amounts of sodium chloride or calcium gluconate are examples of a ready method for study. The effect of respiration, and of anoxia and variation in carbon dioxide, are easily studied by controlling the animal's gaseous respiratory intake.

It is interesting to study the effect of variation of position of the limb. The action potential is known to vary[36,143] as muscle length varies. The effect on the action potential as the muscle is placed at various lengths throughout its length-tension curve is of considerable significance. The geometrical arrangement of the joints of the skeleton, and the antagonists of muscles, restrict the ranges of muscle length.[41] Thus, a muscle in its normal geometrical range occupies a narrow band of the length-tension curve,[143] the lengths of which are possible only in "free" muscles.

Isometric contractions may be studied at various lengths. Isotonic contractions, allowing the muscle to shorten, may be compared.

The possibility of study of the effect of spindle activity may be considered.[78] There is space for additional equipment within the enclosure to study smaller structures. Filaments of the gamma nervous system contained in peripheral nerve may be identified and studied. No attempt has been made to use staining techniques in this study except to stain the preparation for demonstration of the single alpha motor nerves stimulated. This is done at the conclusion of the experimental procedure.

Repetitive stimuli make it very easy to ascertain the refractory period of the motor unit. This has not been done here because of the necessity to spare or to make the most use of all of the axons obtained in order to acquire sufficient data for the studies described above. The experiment, however, casts considerable doubt on the "all-or-none" law. Often action potentials taken in the same environment after an adequate period of time has elapsed have shown a difference in amplitude. One may consider separately the conduction of the impulse along the axon and the behavior of the muscle fibres completing the motor unit. If we consider nervous conduction to be "all-or-none," it is reasonable to consider the possibility that under certain conditions, all of the fibres do not fire.

Katz,[107] in "Quantal Mechanism of Neural

Transmitter Release," refers to his recent work in which he describes "packets" of molecules of acetylcholine. Each packet may contain 1,000 or more molecules; Miledi[132] gave a figure of 100,000 molecules of acetylcholine per packet. Approximately one packet is normally discharged per second. During transmission of the impulse, a few hundred packets are discharged per millesecond. Other factors, such as the concentration of calcium ions, may then influence neuromuscular transmission. Could not other variables in addition to the number of packets of molecules of acetylcholine discharged—such as variation of the percentage of end plates of each axon that is activated—be a factor influencing our concept of the all-or-none law? As demonstrated in Figure 160, the action potential of the single motor unit may vary.

Once some of the basic curves have been obtained for human muscle motor units for certain muscles, it may be desirable to consider a clinical microstimulation technique, using a tiny electrode tip, to stimulate the single axon brought into contact with it. One would have to be in a position to record from the surface of the muscle in the anticipated region. This would be a useful procedure if one knew in advance what form of action potentials to expect. This could be first ascertained by the principal technique described above.

The microelectrode which pierces the cell membrane may cause some artifact. The practice of employing intracellular microelectrodes for selective neuron stimulation and sometimes for action potential recording from both nerve and muscle cells was undoubtedly developed due to the inherent difficulties of the direct method of this study. More recent developments of fine instruments for microtechniques have made the method described possible.

This method has the advantage that the structures involved are not injured. Bipolar, ground-isolated electrical stimulation of carefully dissected single motor neurons leaves no doubt of the identity of the structure from which the recorded action potential results.

CONCLUSIONS

A technique has been presented which provides for observation of the action potential of the single motor unit of the anterior tibial muscle of the experimental animal. The method depends upon microdissection and microstimulation of the single alpha motor neuron close to the myoneural junction of this muscle. The inherent deficiency of *alternating current coupled* recording is demonstrated. The reasonable frequency recording range is suggested. The ideal, but difficult, *direct current coupled* method is used to produce some comparative records. Many characteristic traces, showing remarkable consistency, are presented.

Differential recordings where one electrode is placed at the point of interest and the other into a known isoelectric region make possible measurements and records of absolute potential variation. The valuable artifact-cancellation effect of the differential method is used to great advantage. This may represent the most practical and profitable method of electrodiagnosis (in resting muscle only).

The action potentials of the single motor unit, repeated with remarkable likeness and near identity of form in consecutive experiments, readily lend themselves to the computer analysis reported herein. The modern high-speed computer reveals the frequency spectrum of the single motor unit action potential by rapidly performing the many steps of a modified adaptation of the Fourier analysis.

If this procedure for calculating the frequency spectrum of the single motor unit were applied to human muscle, a spectrum extending into considerably higher frequencies would be expected.

TENSION OF THE MOTOR UNIT

Our concept of the magnitude of tension which muscle is capable of developing has resulted from indirect methods of study. In experimental animals, it is possible to measure the tension of muscle indirectly stimulated, after the tendon has been isolated. Indirect methods, such as calculations of the probable tension developed by the gastrosoleus group attached at the tuber of the calcaneus using the foot as a lever, lead us to believe that human skeletal muscle is capable of developing great amounts of tension. A degree of tension of approximately 500 kg has been suggested as the capacity of this muscle when body motion is accelerated. In Chapter IV, a technique was described by which the tension of normal, voluntarily contracting muscle could be measured directly, as though the muscle end or its tendinous insertion were free (Fig. 66). This was accomplished by immobilizing the tibia with a Steinmann pin which was inserted through the tuberosity of the tibia. Another pin was inserted through the tuber of the calcaneus near the attachment of the triceps surae. The tension developed between these two pins represented the tension in the soleus muscle when the knee was flexed. In this experiment, the subject was trained to suppress the peroneal, posterior tibial and flexor muscles. By attaching a force gauge at the muscle's insertion, the tension developed by the soleus muscle could be recorded. The greatest magnitudes of these voluntary contractions of the soleus muscle were in the order of 35 kg.

In order to record the tension developed by the single motor unit in synchrony with the action potential, the experimental arrangement shown in Figure 175 was used. The nerve to the anterior tibial muscle was dissected as described above and recordings made from the anterior tibial muscle as the single alpha motor neuron was stimulated. The femoral condyles and the distal ends of the tibia and fibula were secured with metallic pins (Fig. 175). The tendon of the anterior tibial muscle was detached (including at its end its bony attachment to the first metatarsal) and sutured securely to the strain-gauge–bearing metal bar. Trauma to the tendon and the Golgi tendon organs was minimized by this procedure. Thus, muscle length was controlled and passive tension regulated and recorded. The work discussed below was based on data recorded from 24 cats.

A simultaneous oscilloscope display of the single motor unit action potential and the tension developed was then possible, using a dual-beam oscilloscope such as the Tektronic 565, which will accept two plug-in units. In this case, a single sweep system was used. Tension was recorded using the carrier amplifier. The tendon of the anterior tibial muscle was attached to a rigid bar upon which were mounted four SR-4 (BLH FAP-25-12, 120 ohm, gauge factor 2.06) very sensitive strain gauges. Thus four external bridge arms were used. The second plug-in amplifier was a high gain differential unit. This was used to record the action potential. Thus, the action potential of the single motor unit and its tension curve were synchronously displayed.

Exposure of the proximal innervation of the anterior tibial muscle[62] was made, as previously described in the studies made of the rabbit single motor unit action potential. Dissection of the single alpha motor neuron was done using a similar technique. Differential recordings of the action potential were made with one electrode in contact with the anterior tibial muscle. The other connection was made to a pin drilled into the calcaneus. The same method of stimulation was used. Six photographs of oscilloscope sweeps are presented in Figure 176. The action potential of a small fasciculus is first demonstrated. It is observed that the onset of tension is close to the point at which the primary spike of the action potential returns to the base line, the termination of repolarization. This, however, is not always the case, as demonstrated in the third photograph. Here, passive tension is increased considerably, and the onset of developed tension is significantly

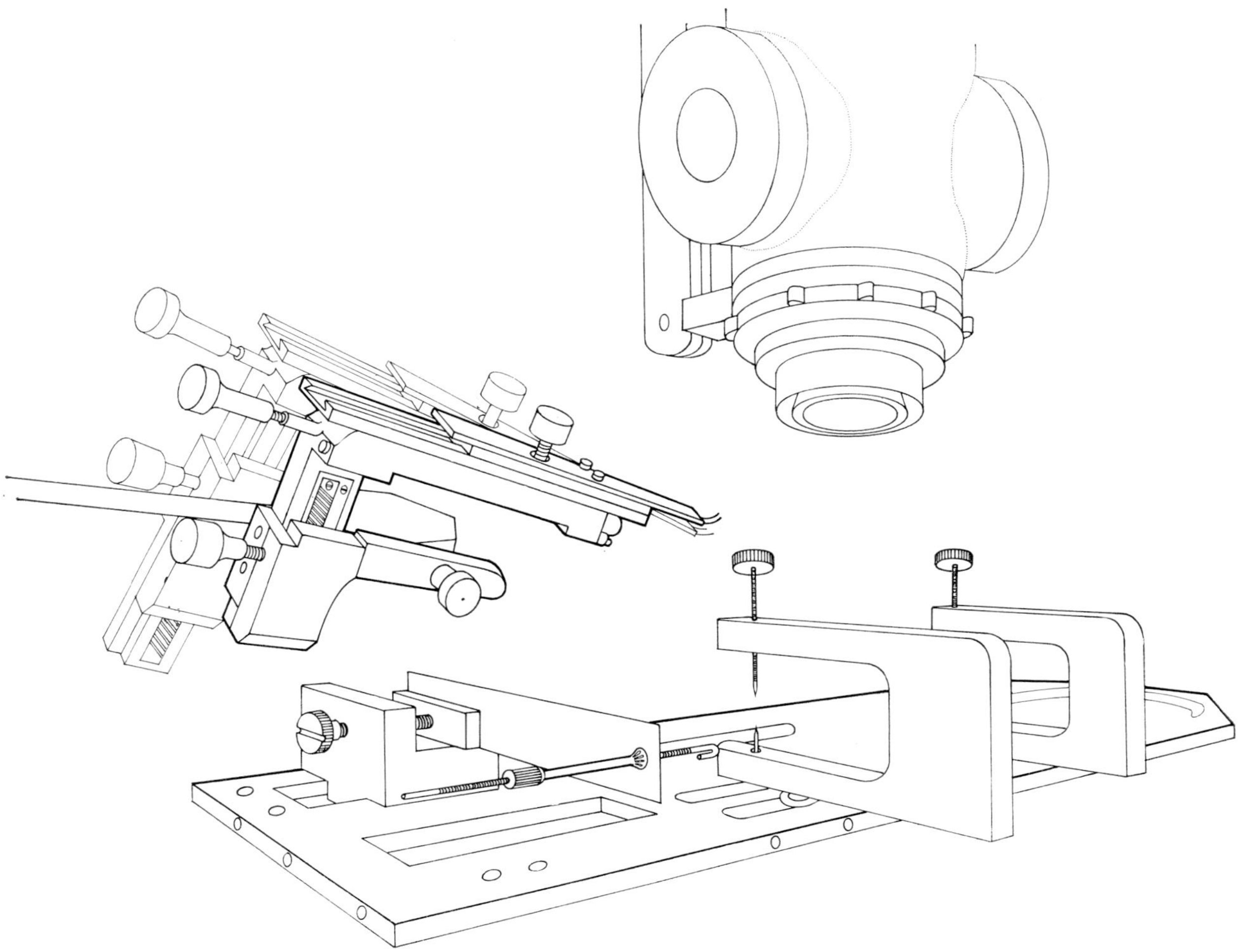

Figure 175. Experimental apparatus added to the apparatus shown in Figure 157 to record the tension as well as the action potentials of single and small groups of motor units.

delayed 5 or 6 msec. A small fasciculus (photograph No. 2, Fig. 176) is stimulated once to produce a single twitch. Superimposed is a tetanic curve produced by stimulus frequency of a little over 60 isolated square waves per second. Number 4 demonstrates the minute tension developed by a single motor unit. The stimulus artifact is followed by the main spike of the action potential. The tension developed in this case by the single twitch is approximately 7 or 8 microstrains, representing approximately 20 to 25 gm. The next photograph (No. 5) is also the stimulation of a single alpha motor neuron, but the tension recorded is much greater. This is undoubtedly due to spreading of the impulse, either by ephapse (artificial communication between neurons) or through the volume conductor, if an excessive amount of Ringer's solution was allowed to remain in the preparation. In a few of the procedures, mineral oil was used instead of Ringer's solution for preservation of tissues. This overcomes some of the difficulties due to the volume conductor but makes microdissection technically more difficult. Photograph Number 6 represents a repetition of Number 5 at a slower sweep speed in order to show the complete classical tension curve.

More recently we have been able to overcome most of the problem associated with spreading of stimulus due to the volume conductor. Improvement in technique has demonstrated unquestionable responses from the single motor unit, and we have acquired addi-

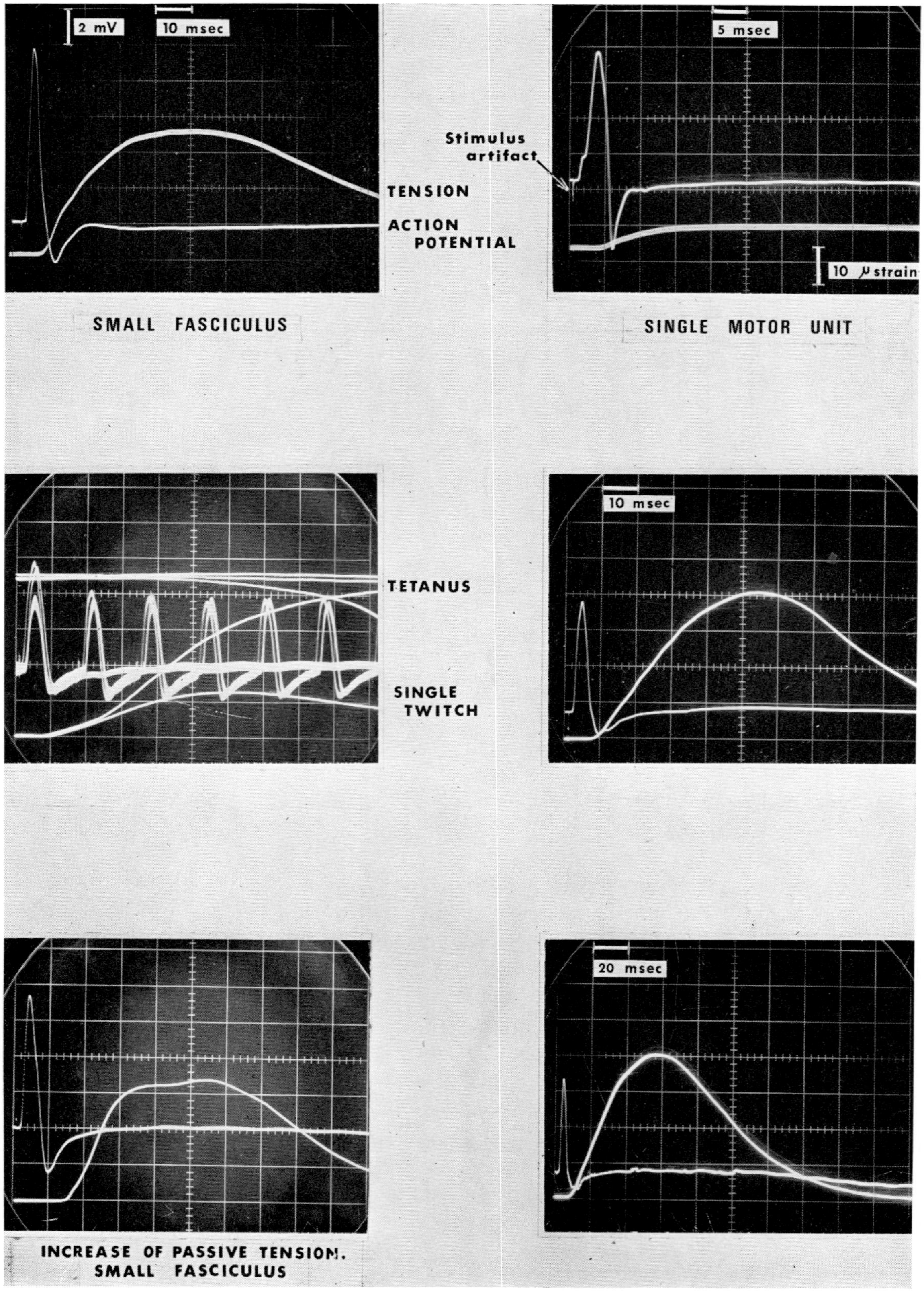

Figure 176. Recordings of muscle action potential and tension. *Left column*, a small group of motor units: single twitch, tetanus, single twitch showing delay of onset of tension when passive tension is increased. *Right column*, muscle action potential produced by stimulation of single alpha motor neuron.

tional knowledge of its action potential and tension, which we have recorded in synchrony at sweep speeds fast enough to demonstrate clearly the minute characteristics of these two waveforms. The tension, however, is subject to a number of variables, such as the state of anesthesia, the time elapsed from the beginning of the experiment to the time of recording and the variation of size of different motor units. The magnitude of tension produced by the single motor unit under these conditions ranges between 8 and 20 gm for single twitches. The tetanic contraction of small fasciculi, as shown in photograph Number 2, Figure 176, is usually approximately three times the magnitude of that of the single twitch of that fasciculus. The single motor unit action potential, however, retains the character of that demonstrated in several of the photographs demonstrated, such as Number 5. This waveform can be subjected to analysis by Fourier transform or complete Fourier analysis, as previously described. Care must be taken to use nonpolarizable recording electrodes. The recording silver electrode used in this work is a small round disc, but recordings have also been made using electrodes of different configurations.

Henneman and co-workers[129,130] report successful stimulation of alpha motor neurons and recording of conduction velocity, electromyograph and tension curve in the soleus muscle of the cat. They were successful in acquiring these data also from the gastrocnemius muscle of the same animal. The criteria for the selection of motor units (97 soleus units) were all-or-none action potentials recorded antidromically from ventral root filaments in response to stimulation of muscle nerve and all-or-none responses of the same unit to stimulation at subthreshold, threshold and suprathreshold intensities.

The value of this procedure, which involves high-fidelity recordings of the single motor unit action potential and its tension curve, is more than basic information. While the basic information acquired consists of further knowledge concerning the action potential and the tension curve and their time relationships, it is of the utmost value to observe the performance of the single motor unit as conditions are changed. These changes may be physical or chemical. There is a wealth of opportunity for profitable experimentation. A polyethylene tube may be inserted into one of the veins and the effect of various injections observed. There is thus opportunity to change drastically the extracellular surroundings of both the alpha motor neuron and the few hundred muscle fibres that it innervates, and to observe changes of the action potential and tension curve of the single motor unit. Studies of this type are in progress at this time. It is hoped that the experimental procedure described herein will lead to more meaningful data concerning the units or quanta of muscle described by Sherrington[158] and by Eccles and Sherrington[58] in a form more comparable to man. Thus it is considered essential to acquire these data from the most highly developed vertebrate possible.

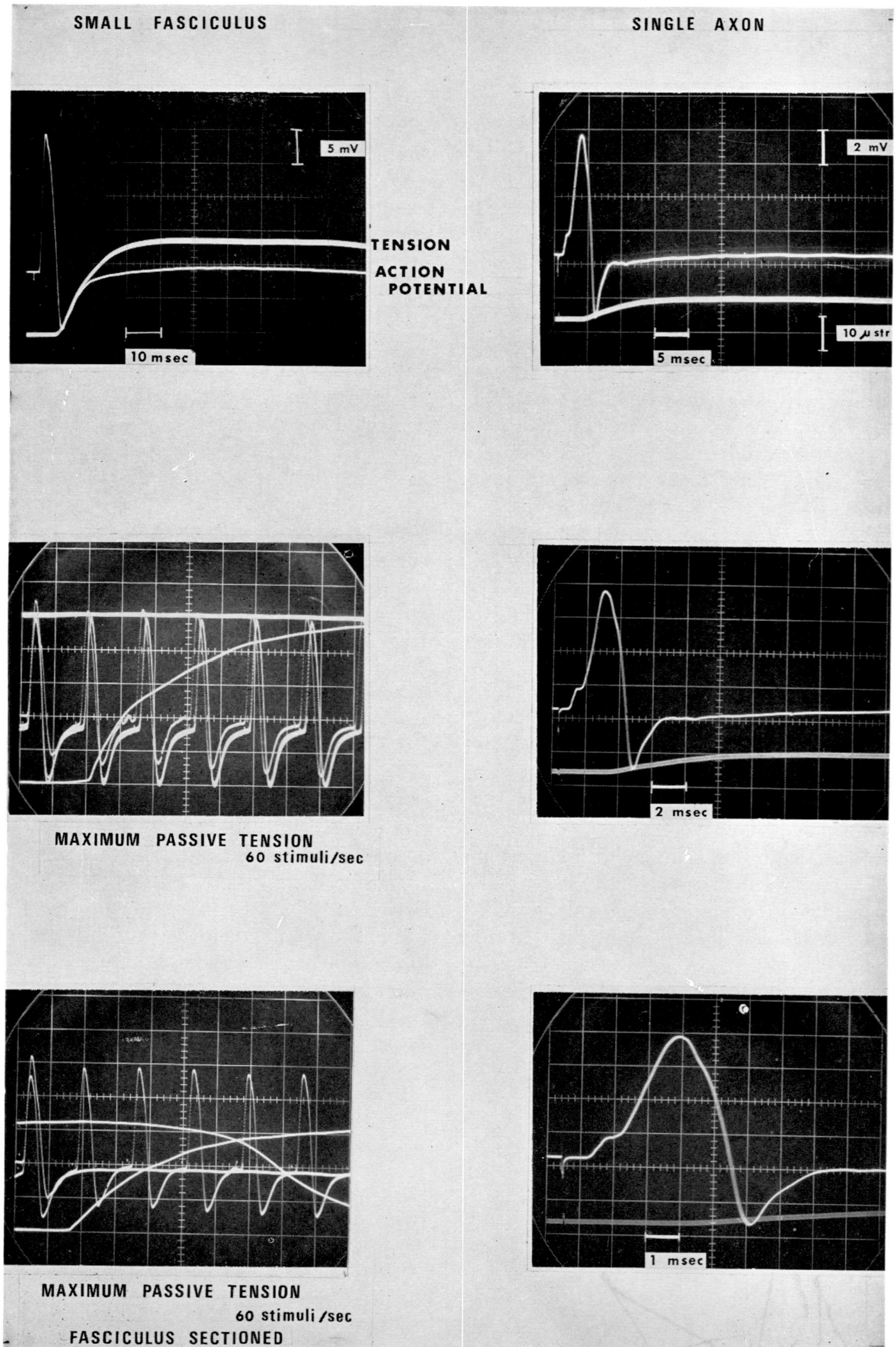

Figure 177. Recordings from a group of motor units. *Left column,* single twitch, tetanus, tetanus after section of the stimulated fasciculus. *Right column,* recordings from a single motor unit produced by stimulation of its (single) axon; end-plate potentials are increasingly apparent as sweep speed is increased.

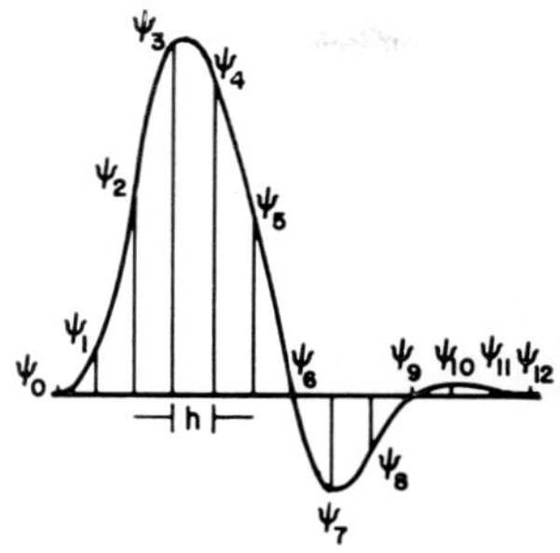

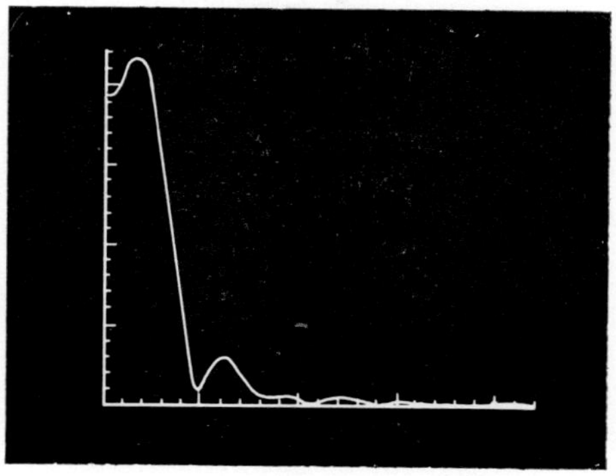

$$A(f) = \frac{1}{\sqrt{2\pi}} \int_{-\infty}^{\infty} \psi(t)\, e^{-2\pi i f t}\, dt$$

$$|A(f)| = \sqrt{\frac{1}{2\pi}\left[\left(\int_{-\infty}^{\infty} \psi(t) \cos \omega t\, dt\right)^2 + \left(\int_{-\infty}^{\infty} \psi(t) \sin \omega t\, dt\right)^2\right]}$$

$$\omega = 2\pi f$$

DIGITIZED WAVEFORM FREQUENCY SPECTRUM

FOURIER TRANSFORM

FOURIER ANALYSIS

BIBLIOGRAPHY

1. Adams, R.D., Eaton, L.M. and Shy, G.M. (Eds.): *Neuromuscular Disorders: The Motor Unit and its Disorders.* Proceedings of the Association for Research in Nervous and Mental Disease. Baltimore, Williams and Wilkins, 1960.
2. Adams, R.D., Denny-Brown, D. and Pearson, C.: *Diseases of Muscle.* New York, Hoeber, 1962.
3. Adrian, E.D. and Bronk, D.W.: The discharge of impulses in motor nerve fibers. *J Physiol (Lond), 67:*119, 1929.
4. Alnaes, E.: Static and dynamic properties of Golgi tendon organs in the anterior tibial and soleus muscles of the cat. *Acta Physiol Scand, 70:* 176–187, 1967.
5. Andrew, B.L.: The sensory innervation of the medial ligament of the knee joint. *J Physiol (Lond), 123:*241–250, 1954.
6. *Annual Report of the Director of the Department of Embryology.* Baltimore, Carnegie Institute of Washington Year Book 59, 1960.
7. Ashworth, B., Grimby, L. and Kugelberg, E.: Comparison of voluntary and reflex activation of motor units. Functional organization of motor neurones. *J Neurol Neurosurg Psychiatry, 30:*91–98, 1967.
8. Bahler, A.S.: Series elastic component of mammalian skeletal muscle. *Am J Physiol, 213:* 1560–1564, 1967.
9. Bahler, A.S., Fales, J.T. and Zierler, K.L.: The dynamii properties of mammalian skeletal muscle. *J Gen Physiol, 51:*369–384, 1968.
10. Bardeen, C.R.: Development and variation of the nerves and the musculature of the inferior extremity and the neighboring regions of the trunk in man. *Am J Anat, 6:*259–390, 1906.
11. Barker, D. (Ed.): *Symposium on Muscle Receptors.* Hong Kong, Hong Kong University Press, 1962.
12. Barnett, C.H. and Napier, J.R.: The axis of rotation at the ankle joint in man. Its influence upon the form of the talus and the mobility of the fibula. *J Anat, 86 (pt 1):*1–9, 1952.
13. Basmajian, J.V.: Electromyography of two-joint muscles. *Anat Rec, 129:*371, 1957.
14. Basmajian, J.V.: Electromyograph of iliopsoas. *Anat Rec, 132:*127–132, 1958.
15. Basmajian, J.V.: A new six-channel electromyograph for studies on muscle. *IRE Transactions on Medical Electronics, PGME, 11:*45–48, July, 1958.
16. Basmajian, J.V.: *Muscles Alive.* Baltimore, Williams and Wilkins, 1962.
17. Becker, R.O.: The electrical response of human skeletal muscle to passive stretch. *J Bone Joint Surg, 42-A:*1091–1103, 1960.
18. Bell, P.O.: Personal communication, 1951.
19. Bessou, P., Laporte, Y. and Pagès, B.: Étude de la relation entre le temps de contraction d'unités motrices et la vitesse de conduction de leurs fibres motrices dans des muscles réinnervés. *C R Acad Sci (Paris), D, 263:*1486–1489, November, 1966.
20. Bigland, B. and Lippold, O.C.J.: The relation between force, velocity and integrated electrical activity in human muscles. *J Physiol (Lond), 123:*214–224, 1954.
21. Bigland, B. and Lippold, O.C.J.: Motor unit activity in the voluntary contraction of human muscle. *J Physiol (Lond), 125:*322–335, 1954.

21a. Bizzi, E., Kalil, R.E. and Tagliasco, V.: Eye-head coordination in monkeys: evidence for centrally patterned organization. *Science, 173:* 452–454, July 30, 1971.

22. Bouman, H.D. and Woolf, A.L. (Eds.): *The Utrecht Symposium on the Innervation of Muscle.* Baltimore, Williams and Wilkins, 1960.
23. Buchtal, F. and Madsen, A.: Synchronous activity in normal and atrophic muscle. *Electroencephalogr Clin Neurophysiol, 2:*425–444, 1950.
24. Buchtal, F., Guld, C. and Rosenfalck, P.: The origin of the motor unit potential. *Acta Physiol Scand, 31 (Suppl. 114):*7–8, 1954.
25. Buchtal, F., Guld, C. and Rosenfalck, P.: Action potential parameters in normal human muscle and their dependence on physical variables. *Acta Physiol Scand, 32:*200–218, 1954.
26. Buchtal, F., Pinelli, P. and Rosenfalck, P.: Action potential parameters in normal human muscle and their physiological determinants. *Acta Physiol Scand, 32:*219–229, 1954.
27. Buchtal, F. and Rosenfalck, P.: Action potential

parameters in different human muscles. *Acta Phychiatr Neurol Scand, 30:*125–131, 1955.

28. Bunnell, Sterling: In Boyes, J.H. (Ed.): *Surgery of the hand,* 5th ed. Philadelphia, Lippincott, 1970.
29. Burke, R.E.: Motor unit types of cat triceps surae muscle. *J Physiol (Lond), 193:*141–160, 1967.
30. Caputo, C.: The role of calcium in the processes of excitation and contraction of skeletal muscle. *J Gen Physiol, 51:*180s–187s, May 1968.
31. Chapman, J.B.: Potentiating effect of potassium on skeletal muscle twitch. *Am J Physiol, 217:* 898–902, 1969.
32. Charpy, A. and Nicolas, A.: *Traité d'Anatomie humaine.* Paris, Masson et Cie, 1911, pp. 824–846.
33. Close, J.R.: Some applications of the functional anatomy of the ankle joint. *J Bone Joint Surg, 38-A:*761–781, 1956.
34. Close, J.R.: *Muscular Phasic Activity in the Lower Extremity and the Effect of Tendon Transfer,* 16 mm sound film, 1958.
35. Close, J.R.: *Electromyography—Its Application in Orthopaedic Surgery.* American Academy of Orthopaedic Surgeons, Instructional Course Lectures, Vol. XVI. St. Louis, Mosby, 1959.
36. Close, J.R.: *Motor Function in the Lower Extremity.* Springfield, Thomas, 1964.
37. Close, J.R. and Inman, V.T.: *The Action of the Ankle Joint.* Report to Advisory Committee on Artificial Limbs, National Research Council. Prosthetic Devices Research Project, Institute of Engineering Research, University of California (Berkeley), Series 11, Issue 22, 1952.
38. Close, J.R. and Inman, V.T.: *The Action of the Subtalar Joint.* Report to Advisory Committee on Artificial Limbs, National Research Council. Prosthetic Devices Research Project, Institute of Engineering Research, University of California (Berkeley), Series 11, Issue 24, May, 1953.
39. Close, J.R., Smart, B.W. and Blosser, J.A.: *Electromyography in the Lower Extremity,* 16 mm sound film, 1955.
40. Close, J.R. and Todd, F.N.: The phasic activity of the muscles of the lower extremity and the effect of tendon transfer. *J Bone Joint Surg. 41-A:*189–209, 1959.
41. Close, J.R., Nickel, E.D. and Todd, F.N.: Motor-unit action-potential counts—their significance in isometric and isotonic contractions. *J Bone Joint Surg. 42-A:*1207–1222, 1960.
42. Close, J.R., Maurer, R.C. and Poor, P.M.: Single motor unit action potentials. *Clin Orthop, 42:* 171–190, 1965.
43. Close, J.R., Inman, V.T., Poor, P.M. and Todd, F.N.: The function of the subtalar joint. *Clin Orthop, 50:*159–179, 1967.
44. Close, J.R. and Kidd, C.C.: The functions of the muscles of the thumb, the index, and long fingers. *J Bone Joint Surg, 51-A:*1601–1620, 1969.
45. Close, R. (Canberra): Properties of motor units in fast and slow skeletal muscles of the rat. *J Physiol (Lond), 193:*45–55, 1967.
46. Close, R. and Hoh, J.F.Y. (Canberra): The after-effects of repetitive stimulation on the isometric twitch contraction of rat fast skeletal muscle. *J Physiol (Lond), 197:*461–477, 1968.
47. Crabb. E.D.: *Principles of Functional Anatomy of the Rabbit.* St. Louis, Swift, 1951.
48. Conrad, J.T. and Glaser, G.H.: Bioelectric properties of dystrophic mammalian muscle. *Arch Neurol, 5:*58–71, 1961.
49. Crafts, L.W., Schneirla, T.C., Robinson, E.E. and Gilbert, R.W.: *Recent Experiments in Psychology.* New York, McGraw-Hill, 1938, p. 371.
50. Crue, B.L., Pudenz, R.H. and Sheldon, C.H.: Observations on the value of clinical electromyography. *J Bone Joint Surg. 39-A:*492, 1957.
51. Cunningham, D.M.: *Components of Floor Reactions during Walking.* Report to Advisory Committee on Artificial Limbs, National Research Council. Prosthetic Devices Research Project, Institute of Engineering Research, University of California (Berkeley), Series 11, Issue 14, 1950.
52. Devanandan, M.S., Eccles, R.M. and Westerman, R.A.: Single motor units of mammalian muscle. *J Physiol (Lond), 178:*359–367, 1965.
53. Donaldson, P.E.K.: *Electronic Apparatus for Biological Research.* London, Butterworths, 1958.
54. Donitz, A.: *Die Mechanik der Fusswurzel.* Inaugural dissertation. Berlin, Gustav Schvade (Otto Fanke), 1903.
55. Dresden, A.: *Solid Analytical Geometry and Determinants.* New York, Dover Publications, 1964.
56. Duchenne, G.B.: *Physiology of Motion.* (Kaplan, E.B., translator.) Philadelphia, Lippincott, 1949.

56a. Eaton, R.G.: *Joint Injuries of the Hand.* Springfield, Thomas, 1971.

57. Eberstein, A. and Goodgold, J.: Slow and fast twitch fibers in human skeletal muscle. *Am J Physiol, 215:*535-541, 1968.
58. Eccles, J.C. and Sherrington, C.S.: Numbers and contraction-values of the individual motor-units examined in some muscles of the limb. *Proc R Soc Lond (Biol), 106:*326–357, 1930.
59. Edshage, S.: Peripheral nerve suture. A technique

for improved intraneural topography. Evaluation of some suture materials. *Acta Chir Scand, suppl. 331:*1–104, 1964.

60. El-Badry, H.M.: *Micromanipulators and Micromanipulation.* New York, Academic Press, 1963.

61. Elftman, H. and Manter, J.: The evolution of the human foot with especial reference to the joints. *J Anat, 70:*56–67, 1938.

62. Elliott, R.: *Anatomy of the Cat.* New York, Holt, Rinehart and Winston, 1963.

63. Eyler, D.L. and Markee, J.E.: The anatomy and function of the intrinsic musculature of the fingers. *J Bone Joint Surg, 36-A:*1–9, 18–20, 1954.

64. Falk, G. and Landa, J.F.: Prolonged response of skeletal muscle in the absence of penetrating anions. *Amer J Physiol, 198:*289–299, 1960.

65. Feinstein, B., Lindegard, B., Nyman, E. and Wohlfart, G.: Studies of action potentials in normal human muscles. *Acta Psychiatr Neurol Scand, 29:*189–195, 1954.

66. Feinstein, B., Lindegard, B., Nyman, E. and Wohlfart, G.: Morphologic studies of motor units in normal human muscles. *Acta Anat, 23:*127–142, 1955.

67. Fernand, V.S.V. and Young, J.Z.: The size of the nerve fibers of muscle nerves. *Proc R Soc Lond (Biol), 139:*38–58, 1951–2.

68. Fick, R.: *Anatomie und Mechanik der Gelenke. Bardelebens Handbuch der Anatomie des Menschen.* Jena, 1911. Band 2, Abteilung 1, Teil 3.

69. Field, J., Magoun, H.W. and Hall, V.E. (Eds.): *Handbook of Physiology, Section 1: Neurophysiology.* American Physiological Society, Washington, D.C., 1959, vol. 1.

70. *Fundamental Studies of Human Locomotion and Other Information Relating to Design of Artificial Limbs.* Subcontractor's Report to Committee on Artificial Limbs, National Research Council. Prosthetic Devices Research Project, College of Engineering, University of California (Berkeley), June, 1947, vols. 1 and 2.

71. Geddes, L.A. and Baker, L.E.: *Principles of Applied Biomedical Instrumentation.* New York, Wiley, 1968.

72. Gelfan, S.: Muscle. *Ann Rev Physiol, 20:*67, 1958.

73. Gellhorn, E.: Patterns of muscular activity in man. *Arch Phys Med Rehab, 28:*568–574, 1947.

74. Gellhorn, E. and Johnson, D.A.: The specificity of the electromyographic method for the investigation of the motor cortex. *Brain, 73:* 275–279, 1950.

75. Glasser, O. (Ed.): *Medical Physics.* Chicago, Year Book Publishers, 1960, vol. III.

76. Golseth, J.G.: Diagnostic contributions of the electromyogram. *Calif Med, 73:*355, 1950.

77. Granit, R.: Systems for control of movement. In *Première Journée Commune: Réunions plénières, rapports et discussions.* Premier Congrès International des Sciences Neurologiques, Bruxelles, juillet 1957. Bruxelles, Editions Acta Medica Belgica, 1957, vol. I, pp. 63–80.

78. Granit, R.: Neuromuscular interaction in postural tone of the cat's isometric soleus muscle. *J Physiol (Lond), 143:*387–402, 1958.

79. Granit, R.: *The Basis of Motor Control.* New York, Academic Press, 1970.

80. Grant, J.C.B.: *Method of Anatomy,* 5th ed. Baltimore, Williams and Wilkins, 1952, pp. 132, 156, 619.

81. Grice, D.S.: An extra-articular arthrodesis of the subastragalar joint for correction of paralytic flat feet in children. *J Bone Joint Surg, 34–A:*924–940, 1952.

82. Gurkow, H.J. and Bast, T.H.: Innervation of striated skeletal muscle. *Am J Phys Med, 37:*269–278, 1958.

83. *Hand, The.* Journal of the British Society for Surgery of the Hand. March, 1969, vol. 1, no. 1.

84. Hawkins, L.G.: Fracture of the lateral process of the talus. *J Bone Joint Surg, 47-A:*1170–1175, 1965.

85. Haxton, H.A.: Absolute muscle force in the ankle flexors of man. *J Physiol (Lond), 103:* 267–273, 1944.

86. Heiland, C.A. and Hrock, M.S.: Review of some sensing and recording applications in human and animal physiology. *Am J Phys Med, 36:*173, 1957.

87. Henke, W.: Die Bewegung des Fusses am Sprungbein. *Zeitschrift für Rational Medizin,* Band 7, S.225–234, 1859.

88. Henry, A.K.: *Extensile Exposure,* 2nd ed. Baltimore, Williams and Wilkins, 1957, p. 97.

89. Hertz, H., Madsen, A. and Buchtal, F.: Prognostic implications of electromyography in acute anterior poliomyelitis. *J Bone Joint Surg, 36–A:*902, 1954.

90. Hettinger, T.: *Physiology of Strength.* Springfield, Thomas, 1961.

91. Hill, A.V.: The mechanics of active muscle. *Proc R Soc Lond (Biol), 141:*104–117, 1953.

92. Hill, A.V.: Production and absorption of work by muscle. *Science, 131:*897–903, March 25, 1960.

93. Hill, A.V.: *First and Last Experiments in Muscle Mechanics.* Cambridge, Cambridge University Press, 1970.

94. Hirschberg, G.C. and Abramson, A.S.: Clinical electromyography. *Arch Phys Med Rehabil, 31:*576, 1950.

95. Houk, J. and Henneman, E.: Responses of Golgi tendon organs to active contractions of the soleus muscle of the cat. *J Neurophysiol 30:*466–481, 1967.

96. Hunt, C.C. and Kuffler, S.W.: Motor innervation of skeletal muscle: multiple innervation of individual muscle fibres and motor unit function. *J Physiol (Lond), 126:*293–303, 1954.

97. Huxley, A.F. *et al,*: Second Conference of Physicochemical Mechanism of Nerve Activity and Second Conference on Muscular Contraction. Nachmanson, D.A. (Ed.) *Ann N Y Acad Sci, 81:*215–510, 1959.

98. Huxley, H.E.: The mechanism of muscular contraction. *Science, 164:*1356–1365, June 20, 1969.

99. Ikai, M. and Fukunaga, T.: Calculation of muscle strength per unit cross-sectional area of human muscle by means of ultrasonic measurement. *Int Z Angew Physiol, 26:*26–32, 1968.

100. Inman, V.T. Ralston, H.J., Saunders, J.B. deC.M., Feinstein, B. and Wright, E.W., Jr.: *Relations of Human Electromyogram to Muscular Tension.* Report to Advisory Committee on Artificial Limbs, National Research Council. Prosthetics Devices Research Project, Institute of Engineering Research, University of California (Berkeley), Series 11, Issue 18, November, 1951.

101. Inman, V.T., Ralston, H.J., Saunders, J.B. deC.M., Feinstein, B. and Wright, E.W., Jr., Relations of human electromyogram to muscular tension. *Electroencephalogr Clin Neurophysiol, 4:*187–194, 1952.

102. Irwin, R.L. and Oliver, K.L.: Sodium-induced tension in slow skeletal muscle deprived of external calcium. *Am J. Physiol, 217:*13–18, 1969.

103. Jacobson, L.E.: The electrophysiology of mental activities. *Am J Physiol, 44:*677–694, 1932.

104. Jampolsky, A., Tamler, E. and Marg, E.: Elements in the technique of extraocular electromyography. *Arch Ophthalmol, 61:*258, 1959.

105. Joseph, J.: *Man's Posture.* Springfield, Thomas, 1960.

106. Kaplan, E.B.: *Functional and Surgical Anatomy of the Hand,* 2nd ed. Philadelphia, Lippincott, 1965.

107. Katz, B.: Quantal mechanism of neural transmitter release. *Science, 173:*123–126, July 9, 1971.

108. Katz, B. and Schmitt, O.H.: Electrical interaction between two adjacent nerve fibres. *J Physiol (Lond), 97:*471–488, 1940.

109. Keibel, F. and Mall, F.P. (Eds.): *Manual of Human Embryology.* Philadelphia, Lippincott, 1910.

110. Krnjevic, K. and Miledi, R.: Motor units in the rat diaphragm. *J Physiol (Lond), 140:*427–439, 1958.

111. Kugelberg, E. and Taverner, D.: Comparison between voluntary and electrical activation of motor units in anterior horn cell diseases; on "central synchronization" of motor units. *Electroencephalogr Clin Neurophysiol, 2:*125–132, 1950.

112. Landsmeer, J.M.F.: The anatomy of the dorsal aponeurosis of the human finger and its functional significance. *Anat Rec, 104:*31–44, 1949.

113. Landsmeer, J.M.F.: Anatomical and functional investigations on the articulation of the human fingers. *Acta Anat, 25 (suppl. 24):*1–69 1955.

114. Last, J.R.: *Anatomy, Regional and Applied.* Boston, Little, Brown, 1954.

114a. Lennerstrand, G. and Thoden, U.: Muscle spindle responses to concomitant variations in length and in fusimotor activation. *Acta Physiol Scand, 74:*153–165, 1968.

115. Levens, A.S., Inman, V.T. and Blosser, J.A.: Transverse rotation of the segments of the lower extremity in locomotion. *J Bone Joint Surg, 30–A:* 859–872, 1948.

116. Lewin, P. and Boshes, L.D.: The skeletal muscles and tendon apparatus, sesamoids, bursae and fasciae. In McQuarrie, I. (Ed.): *Brennemann's Practice of Pediatrics.* Hagerstown, W.F. Prior, 1959, vol. 4, Chap. 37.

117. Libet, B., Feinstein, B. and Wright, E.W., Jr.: Tendon afferents in autogenetic inhibition in man. *Electroencephalogr Clin Neurophysiol, 11:*129–140, 1959.

118. Licht, Sidney (Ed.): *Electrodiagnosis and Electromyography,* 3rd ed. New Haven, Elizabeth Licht, 1971.

119. Lippold, O.C.J.: The relation between integrated action potentials in a human muscle and its isometric tension. *J Physiol (Lond), 117:*492–499, 1952.

120. Long, C., II: Intrinsic-extrinsic muscle control of the fingers. Electromyographic studies. *J Joint Surg, 50–A:*973–984, 1968.

121. Long, C., II, and Brown, M.E.: *The Moving Finger Writes: An Electromyographic Study of Muscles Driving the Middle Finger,* 16 mm sound film. Highland View Hospital, Cleveland, Ohio, 1962.

122. Lorente de Nó, R.: *A Study of Nerve Physiology.* Studies from the Rockefeller Institute

for Medical Research, *131:*1–496, *132:*1–548, 1947.

123. Lucas, D.B. and Bresler, B.: *Stability of the Human Spine.* Biomechanics Laboratory, University of California, San Francisco and Berkeley, March, 1961.
124. Magun, R. and Esslen, E.: Electromyographic study of reinnervated muscle and of hemifacial spasm. *Am J Phys Med, 38:*79, 1959.
125. Manter, J.T.: Movements of the subtalar and transverse tarsal joints. *Anat Rec, 80:*397–410, 1941.
126. Marinacci, A.A.: *Applied Electromyography.* Philadelphia, Lea and Febiger, 1968.
127. Max, L.W.: An experimental study of the motor theory of consciousness, Part III. Action-current responses in deaf mutes during sleep, sensory stimulation and dreams. *J Comp Psychol, 19:*469–486, 1935.
128. Max, L.W.: An experimental study of the motor theory of consciousness, Part IV. Action-current responses in the deaf during awakening, kinaesthetic imagery and abstract thinking. *J Comp Psychol, 24:*301–344, 1937.
129. McPhedran, A.M., Wuerker, R.D. and Henneman, E.: Properties of motor units in a homogeneous red muscle (soleus) of the cat. *J Neurophysiol, 28:*71–84, 1965.
130. McPhedran, A.M., Wuerker, R.D. and Henneman, E.: Properties of motor units in a heterogeneous pale muscle (m. gastrocnemius) of the cat. *J Neurophysiol, 28:*85–99, 1965.
131. Meyer, H.: *Statik und Mechanik des Menschlichen Fusses.* Jena, 1886.
132. Miledi, R.: From nerve to muscle. *Discovery (London), 22:*442–450, 1961.
133. Miller, H.C. and Darrow, D.C.: The effect of changes in muscle electrolyte on the response of skeletal muscle to tetanic stimulation with particular reference to potassium. *Am J Physiol, 129:*264–270, 1940.
134. Morris, C.B.: The measurement of the strength of muscle relative to the cross section. *Res Q Am Assoc Health Phys Ed, 19:*295–303, December, 1948.
135. Mustard, W.T.: Iliopsoas transfer for weakness of the hip abductors. *J Bone Joint Surg, 34–A:*647–650, 1952.
136. Nastuk, W.L. (Ed.): *Physical Techniques in Biological Research.* New York, Academic Press, 1963, 1964, Vols. V (1964) and VI (1963).
137. Oakley, C.O.: *Analytic Geometry.* New York, Barnes and Noble, 1957.
138. Olson, C.B. and Swett, C.P., Jr.: A functional histochemical characterization of motor units in a heterogeneous muscle (flexor digitorum longus) of the cat. *J Comp Neurol, 128:*475–498, 1966.
139. Olson, C.B., Carpenter, D.O. and Henneman, E.: Orderly recruitment of muscle action potentials. Motor unit threshold and EMG amplitude. *Arch Neurol, 19:*591-597, 1968.
140. *Pattern of Muscular Activity in the Lower Extremity during Walking.* Report to Advisory Committee on Artificial Limbs, National Research Council. Prosthetics Devices Research Project, Institute of Engineering Research, University of California (Berkeley), Series 11, Issue 25, September, 1953.
141. Pierce, D.S. and Wagman, I.H.: A method of recording from single muscle fibers or motor units in human skeletal muscle. *J Appl Physiol, 19:*366–368, 1964.
142. Ralston, H.J.: Recent advances in neuromuscular physiology. *Am J Phys Med, 36:*94, 1957.
143. Ralston, H.J., Inman, V.T., Strait, L.A. and Shaffrath, M.D.: Mechanics of human isolated voluntary muscle. *Am J Physiol, 151:*612–620, 1947.
144. Ralston, H.J., Polissar, M.J., Inman, V.T., Close, J.R. and Feinstein, B.: Dynamic features of human isolated voluntary muscle in isometric and free contractions. *J Appl Physiol, 1:*526–533, 1949.
145. Reighard, J. and Jennings, H.S.: *Anatomy of the Cat,* 3rd ed. New York, Holt, Rinehart and Winston, c. 1901, 1963.
146. Rodriquez, M.J.: Therapeutic exercise in general practice. *Calif Med, 95:*378–381, 1961.
147. Rosenblueth, A., Alanis, J. and Rubio, R.: A comparative study of the isometric and isotonic contractions of striated muscles. *Arch Int Physiol Biochim (Liège), 66:*330–353, July, 1958.
148. Rosenblueth, A., Alanis, J. and Rubio, R.: The influence of initial load and length on the amplitude of muscular contractions. *Arch Int Physiol Biochim* (Liège), *66:*354–372, July, 1958.
149. Ruch, T.C. and Fulton, J.F. (Eds.): *Medical Physiology and Biophysics,* 18th ed. Towell-Fulton Textbook of Physiology. Philadelphia, Saunders, 1960.
150. Ruch, T.C. and Patton, H.D. (Eds.): *Medical Physiology and Biophysics,* 19th ed. Towell-Fulton Textbook of Physiology. Philadelphia, Saunders, 1965.
151. Ryker, N.J., Jr.: *Glass Walkway Studies of Normal Subjects during Normal Level Walking.* Report to Advisory Committee on Artificial Limbs, National Research Council. Prosthetics Devices Research Project, Institute of

Engineering Research, University of California (Berkeley), Series 11, Issue 20, January, 1952.

152. Salisbury, C.R.: The interosseous muscles of the hand. *J Anat (Lond), 71:*395–403, 1936.

153. Saunders, J.B. deC. M., Inman, V.T. and Eberhart, H.D.: The major determinants in normal and pathological gait. *J Bone Joint Surg, 35–A:*543–558, 1953.

154. Schaeffer, B.: The morphological and functional evolution of the tarsus in amphibians and reptiles. *Bull Amer Museum Nat Hist, 78:*395–472, 1941.

155. Scherb, R.: Kinetisch-diagnostische Analyse von Gehstörungen: Technik und Resultate der Myokinesigraphie. *Z Orthop, 82 (suppl.):* 1–199, 1952.

156. Schwartz, R.P. and Bouman, H.D.: Muscle spasm in acute stage of infantile paralysis as indicated by recorded action current potentials. *JAMA, 119:*923, 1942.

157. Schwartz, R.P., Bouman, H.D. and Smith, W.K.: Significance of muscle spasm in acute stage of infantile paralysis based on action current records. *JAMA, 126:*695, 1944.

158. Sherrington, C.S.: *The Integrative Action of the Nervous System.* New York, Scribner, 1906.

159. Shieltron Inc., Riverton, N.J.: *An Introduction to R.F. [Radio Frequency] Shielded Enclosures.* 6 pp., 1962.

160. Stack, H.G.: Muscle function in the fingers. *J Bone Joint Surg, 44–B:*899–909, 1962.

161. Stacy, R.W.: *Biological and Medical Electronics.* New York, McGraw-Hill, 1960.

162. Stilwell, D.L.: The innervation of tendons and aponeuroses. *Am J Anat, 100:*289–317, 1957.

163. Szent-Gyorgi, A.: Muscle research. *Science, 127:*699–702, September 26, 1958.

164. Tasaki, I. and Tasaki, N.: The electrical field which a transmitting nerve fibre produces in the fluid medium. *Biochim Biophys Acta, 5:*335–342, 1950.

165. Viets, H.R. (Ed.): *Myasthenia Gravis.* Springfield, Thomas, 1961.

166. Walton, J.N.: The electromyogram in myopathy: analysis with the audiofrequency spectrometer. *J Neurol Neurosurg Psychiat, 15:*219, 1952.

167. Waring, W. and Nickel, V.L.: *Investigation of Myoelectric Control of Functional Braces.* Final Report on Planning Grant RD–1655–M–64, March, 1965.

168. Weddell, G., Feinstein, B. and Pattle, R.E.: The electrical activity of voluntary muscle in man under normal and pathological conditions. *Brain, 67:*178, 1944.

169. Whitfield, I.C.: *Introduction to Electronics for Physiological Workers.* London, Macmillan, 1960.

170. Williams, M. and Lissner, H.R.: *Biomechanics of Human Motion.* Philadelphia, Saunders, 1962.

171. Wright, D.G., Desai, S.M. and Henderson, W.H.: Action of the subtalar and ankle-joint complex during the stance phase of walking. *J Bone Joint Surg, 46–A:*361–382, 1964.

172. Wynn Parry, C.B.: Electrodiagnosis. *J Bone Joint Surg, 43–B:*222–236, 1961.

INDEX

M

N

P

Q

R

S

T